INTRODUCTION TO
MODERN LIQUID CHROMATOGRAPHY

L. R. SNYDER

Technicon Instruments Corporation
Tarrytown, New York

J. J. KIRKLAND

E. I. DuPont de Nemours & Company
Wilmington, Delaware

A WILEY-INTERSCIENCE PUBLICATION

JOHN WILEY & SONS, New York • London • Sydney • Toronto

Library of Congress Cataloging in Publication Data:

Snyder, Lloyd R.
   Introduction to modern liquid chromatography.

  "A Wiley-Interscience publication."
  1. Liquid chromatography. I. Kirkland, Joseph
Jack, joint author. II. Title.

QD79.C454S58   544'.924   73–17453
ISBN 0–471–81019–3

Printed in the United States of America

10 9 8 7 6 5 4 3 2 1

And thus the native hue of resolution
Is sicklied o'er with the pale cast of
thought.

Shakespeare
Hamlet, Scene I, Act 3

PREFACE

This book is about modern liquid chromatography. By
this we mean automated, high-pressure liquid chromatography in columns, with a capability for the high-
resolution separation of a wide range of sample types,
within times of a few minutes to perhaps an hour. Modern liquid chromatography (LC) is now about five years
old. By early 1969 it was possible to purchase equipment and high-performance column packings which together largely bridged the gap between classical liquid
chromatography and gas chromatography. Since that time
there has been a flurry of activity on the part of companies that supply equipment and materials for LC.
Within the past few years there have been further major advances in the theory and practice of LC. Finally, numerous applications of modern LC to a wide range
of problems are now being reported. The technique has
reached the point where the average chromatographer
can achieve--by yesterday's standards--consistently
spectacular results.

To get the most out of modern LC, some care is required in choosing the right technique, selecting the
best separation conditions, and using the proper equipment to best advantage. In short, the practical worker
must know what he is doing. Moreover, his knowledge
must be a balance of theory and experience; it must

include both principles and practice. Unfortunately,
there has been a tendency for those in chromatography
to stress either the theoretical or the "practical"
side of the subject. Also, the theory of chromatog-
raphy--and of modern LC--has often been represented as
highly complex, with its application to real separation
problems either not obvious, or impossibly tedious.
We think there is a better approach.

An effective presentation of what modern LC is all
about must be a blend of practical details plus down-
to-earth theory. This conviction led us in 1971 to
develop the American Chemical Society short course
"Modern Liquid Chromatography." Within the next two
years we presented the course to about 800 students,
with a highly enthusiastic response. The course itself continued to evolve during this time, largely in
response to the questions and comments of the students.
By late 1972 it appeared worthwhile to reduce our approach to textbook form, and the present book is the
result.

Our goal for this book was to retain the essential
elements of the short course, and to add certain materials that could not be included in a two-day series
of lectures. We did not hope to present everything of
conceivable interest to LC in a book of this size, yet
we were determined not to slight any area of significant practical importance. Where compromise between
these two objectives eventually proved necessary, it
has been handled by referencing other sources. The
book was written to be self-sufficient in terms of
the needs of the average professional or technician
who plans to work with modern LC. We believe this book
will prove useful in most laboratories where modern LC
is practiced.

In conclusion, we would like to express our appre-
ciation to numerous professional associates who have
shaped our thinking as set down in the present book,
to the hundreds of students who took our course--and

helped make it better for the next class, and to the staff of the Continuing Education Department of the American Chemical Society for their help with and sponsorship of the short course. We are especially grateful to the special contributors, Dr. J. J. DeStefano of the Biochemicals Department, E. I. du Pont de Nemours & Co., and Mr. H. J. Adler of the Research Department, Technicon Instrument Corporation, who wrote material in areas of their expertise to fill in gaps in the original ACS short course. We are also grateful to several of our friends for agreeing to review the initial manuscript, and who much improved it: Dr. D. L. Saunders of the Union Oil Company, Dr. H. M. McNair and Mr. Daniel Marsh of the Virginia Polytechnic Institute, Dr. B. L. Karger of Northeastern University, Dr. C. Horvath of Yale University, and Dr. J. J. DeStefano. However, we should hasten to add that we did not always heed their advice, and we must take responsibility for any errors that remain.

The efforts of our dedicated secretaries, Mrs. Elizabeth Gill and Mrs. Ester Gerber of the Technicon Instrument Corporation, and Mrs. Mary Lynn Edwards of Du Pont are much appreciated. We are particularly indebted to Miss Mildred Syvertsen of Du Pont for her preparation of the final manuscript copy. The support of our respective departments in this connection was also of great assistance. Finally, we owe much to our families for their forbearance during the several months that were required to complete this book.

L. R. Snyder

J. J. Kirkland

October 1973

CONTENTS

INTRODUCTION TO
MODERN LIQUID CHROMATOGRAPHY

CHAPTER ONE

INTRODUCTION

Over the past forty years the practice of chromatography has witnessed a continuing growth in almost every respect: the number of chromatographers, the amount of published work, the variety and complexity of samples being separated, separation speed and convenience, and so on. However, this growth curve has not moved smoothly upward from year to year. Rather the history of chromatography is one of periodic upward spurts that have followed some major innovation. It is easy to recognize several major milestones of this type: the introduction of partition and paper chromatography in the 1940s, gas and thin-layer chromatography in the 1950s, and the various gel or exclusion methods in the early 1960s. A few years later it was possible to foresee still another of those major developments which were destined to revolutionize the practice of chromatography: a technique which we will call <u>modern liquid chromatography</u>.

What do we mean by "modern liquid chromatography"? Liquid chromatography (LC) refers to any chromatographic process in which the moving phase is a <u>liquid</u>, in contrast to the moving <u>gas</u> phase of gas chromatography. Traditional column chromatography (whether adsorption, partition, or ion exchange), thin-layer and paper chromatography, and modern LC are all forms of liquid chromatography. The difference between modern LC and these older procedures includes details of equipment, materials, technique, and theory. But mainly, modern LC offers major advantages in convenience, accuracy, speed, and the ability to carry out difficult separations. To appreciate the unique value of modern LC it will help to draw two comparisons:

- Liquid chromatography versus gas chromatography;
- Modern LC versus traditional LC procedures.

1.1   LIQUID VERSUS GAS CHROMATOGRAPHY

The tremendous ability of gas chromatography (GC) to separate and analyze complex mixtures is now widely appreciated. Compared to previous chromatographic methods, GC features separations that are much faster and better. Moreover, automatic equipment for GC was soon available for convenient, unattended operation. However, many samples cannot be handled by GC, either because they are insufficiently volatile and cannot pass through the column, or because they are thermally unstable and decompose under the conditions of separation. It has been estimated that only 20% of known organic compounds can be handled satisfactorily by GC, without prior chemical modification of the sample.

    LC on the other hand is not limited by sample volatility or thermal stability. Thus, LC is ideally suited for the separation of macromolecules and ionic species of biomedical interest, labile natural products, and a wide variety of other high-molecular-weight and/or less stable compounds; for example,

proteins	polysaccharides	pharmaceuticals
nucleic acids	plant pigments	dyes
amino acids	polar lipids	surfactants

drug and pesticide   steroids   synthetic polymers
   conjugates
purines   vitamins   antioxidants

Liquid chromatography also enjoys certain other
advantages with respect to GC. Very difficult separations are often more readily achieved by liquid than by
gas chromatography. The reasons for this include:

- Two chromatographic phases in LC for selective interaction with sample molecules,
  versus only one in GC;
- A greater variety of uniquely useful column
  packings (stationary phases) in LC;
- Lower separation temperatures in LC.

Chromatographic separation is the result of specific
interactions between sample molecules and the stationary and moving phases. These interactions are essentially absent in the moving gas phase of GC, but are
present in the liquid phase of LC--thus providing an
additional variable for controlling and improving separation. Furthermore, a greater variety of stationary
phases has been found useful in LC, which again allows
a wider variation of these selective interactions, and
greater possibilities for separation. Finally, chromatographic separation is generally enhanced as the temperature is lowered, because intermolecular interactions
then become more effective. This favors procedures
such as LC which are usually used at temperatures near
ambient.
      Liquid chromatography also offers a number of
unique detectors which have so far found little or no
application in GC:

- Colorimeters combined with color-forming
  reactions of separated sample components.
- UV absorption and fluorescence detectors.

- Radiometric detectors.
- Conductivity detectors.
- Polarographic detectors.
- Refractive index detectors.

In certain LC analyses that are illustrated later, the
availability of the right selective detector can eliminate the need for complete separation.
   A final advantage of liquid versus gas chromatography is in the relative ease of sample recovery. Separated fractions are easily collected in LC, simply by
placing an open vessel at the end of the column. Recovery is quantitative, and separated sample components
are readily isolated. The recovery of separated sample
components in GC is also possible--but is generally
less convenient and quantitative.
   Despite these advantages of LC relative to GC, the
latter is usually the method of choice when no special
problems are expected for a given sample. Separations
by GC are often faster and more sensitive, as well as
more convenient. Also, most laboratories are already
well equipped for GC, in terms of both equipment and
operating personnel. The time is rapidly passing, however, when GC is automatically a first choice for the
separation of most samples. Before long we can expect
to see more samples being analyzed by modern LC than
by GC.

1.2   MODERN VERSUS TRADITIONAL LC PROCEDURES

Consider now the difference between modern LC and classical column or open-bed chromatography. These three
general procedures are illustrated in Figure 1.1.   In
classical LC a column was generally used only once,
then discarded. Therefore, packing a column (step 1
of Figure 1.1, "Bed Preparation") had to be repeated
for each separation, and this represented a significant expense of both manpower and material. Sample

application in classical LC (step 2), if it was done right, required some skill and time on the part of the operator. Solvent flow in classical LC (step 3) was achieved by gravity feeding of the column, and individual sample fractions were collected manually. Since typical separations required several hours in classical LC, this was a tedious, time-consuming operation. Detection and quantitation (step 4) were achieved by the manual analysis of individual fractions. Many fractions were normally collected, and their processing again required much time and effort. Finally, the results of the separation were recorded in the form of a chromatogram: a bar graph of sample concentration versus fraction number.

The advent of paper chromatography in the 1940s and thin-layer chromatography (TLC) in the 1950s greatly simplified the practice of analytical liquid chromatography. This is also illustrated in Figure 1.1. Bed preparation in TLC or paper chromatography (step 1) was much simpler and cheaper than in classical LC. The paper or adsorbent-covered plates could be purchased in ready-to-use form, at nominal expense. Sample application was achieved rather easily, and solvent flow was accomplished by placing the spotted paper or plate into a closed vessel with a small amount of solvent. Solvent flow up the paper or plate proceeded by capillary action, without the need for operator intervention. Finally, detection and quantitation could be achieved by spraying the dried paper or plate with some chromogenic reactant to provide a visible spot for each separated sample component.

The techniques of paper and thin-layer chromatography greatly simplified liquid chromatography and made it much more convenient. A further advantage, particularly for TLC, was that the resulting separations were much better than in classical LC, and required much less time--typically 30-60 min versus several hours. However, certain limitations were still apparent in these open-bed methods:

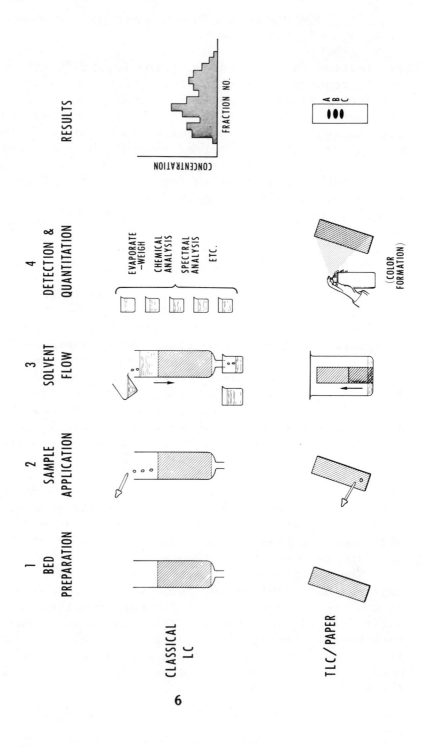

6

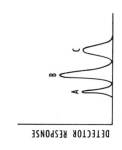

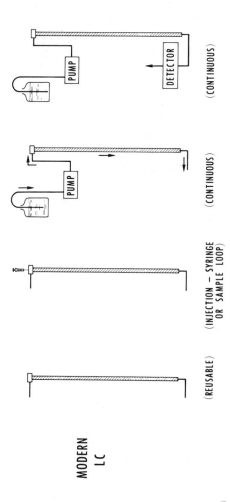

Figure 1.1 Different forms of liquid chromatography.

7

- Difficult quantitation and poor reproducibility.
- Difficult automation.
- Longer separation times and poorer separation than in GC.
- Limited capacity for preparative separation (i.e., maximum sample sizes of a few milligrams).

Despite these limitations, TLC and paper chromatography became the techniques of choice for carrying out most LC separations.

Let us look now at modern LC. Closed, reusable columns are employed (step 1, Figure 1.1), so that hundreds of individual separations can be carried out on a given column. Since the cost of an individual column can now be prorated over a large number of samples, it is possible to use more expensive column packings for high performance, and to spend more time on the careful packing of a column for best results. Precise sample injection (step 2) is achieved easily and rapidly in modern LC, using either syringe injection or a sample valve. Solvent flow (step 3) is achieved by means of high-pressure pumps. This has a decided advantage: controlled, rapid flow of solvent through relatively impermeable columns. Controlled flow results in more reproducible operation, which means greater accuracy in LC analysis. High-pressure operation leads to better, faster separation--as will be seen in Chapter 3. Detection and quantitation in modern LC are achieved with continuous detectors of various types. These yield a continuous chromatogram without intervention by the operator. The result is an accurate record of the separation, with minimum effort.

Repetitive separation by modern LC can be reduced to a simple sample injection and final data reduction, although the column or solvent usually must be changed for each new application. From this it is obvious that modern LC is considerably more convenient and requires

less work than either classical LC or TLC. The greater reproducibility and continuous, quantitative detection in modern LC also lead to higher accuracy and
precision in both qualitative and quantitative analysis.
As discussed in Chapter 13, quantitative analysis by
modern LC can achieve accuracies of better than ±1%.
Finally, preparative separations of gram quantities of
sample are proving relatively simple by modern LC.

Modern LC also provides a major advance over the
older LC methods, with respect to speed and separation
power. In fact, it now rivals GC in this respect.
Figure 1.2 shows an example of the speed of modern LC;
the separation of seven hydroxylated benzene derivatives in about 1 min, using a small-particle silica
column. Figure 1.3 shows the separation of an 18-
component sample in 20 min, using gradient elution
(Section 3.4) from another silica column. Modern LC
also features a number of new column packings which
provide separations that were previously impossible.
Figure 1.4 shows the partial separation of a synthetic
polymer sample by molecular weight. The resulting
chromatogram can then be translated into a molecular
weight distribution for the polymer sample. Most important, all of these advantages of modern LC are now
routinely obtainable with commercial equipment and
column packings. In most cases, even the packed columns can be purchased.

1.3   HOW DID MODERN LC ARISE?

Modern LC is based on developments in several
areas: equipment, special columns and column packings,
and theory. High-pressure, low dead-volume equipment
with sensitive detectors plays a vital role. The new
column packings that have been developed specifically
for modern LC are also essential for rapid, high-
performance separations. Theory has played a much

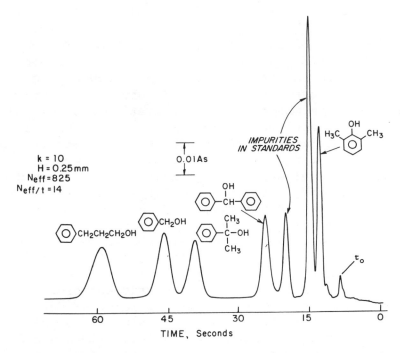

Figure 1.2   High-speed separation of hydroxylated aromatics (13).   Reprinted by permission of editor.
Column, 250 x 3.2 mm i.d., 8-9 μ porous silica microspheres (~ 75 Å); carrier, dichloromethane (50% H_2O-
saturated); pressure, 2000 psi; flow, 10.5 ml/min; temperature, 27°C; sample, 4 μl total in hexane solution.

more important role in the development of modern LC
than for preceding chromatographic innovations.   Fortunately, the theory of LC is not very different from
that of GC, and a good understanding of the fundamentals of GC had evolved by the early 1960s (see e.g.,
ref. 1.).   This GC theory was readily extended to include LC, and this in turn led directly to the development of the first high-performance column packings
for LC, and the design of the first modern LC units.
A proper understanding of how the different separation
variables are selected for optimum performance is

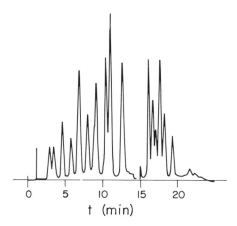

t (min)

Figure 1.3   Separation of an 18-component mixture by
gradient elution from silica (14).   Reprinted with
permission.

particularly important in modern LC; theory has been
most useful in providing the necessary insights.
     The potential advantages of modern LC first came
to the attention of a wide audience in early 1969,
when a special session on liquid chromatography was
organized as part of the Fifth International Symposium
on Advances in Chromatography (2).   However, modern LC
had its beginnings in the late 1950s, with the introduction of automated amino acid analysis by Spackman,
Stein, and Moore (3).   This was followed by the pio-
neering work of Hamilton (4) and Giddings (1) on the
fundamental theory of high-performance LC in columns,
the work of C. D. Scott at Oak Ridge on high-pressure
ion exchange chromatography, and the introduction of
gel permeation chromatography by J. C. Moore (5) and
Waters Associates in the mid-1960s.   At this point a
number of workers began active research into what was
to become modern LC, and their combined efforts (see
ref. 2 for a review) led to the 1969 breakthrough.

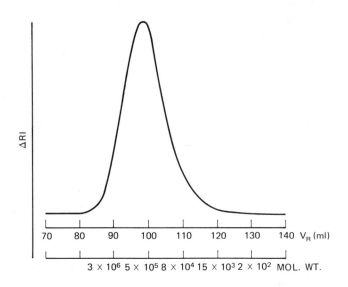

Figure 1.4   Separation of polyvinyl chloride by gel
permeation chromatography.
Reprinted by permission of Waters Associates.

Since 1969 there has been a tremendous activity aimed
at the development of better equipment and columns and
further improvements in our understanding of modern LC.
The technique and theory of modern LC have now matured
to the point where a reasonably complete treatment is
possible, as is attempted in the present book.

1.4   THE LITERATURE OF LC

The useful literature of modern LC extends back into
the 1950s, since much of the work in classical LC is
equally applicable to modern LC. A comprehensive review of all books on chromatography through 1969 is
given (6), and reference to this list will prove useful. Since then several books on modern LC have
appeared (7-11).

The primary literature of modern LC has so far appeared mainly in the various chromatographic and analytical journals (e.g., <u>Analytical Chemistry</u>, <u>Journal of Chromatographic Science</u>, etc.), but work is now being described in the general scientific literature. Three abstract services which review the recent LC literature are now available: a combined GC/LC summary from the Institute of Petroleum* ($10/yr), and LC Abstracts from the Preston Technical Abstracts Co.** ($180/yr), or the Science and Technology Agency*** $78/yr). Reviews of the current LC literature can also be found in the biennial reviews of <u>Analytical Chemistry</u> (see, e.g., ref. 12), as well as the bibliography section of each volume of the <u>Journal of Chromatography</u>.

The commercial literature offered by companies supplying LC equipment and materials is an important primary source of information, particularly for applications of modern LC. Table 1.1 summarizes some of these companies and specific items of interest. This literature is free and often uniquely useful.

TABLE 1.1   SOME COMMERCIAL LC LITERATURE

Applied Science	"Gas-Chrom Newsletter"
Bio-Rad	Price List W, "Gel Chromatography"
Chromatronix	LC Catalog, "LC Applications," "Lab Notes"
Du Pont	"Chromatographic Methods" (LC application sheets)

*   Applied Science Publishers, Ltd., Ripple Road, Barking, Essex, England.
** Post Office Box 312, Niles, Illinois (60648)
***3 Harrington Rd., S. Kensington, London, SW73ES

Table 1.1 - Cont'd.

Pharmacia	"Sephadex, Gel Filtration in Theory and Practice";"Sephadex Ion Exchangers. A Guide to Ion Exchange Chromatography"
Varian	"1971 Instruments & Accessories" "Nucleic Acid Constituents by LC"
Waters Associates	"Chromatography" (cat.), "Chromatography Notes"

Addresses

Applied Science Labs, Inc.
State College, Pa. 16801

Bio-Rad Laboratories
32nd and Griffin Avenue
Richmond, Calif. 94804
(415) 234-4130

Chromatronix, Inc.
2743 Ninth Street
Berkeley, Calif. 94710
(415) 841-7221

E. I. du Pont de Nemours
& Co., Instrument Products
Division
1007 Market Street
Wilmington, Del. 19898
(302) 774-2421

Pharmacia Fine Chemicals,
Inc., 800 Centennial Ave.
Piscataway, N.J. 08854
(201) 469-1222

Varian Aerograph
2700 Mitchell Drive
Walnut Creek, Calif. 94598
(415) 939-2400

Waters Associates, Inc.
61 Fountain Street
Framingham, Mass. 01701
(617) 879-2000

REFERENCES

1.   J. C. Giddings, Dynamics of Chromatography, Marcel Dekker, New York, 1965.
2.   A. Zlatkis, ed., Advances in Chromatography, 1969, Preston Tech. Abst. Co., 1969.
3.   D. H. Spackman, W. N. Stein, and S. Moore, Anal. Chem., 30, 1190 (1958).
4.   P. B. Hamilton, Advan. Chromatog. 2, 3 (1966).
5.   J. C. Moore, J. Polymer Sci., A2, 835 (1964).
6.   J. Chromatog. Sci., 8 (July), D2 (1970).
7.   N. Hadden, et al, Basic Liquid Chromatography, Varian Aerograph, Walnut Creek, Calif., 1972.
8.   J. J. Kirkland, ed., Modern Practice of Liquid Chromatography, Wiley-Interscience, New York, 1971.
9.   J. Q. Walker, M. T. Jackson, Jr., and J. B. Maynard, Chromatographic Systems. Maintenance and Troubleshooting, Academic Press, New York, 1972.
10. S. G. Perry, R. Amos, and P. I. Brewer, Practical Liquid Chromatography, Plenum Press, New York, 1972.
11. P. R. Brown, High Pressure Liquid Chromatography: Biochemical and Biomedical Applications, Academic Press, New York, 1973.
12. G. Zweig and J. Sherma, Anal. Chem., 44 (5), 47-49 (1972).
13. J. J. Kirkland, J. Chromatog. Sci., 10, 593 (1972).
14. L. R. Snyder and D. L. Saunders, J. Chromatog. Sci., 7, 195 (1969).

CHAPTER TWO

BASIC CONCEPTS

The successful use of LC for a given problem depends
on our ability to choose the right combination of operating conditions: the type of column packing and
solvent, the length, diameter, and operating pressure
of the column, separation temperature, sample size,
and so on. This in turn requires a basic understanding of the various factors that control LC separation.
In this chapter we will review the essential features
of liquid chromatography. Then in Chapter 3 we will
apply this theory to the control of LC separation. In
most cases this discussion can be followed qualitatively. The few important mathematical relationships are
indicated by enclosure within a box (e.g., Eq. 2.3).

2.1 THE CHROMATOGRAPHIC PROCESS

Figure 2.1 shows the hypothetical separation of a three-
component mixture in an LC column. Individual molecules are shown as triangles for compound A, dark
squares for compound B, and circles for compound C.
Four successive stages of the separation are shown,
beginning in (a) with application of the sample to a
dry column. In modern LC the column will always be
pre-wet prior to sample injection; however, a dry column is shown here to better illustrate the separation

17

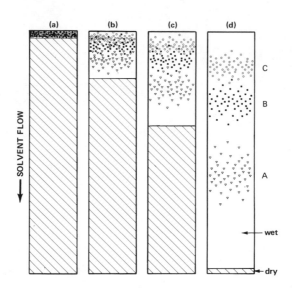

Figure 2.1   Hypothetical separation of a 3-component
sample; △ compound A, ■ compound B, ○ compound C.

process (separation proceeds in the same manner for
both wet and dry columns). In step (b) solvent or
mobile phase begins to flow through the column, resulting in movement of sample molecules along the column,
and a partial separation of components A, B, and C.
The movement of solvent through the column has proceeded further in step (c), and by step (d) the three
compounds are essentially separated from each other.
        In step (d) of Figure 2.1 we can recognize two
characteristic features of chromatographic separation:
differential migration of various compounds in the
original sample, and a spreading along the column of
molecules of a given compound. Differential migration
in LC refers to the varying rates of movement of different compounds through a column. In Figure 2.1, compound A moves most rapidly and leaves the column first;
compound C moves the slowest and leaves the column
last. As a result compounds A and C gradually become

separated as they move through the column. Differential migration is the basis of separation in LC; without a difference in migration rates for two compounds,
no separation is possible.

Differential migration in LC results from the
equilibrium distribution of different compounds--such
as A, B, and C--between particles or <u>stationary phase</u>,
and the flowing solvent or <u>moving phase</u>. This is illustrated in Figure 2.2, for a single particle of column
packing and compounds A and C. Compound C at equilibrium is present mainly in the stationary phase or particle, with only a small fraction of its molecules in
the moving phase at any given time. The speed with
which each compound moves through the column is determined by the number of molecules of that compound in
the moving phase at any instant, since sample molecules
do not move through the column while they are in the
stationary phase. Therefore compounds such as C, whose
molecules spend most of the time in the stationary
phase, move through the column rather slowly. Compounds such as A, whose molecules are found in the moving phase most of the time, move through the column

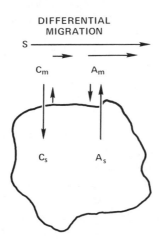

Figure 2.2   The basis of retention in LC.

more rapidly. Molecules of solvent or moving phase (S
in Fig. 2.2) usually move through the column at the
fastest possible rate (except in gel chromatography;
see Chapter 10).

Differential migration or the movement of individual compounds through the column depends upon the
equilibrium distribution of each compound between stationary and moving phases. Therefore, differential
migration is determined by those experimental variables
which affect this distribution: the composition of the
moving phase, the composition of the stationary phase,
and the separation temperature. When we want to change
differential migration to improve separation, we must
change one of these three variables. In principle,
the pressure within the column should also affect equilibrium distribution and differential migration, but
in fact such effects are normally negligible at the
usual column pressures (i.e., less than 10,000 psi);
see ref. 1.

Consider next the second characteristic of chromatographic separation: the spreading of molecules
along the column for a given compound, such as C in
Figure 2.1. In Figure 2.1(a) molecules of C (triangles) begin as a narrow line at the top of the column. As these molecules move through the column the
initial narrow line gradually broadens, until in (d)
molecules of C are spread over a much wider portion of
the column. It is apparent, therefore, that the average migration rates of individual molecules of C are
not identical. These differences in molecular migration rate for molecules C do not arise from differences
in equilibrium distribution--as in Figure 2.2. Rather,
this spreading of molecules C along the column is
caused by physical or rate processes. The more important of these physical processes are illustrated in
Figure 2.3. In A of Figure 2.3 we show a cross section
of the top of the column (with individual particles
numbered from 1 to 10). Sample molecules are shown as
x's at the top of the column, that is, just after

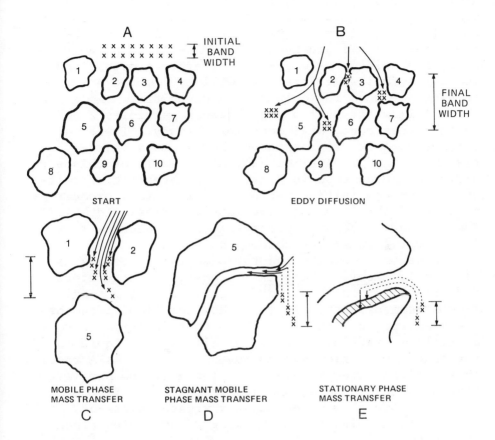

Figure 2.3   Contributions to molecular spreading in LC.

injection.   At this point (A in Figure 2.3) these mole-
molecules form a fairly narrow line, as measured by
the vertical, double-tipped arrow alongside the x's
(labeled "initial band width").

    In Figure 2.3B we illustrate one of the four pro-
cesses leading to molecular spreading: eddy diffusion.
Eddy diffusion arises from the different solvent flow
streams within the column.   As a result, sample molecules take different paths through the packed bed, depending upon what flow streams they follow.   These

different flow paths are illustrated in B by the various arrows. Liquid moves faster in wide flow paths,
and slower in narrow paths. Thus, the flow path between particles 1 and 2--and 5 and 6--is relatively
wide, the solvent velocity is therefore greater, and
molecules following this path will have moved a greater distance down the column in a given time. Molecules
which follow the narrow path between particles 2 and 3,
on the other hand, will move more slowly. These molecules therefore progress down the column a shorter
distance in a given time. So, as a result of this eddy
diffusion phenomenon in B, we see a spreading of molecules from the initial narrow line in A to a broader
portion of the column (arrow labeled "final width").
This spreading becomes progressively greater as flow
of solvent through the column continues.

   A second contribution to molecular spreading is
seen in C: mobile phase mass transfer. This refers to
differing flow rates for different parts of a single
flow stream or path. In C we show the flow stream between particles 1 and 2. Liquid that is adjacent to a
particle moves slowly or not at all, while liquid in
the center of a flow stream moves fastest. As a result,
in any given time, sample molecules near the particle
move a short distance, and sample molecules in the
middle of the flow stream move a greater distance.
Again this results in a spreading of molecules along
the column.

   Figure 2.3D shows the contribution of stagnant
mobile phase mass transfer to molecular spreading. In
the case of porous column-packing particles, the mobile
phase contained within the pores of the particle is
stagnant or unmoving. In D we show one such pore, for
particle 5. Sample molecules move in and out of these
pores by diffusion. Those molecules that happen to
diffuse a short distance into the pore, then diffuse
out, quickly return to the moving phase and move a certain distance down the column. Molecules that diffuse
further into the pore spend more time in the pore, and

less time in the external moving phase. As a result
these molecules move a shorter distance down the column. Again there is an increase in molecular spreading.
    Finally, in E is shown the effect of stationary
phase mass transfer. After molecules diffuse into a
pore, they then penetrate the stationary phase (cross-
hatched region), or become attached to it in some fashion. If a molecule penetrates deep into the stationary
phase it spends a longer time in the particle, and
travels a shorter distance down the column--just as in
D. Molecules which spend only a little time moving
into and out of the stationary phase return to the moving phase sooner, and move further down the column. In
Section 2.3 we will see how each of these various contributions to molecular spreading depends upon experimental conditions. Then in Chapter 6 we will discuss
how molecular spreading can be minimized for improved
separation.
    Eventually the various compounds reach the end of
the column and are carried off to the detector, where
their concentrations are recorded as a function of separation time. The resulting chromatogram for our hypothetical separation of Figure 2.1 is shown in Figure
2.4. Such a chromatogram can be characterized by four
features which are important in describing the resulting separation. First, each compound leaves the column in the form of a symmetrical, bell-shaped band or
Gaussian (standard error) curve. Second, each band
emerges from the column at a characteristic time that
can be used to identify that compound. This retention
time, t_R, is measured from the time of sample injection
to the time the band maximum leaves the column. A
third characteristic feature is the difference in retention times between adjacent bands; for example, t_R
for compound C minus t_R for compound B. The larger
this difference is, the easier is the separation of
the two bands. Finally, each band is characterized by
a bandwidth t_w, as shown for band B in Figure 2.4.
Tangents are drawn to each side of the band and

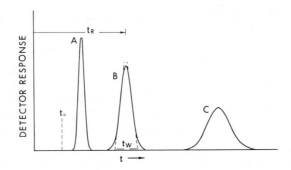

Figure 2.4   The resulting chromatogram.

extended to touch the <u>baseline</u> (detector line for zero sample concentration). The smaller is t_w and the narrower are these bands, the better is separation.

Four separate mechanisms or processes exist for retention of sample molecules by the stationary phase. These in turn give rise to the four basic LC methods: liquid-liquid, liquid-solid, ion exchange, and gel chromatography. Liquid-liquid or <u>partition</u> chromatography involves a liquid stationary phase whose composition is different from that of the moving liquid phase. Sample molecules distribute between the moving and stationary liquid phases, just as in liquid-liquid extraction within a separatory funnel. The moving and stationary phase liquids may be immiscible, or the stationary phase can be chemically bonded to the particle or <u>support</u>. Liquid-solid or <u>adsorption</u> chromatography involves high-surface-area particles, with retention of sample molecules occurring by attraction to the surface of the particle. In ion-exchange chromatography the column packing contains fixed ionic groups such as $-SO_3^-$ along with counter ions of opposite charge (e.g., Na^+). The counter ions are also present in the moving phase in the form of a salt (e.g., $NaCl$). Ionic sample molecules of the same charge as the counter ion (e.g., X^+) are retained by ion exchange:

$$X^+ + -SO_3^-Na^+ \rightleftharpoons Na^+ + -SO_3^-X^+.$$

Finally, in gel or <u>exclusion</u> chromatography, the column packing is a porous material with pores of different sizes. Large molecules are excluded from all the pores, because they are too large to enter, while small molecules penetrate most of the pores. Thus, large molecules move through the column quickly, and small molecules are retained by the packing. Usually separation in gel chromatography is determined strictly by molecular size.

Chapters 7-10 describe the characteristic features and use of each of these four LC methods. Each method provides a unique basis for sample retention and differential migration, and therefore each method is particularly useful for certain kinds of samples or separation problems. Chapter 11 describes how we select one or more of the four LC methods for a given problem.

2.2   RETENTION IN LC

Up to this point our discussion of LC has been qualitative, rather than quantitative. Now it is time to reduce some of the concepts we have been describing to mathematical relationships. We will begin with band migration rate and differential migration. Let the solvent velocity in the column be u (cm/sec), and let the average velocity of a sample band X be u_X. From our discussion of Figure 2.2, it should be clear that u_X depends upon the fraction R of molecules X in the moving phase, and upon u. Specifically,

$$u_X = uR \qquad (2.1)$$

If the fraction of molecules X in the moving phase is zero, no migration can occur, and u_X is zero. If the fraction of molecules X in the moving phase is one (i.e., all molecules X are in the moving phase), then molecules X move through the column at the same rate as solvent molecules, and $u_X = u$.

Let us next derive an expression for R, the fraction of molecules X in the moving phase. We begin by defining a fundamental LC parameter: the capacity fac-tor k', which is equal to n_S/n_m, where n_S is the total moles of X in the stationary phase, and n_m is the total moles of X in the moving phase.   We can now write

$$k' + 1 = \frac{n_S}{n_m} + \frac{n_m}{n_m}$$

$$= \frac{n_S + n_m}{n_m} ,$$

and

$$R = \frac{n_m}{n_S + n_m} = \frac{1}{1 + k'} \qquad (2.2)$$

Therefore from Eq. 2.1

$$u_X = \frac{u}{1 + k'} . \qquad (2.2a)$$

The quantity u_X can be related to retention time t_R and column length L in terms of the relationship: time equals distance divided by velocity.   In this case the time is that required for band X to traverse the column, t_R (sec), the distance is the column length L (cm), and the velocity is that of band X, u_X:

$$t_R = L/u_X$$

Similarly, the time t_o for solvent molecules (or other nonretained compounds) to move from one end of the column to the other is,

$$t_o = L/u.$$

Eliminating L between these last two equations then gives

$$t_R = \frac{u \, t_o}{u_X}$$

which with Eq. 2.2a gives

$$\boxed{t_R = t_o (1 + k')} \qquad (2.3)$$

Here we have t_R expressed as a function of the funda-mental column parameters t_o and k'; t_R can vary between

t_O (for k' = 0) and any larger value (for k' large).
Since t_O varies inversely with solvent velocity u, so
does t_R.   For a given column, solvent, separation temperature, and sample component X, k' is normally constant for sufficiently small samples (see Section 2.6).
Thus t_R is defined for a given compound X by the chromatographic system, and t_R can be used to tentatively
identify a compound in that LC system (by comparison
with a t_R value for a known compound in the same system).

Rearrangement of Eq. 2.3 gives an expression for
k':

$$k' = \frac{t_R - t_O}{t_O} \qquad (2.3a)$$

To plan a strategy for improving separation, it will
often be necessary to determine k' for one or more
bands in a chromatogram.   Equation 2.3a provides a
simple, rapid basis for estimating k' values in these
cases; k' is seen to be equal to the distance between
t_O and the band center, divided by the distance between
injection and t_O.   For example, for band B in Figure
2.4, k' = 2.2.   In the same way values for bands A and
C in this chromatogram can also be estimated (see
below* for results).

The important column parameter t_O can be determined in several different ways.   In most cases the
center of the first band or baseline disturbance (following sample injection; e.g., Fig. 1.2) marks t_O.   If
there is any doubt on the position of t_O, a weaker
solvent (or other nonretained compound) can be injected
as sample, and its t_R value will equal t_O.   A weaker
solvent is one that gives larger k' values and stronger
sample retention than the solvent used as mobile phase
(see Chapters 7-9 for lists of solvents according to
strength).   For example, if chloroform is used as solvent in liquid-solid chromatography, hexane might be

---

*Band A, k' = 0.7; band C, k' = 5.7.

injected as sample to measure t_o.

Retention in LC is sometimes measured in volume units, rather than in time units. Thus the retention volume V_R is the total volume of mobile phase required to elute the center of a given band X; that is, the total solvent flow between sample injection and the appearance of the band center at the detector. The retention volume V_R (ml) is seen to be equal to the retention time t_R (sec) times the volumetric flow rate F (ml/sec) of solvent through the column:

$$V_R = t_R F .$$

Similarly, the total volume of solvent within the column V_m is equal to F times t_o:

$$V_m = t_o F .$$

Eliminating F between the above two equations gives

$$V_R = V_m \frac{t_R}{t_o}$$
$$= V_m (1 + k') \qquad (2.4)$$

Values of V_R are sometimes preferred to values of t_R. The reason is that t_R varies with flow rate F, while V_R is independent of F. V_R values are thus more suitable for permanent column calibrations (as in gel chromatography; see Chapter 10), particularly when flow rates can vary or are not precisely known. This might arise with constant-pressure pumps (see Chapter 4) and columns of varying permeability.

So far we have ignored the effect on retention volume of the volume V_s of stationary phase within the column. The quantity k' is given as n_s/n_m, where $n_s = (X)_s V_s$ and $n_m = (X)_m V_m$. Here $(X)_s$ and $(X)_m$ refer to the concentrations (moles/ml) of X in the stationary and mobile phases, respectively. Therefore

$$k' = \frac{(X)_s V_s}{(X)_m V_m}$$

$$= K \frac{V_s}{V_m} . \tag{2.5}$$

Here $K = (X)_s/(X)_m$ is the well-known <u>distribution constant</u>, which measures the equilibrium distribution of X between the stationary and moving phases. It is seen in Eq. 2.5 that k' is proportional to V_s. Thus the k' values of all sample bands change with the relative loading of support by stationary phase in liquid-liquid chromatography. It follows then that <u>pellicular</u> or surface-coated glass beads (see Fig. 6.1a) have smaller values of V_s than porous particles and therefore provide smaller values of k', other factors being equal.

2.3   BAND BROADENING

Band width t_w in LC is commonly expressed in terms of the <u>theoretical plate number</u> N of the column:

$$N = 16 \left(\frac{t_R}{t_w}\right)^2 \tag{2.6}$$

The quantity N is approximately constant for different bands in a chromatogram, for a given set of operating conditions (a particular column and solvent, with solvent velocity and temperature fixed). Therefore N is a useful measure of <u>column efficiency</u>: the relative ability of a given column to provide narrow bands (small values of t_w) and improved separations. In some cases (see discussion of 2, 3) N varies somewhat with k', often being larger for $k' \approx 0$; this is not particularly important here. In gel chromatography (see Chapter 10) t_w is roughly constant as k' changes, which means that N increases with k'.

With N constant, Eq. 2.6 predicts that band widths widen proportionately with t_R, and this is generally observed; for example, see Figure 2.4. As a result of the increased width of later eluting bands, band height tends to decrease, and at sufficiently large values of k', all bands become indistinguishable from the baseline.

The quantity N is proportional to column length L, or

$$N = \frac{L}{H}$$

(2.6a)

The proportionality constant H is the so-called height equivalent of a theoretical plate, or HETP value; it measures the efficiency of the column per unit length. Small H values mean more efficient columns or larger N values. A central goal in LC practice is the attainment of small H values and maximum N values.

Values of H for a given column are the result of the various band broadening processes outlined in Figure 2.2, plus (very occasionally) one additional effect: longitudinal molecular diffusion. The latter refers to the tendency of individual molecules to diffuse randomly along the column away from the band center. The contributions H_i of each of these processes to H have been derived (2) as functions of (a) the diameter d_p of column-packing particles, (b) solvent velocity u, and (c) the molecular diffusion coefficients D_m and D_s of sample in the moving and stationary phases, respectively:

Process	H_i
Eddy diffusion	$C_e d_p$
Mobile phase mass transfer	$\dfrac{C_m d_p^2 u}{D_m}$
Longitudinal diffusion	$\dfrac{C_d D_m}{u}$

Stagnant mobile phase mass transfer	$\dfrac{C_{sm}d_p{}^2u}{D_m}$
Stationary phase mass transfer	$\dfrac{C_s d_f{}^2u}{D_s}$

The quantity d_f refers to the thickness of the stationary phase film. When the stationary phase is the same as the support or particle (as in porous ion exchange resins), d_f is replaced by d_p. The values of the coefficients C_e, C_m, C_{sm}, C_s, and C_d are given in ref. 2. These constants remain unchanged for particles of given type; e.g., porous or superficially porous particles, as in Figure 6.1.

The combined contribution of these various terms to H (see ref. 2) is

$$H = \underbrace{\frac{1}{(1/C_e d_p) + (1/C_m d_p{}^2 u/D_m)}}_{(i)} + \underbrace{\frac{C_d D_m}{u}}_{(ii)}$$

$$+ \underbrace{\frac{C_{sm}d_p{}^2 u}{D_m}}_{(iii)} + \underbrace{\frac{C_s d_f{}^2 u}{D_s}}_{(iv)} . \qquad (2.7)$$

Term (iv) is generally small in LC, except for porous ion-exchange resins or heavily loaded liquid-liquid supports. We can therefore omit term (iv). If we also define the reduced plate height h and reduced velocity ν ,

$$h = \frac{H}{d_p} \quad \text{and} \quad \nu = \frac{u d_p}{D_m}$$

Eq. 2.7 then becomes

$$h = \frac{1}{(1/C_e) + (1/C_m \nu)} + \frac{C_d}{\nu} + C_{sm}\nu . \qquad (2.7a)$$

The significance of Eqs. 2.7 and 2.7a is examined more

fully in Appendix I.   We can summarize these conclusions as follows:

- H is smaller for small values of d_p and u.

- H is smaller for less viscous solvents.

- H is smaller for higher separation temperatures.

- H is smaller for small sample molecules X.

Therefore large N values and <u>improved separations</u> are favored by <u>small column-packing particles, slow movement of solvent</u> through the column, <u>solvents that are as nonviscous as possible</u> (e.g., viscosities less than 0.5 cP, at the temperature of separation), <u>higher separation temperatures</u>, and (Eq. 2.6a) <u>long columns.</u> Furthermore the separation of macromolecular species such as proteins will generally be more difficult, compared to compounds with molecular weights in the 200-2000 range.

Over a practical range in operating conditions, terms (ii) and (iv) of Eq. 2.7 are normally negligible, which leads to a characteristic dependence of H on u: this is illustrated by the representative data of Figure 2.5.   The H versus u plot is almost invariably convex to the u axis, and this has some practical consequences that we will explore in Chapter 3.   In most cases such plots are well approximated by the empirical function

$$\boxed{H = Du^n}\qquad(2.8)$$

as can be seen in Figure I-i (Appendix I) for practical values of u ($\nu > 10$).   Here D and n are constants for a given column and solvent.   Normally $0.3 \geq n \geq 0.6$, but values of n as low as 0.2 and as high as 0.8 have been reported.   The quantity D is a good measure of overall efficiency (since D = H, when u = 1), and n is an

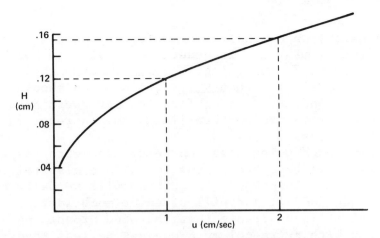

Figure 2.5   Typical plot of plate height H versus solvent velocity u (8); 20 μ silica column.

important parameter for optimizing the use of a given
column (see Section 3.5).   A plot of log H versus log
u gives n directly, as the slope of the resulting plot.
It is generally useful to determine a plot of H versus
u for every column that will be used.

   No discussion of band broadening in LC is complete
without considering the additional broadening that can
occur <u>outside</u> the column.   This so-called <u>extra-column</u>
band broadening results from the appreciable volumes
of the injected sample V_x and of the various elements
of an LC unit: injection system, connecting lines,
detector cell, column end-fittings, and so on.   Thus a
sample band spreads to a volume V_w ($V_w = t_w F$) as a result of separation within the column, but a final band
width V_w' is recorded by the detector.   The quantity
V_w' is related to V_w, V_x, and the extra-column band
broadening V_i, V_j, and so on that occurs in elements
i, j, and so on of the LC unit:

$$V_w'^2 = V_w^2 + V_x^2 + V_i^2 + V_j^2 + \cdots . \qquad (2.9)$$

The quantity V_i is the width of the resulting band

(volume units) when a narrow band ($V_w \approx 0$) passes
through the element i.   In general we desire to keep
extra-column band broadening to a minimum, so that
$V_w \approx V_w'$.   This means that V_x, V_i, and so on should
be less than one-third of V_w (for the narrowest band in
the chromatogram), which then limits the increase in
V_w to 10% or less.

A few rough guidelines can be offered with respect
to extra-column band broadening in LC (see also refs.
2, 4-6).   The volume V_w of an LC band will normally be
greater than 100 μl, and will often be much larger.
In general it is desirable that the band broadening
associated with extra-column processes be less than
100 μl, and ideally less than 30 μl.   This also means
that the volume of sample V_x charged to a column should
be less than one-third of V_w for the first band in the
chromatogram, which means that V_x can still be as large
as 30-100 μl.   If the sample is dissolved in a solvent
that is much weaker than the moving phase used in the
separation, even larger samples can be charged.   In
this case the sample concentrates near the inlet of
the column, because the k' values of all sample components (at the inlet) have been temporarily increased
by the weak solvent.

The band broadening V_i that occurs in a length L
of connecting tubing of internal diameter d is given
(4) as

$$V_i^2 = \frac{\pi d^4 \; FL}{24 \; D_m} \qquad .$$ (2.9a)

All connecting lines should therefore be short, with
internal diameters as small as possible.   Tubing with
an i.d. of 0.010-0.012 in. is often used for such
lines, since larger diameter tubing gives much larger
V_i values, while smaller diameter tubing tends to plug
frequently.   For typical conditions ($D_m \approx 10^{-5}$, F =
1 ml/min. and 0.01 in. tubing),Eq. 2.9a suggests that
up to 30 cm of tubing can be used without degrading
column performance.

The detector cell volume should not be larger than about 10 μl, for which $V_i \approx 100$ μl. Ideally the detector cell should have an even smaller volume, for example, for highly efficient columns packed with small (5-10 μ) particles.

When larger and/or less efficient columns are used (e.g., as is commonly the case in gel chromatography), the above requirements ($V_i < 100$ μl) can be relaxed considerably; relatively large samples (1-2 ml) can be charged, longer connecting lines can be used, and larger detector cell volumes are possible.

2.4 RESOLUTION

The usual goal in LC is the adequate separation of a given sample mixture. In approaching this goal we must have some quantitative measure of relative sep-aration or resolution. The resolution R_S of two adja-cent bands 1 and 2 is defined equal to the distance between the two band centers, divided by average band width (see Fig. 2.6):

$$R_S = \frac{t_2 - t_1}{(1/2)(t_{w1} + t_{w2})} \qquad (2.10)$$

The quantities t_1 and t_2 refer to the t_R values of bands 1 and 2, and t_{w1} and t_{w2} are their t_w values. When $R_S = 1$, as in Figure 2.6, the two bands are rea-sonably well separated; that is, there is only 2% overlap of one band on the other. Larger values of R_S mean better separation, and smaller values of R_S show poorer separation, as illustrated in Figure 2.7. For a given value of R_S, band overlap becomes more severe when one of the two bands is much smaller than the other (Figure 2.7). The significance and use of the R_S function are discussed in detail in Sections 3.1-3.2.

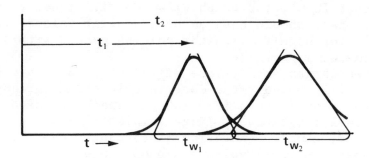

Figure 2.6   Resolution in LC.

The parameter R_s of Eq. 2.10 serves to <u>define</u> separation. To <u>control</u> resolution, we must know how R_s varies with experimental parameters such as k' and N. We will next derive such a relationship for two closely spaced bands. Equation 2.3 gives $t_1 = t_0(1 + k_1)$ and $t_2 = t_0(1 + k_2)$, where k_1 and k_2 are the k' values of bands 1 and 2. Since $t_1 \approx t_2$, from Eq. 2.6 we have $t_{w1} \approx t_{w2}$ (N is assumed constant for bands 1 and 2). Inserting these relationships into Eq. 2.10 then gives

$$R_s = \frac{t_0 \ (k_2 - k_1)}{t_{w1}} \qquad (2.10a)$$

Similarly, Eq. 2.6 for band 1 gives $t_{w1} = 4 \ t_1/\sqrt{N} = 4 \ t_0(1 + k_1)/\sqrt{N}$. Inserting this expression for t_{w1} into Eq. 2.10a gives

$$R_s = \frac{(k_2 - k_1) \ \sqrt{N}}{4 \ (1 + k_1)}$$
$$= (\frac{1}{4})\left[(\frac{k_2}{k_1}) - 1\right] \sqrt{N} \left[\frac{k_1}{1 + k_1}\right] \qquad (2.10b)$$

If we define the <u>separation factor</u> $\alpha = k_2/k_1$ for bands 1 and 2, and recognize that $k_1 \approx k_2 \approx$ <u>the average value k'</u> for the two bands, Eq. 2.10b becomes

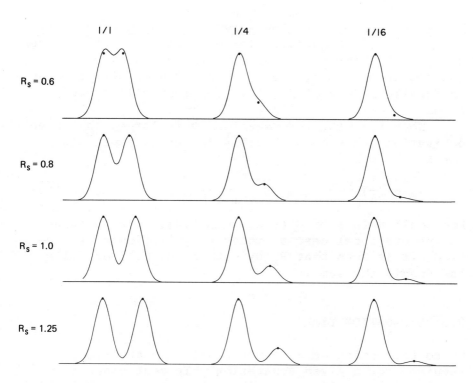

Figure 2.7   Separation as a function of R_S and relative band concentration.

$$R_S = (\frac{1}{4})(\alpha - 1) \sqrt{N} \left[\frac{k'}{(1 + k')}\right] \qquad (2.11)$$
$$\text{(i)} \qquad \text{(ii)} \quad \text{(iii)}$$

Equation 2.11 is a fundamental relationship in LC which
allows us to control resolution by varying α, N, or k'.
The three terms (i)-(iii) of Eq. 2.11 are essentially
independent, so that we can optimize first one term,
then another.   Separation selectivity as measured by
α term $[(i)]$ is varied by changing the composition of
the moving and/or stationary phases. Separation efficiency as measured by N term $[(ii)]$ is varied by changing column length L or solvent velocity u.   Term (iii),

k', is varied by changing <u>solvent strength</u>, the ability
of the solvent to provide large or small k' values (see
Chapters 7-9 for a listing of different solvents according to strength).   In Sections 3.4-3.6 we will examine
in detail the use of Eq. 2.11 for controlling resolution.

When $t_1 << t_2$, and therefore α is large, the above
derivation leads to a slightly different expression
for R_s:

$$R_s = (\frac{1}{4})\left[\frac{\alpha - 1}{\alpha}\right] \sqrt{N} \left[\frac{k'}{1 + k'}\right] . \qquad (2.11a)$$

For small values of α (e.g., $\alpha \leq 1.2$), corresponding
to the practical case of moderately difficult separation, it is seen that R_s by either Eq. 2.11 or 2.11a
is roughly the same.

2.5   SEPARATION TIME

Up to this point we have not considered the time, t,
required for a given separation.   In most cases t
should be as small as possible, both for convenience
and to allow a maximum number of separations or analyses per day.   In Chapter 3 we will see that separation
time and resolution are interrelated, so that difficult
separations require longer times.   But for the moment
let us concentrate on t, apart from sample resolution.
The quantity t is approximately equal to the retention
time t_R for the last band off the column.   For a two-
component separation as in Figure 2.6, therefore,

$$t = t_2 \doteq t_o(1+ k') . \qquad (2.12)$$

As discussed in Appendix II, t_o can in turn be related
to solvent viscosity η (Poise), total column porosity
f (ranging from f = 0.4 for pellicular packings to f =
0.8 for typical porous packings), column length L (cm)
and the pressure drop P across the column (dynes/cm^2):

$$t_0 = \frac{1000 \, \eta \, fL^2}{Pd_p^2} . \qquad (2.13)$$

Equation 2.13 is in units of the cgs system. For porous packings ($f = 0.8$) and more commonly used units (t_0 in sec, η in cP, L in cm, P in psi, d_p in μ, Eq. 2.13 becomes

$$t_0 = \frac{1.2 \times 10^4 \, L^2 \eta}{Pd_p^2} . \qquad (2.13a)$$

Figure 2.8 plots values of t_0 from Eq. 2.13a versus L for different particle sizes d_p and fixed values of P (1000 psi) and η (0.3 cP). Equation 2.13a and Figure

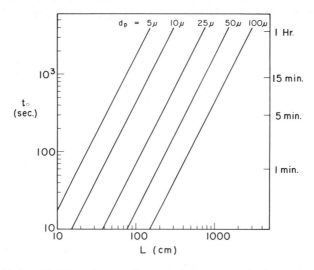

Figure 2.8   Variation of t_0 with particle size (d_p) and column length L (P = 1000 psi, = 0.3 cP). Values for porous irregular particles; t_0 values are one-half as large for porous, spherical particles, and one-fourth as large for pellicular particles.

2.8 apply reasonably well for well-packed columns of porous, irregular particles. Spherical particles give

higher permeabilities, so that t_O values are about
half as great as those of Figure 2.9. Pellicular particles (see Figure 6.1a), which are normally spherical,
give t_O values about one-fourth the values shown in
Figure 2.8 and Eq. 2.13a, because of the smaller value
of f. Columns of soft gels, on the other hand, can
give larger t_O values because of deformation of the gel
particles within the column (see Chapter 6).

Figure 2.8 and Eq. 2.13a are useful in checking
the permeability of newly prepared columns. If t_O is
found to be much larger than the value predicted by
Eq. 2.13a, this suggests a poor column or a blockage
in some part of the system. For example, small particles may have partially clogged the column inlet or
outlet, or the lines leading from the column. Figure

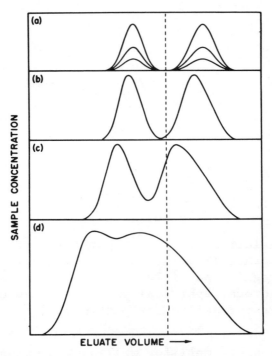

Figure 2.9   Sample size effects in LC.

2.8 is also useful for selecting initial separation
conditions (L, P, etc.) to give reasonable values of
the estimated separation time. Thus assume that t is
desired to be 10 min, and a maximum value of k' = 10
is chosen. This means t_0 = 600/11 = 55 sec (Eq. 2.12).
Further assume a solvent of viscosity 0.3 cP and particles of d_p = 25 μ. For P = 1000 psi, Figure 2.8
indicates that L should be about 100 cm.

2.6   SAMPLE SIZE EFFECTS

The effect of varying sample size in LC is illustrated
in Figure 2.9. For sufficiently small samples (a),
band height increases with sample size, but retention
times are not affected and resolution remains the same
(so-called linear-isotherm sorption). At some critical sample size (b) a noticeable decrease in retention
time occurs for one or more sample bands, and for further increase in sample size (c and d) there is usually a rapid degradation of separation and a further
decrease in all retention time values.

The same effects are illustrated for a single
sample component in Figure 2.10, where corrected retention volumes R (≈ k') and H values are plotted
versus sample size (wt. sample/wt. stationary phase).
At a certain sample size, $\theta_{0.1}$, the so-called linear
capacity of the stationary phase or column, a 10% reduction in k' (and R) has occurred, relative to the
constant k' value observed for sufficiently small
samples. In the example of Figure 2.10, this adsorbent has a linear capacity of 0.5 mg sample/gram of
adsorbent. When the linear capacity of the column is
exceeded, H values rapidly increase and separation is
quickly degraded.

For analytical separations it is almost always
preferable to work within a sample size range where k'
values are constant; that is, sample sizes should be

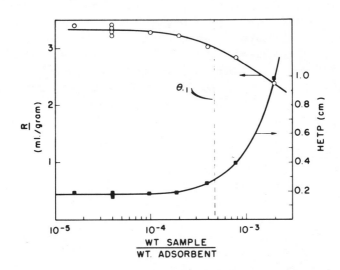

Figure 2.10   Linear capacity in LC (data for 4% water-silica) (9).

less than the linear capacity of the column. In the
case of preparative separations (see Chapter 12), the
linear capacity of the column can be exceeded somewhat,
but resolution inevitably becomes poorer for sufficiently large samples. Chapters 7-10 discuss the maximum
allowable sample sizes for the individual methods.
Because pellicular packings have a lower concentration
of stationary phase per particle, their linear capacities are normally lower by a factor of 5 or more,
relative to porous particles.

2.7   BAND TAILING

In well-designed LC systems, individual bands are normally Gaussian, as shown in Figure 2.4. However, band
<u>tailing</u> as illustrated in Figure 2.11 can occasionally

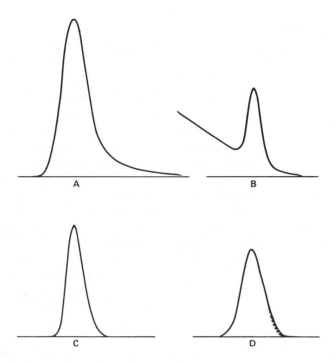

Figure 2.11   Band tailing in LC.   Reprinted with permission from Analytical Chemistry

arise.   Four different types of tailing can be dis-
tinguished.   Chemical tailing (A) occurs because of a
mismatch between the sample and the stationary or
(less commonly) moving phases.   It is noted by a slow
return of the band tail to baseline, poor separation
of adjacent bands, and frequent low recovery of the
sample.   When chemical tailing occurs it is necessary
to try a different stationary phase or solvent--or
even a different LC method (e.g., liquid-liquid, rather than liquid-solid chromatography).
     Solvent tailing (B) is the result of trying to
separate a small band in the presence of an earlier

band of much greater size (e.g., the solvent in which
the sample occurs, if this solvent is different from
the moving phase). Little can be done about solvent
tailing except to use smaller sample sizes or increase
resolution.

Poisson tailing (C) is not often observed in LC.
It is the result of inefficient columns--which we try
to avoid--and results in mildly asymmetric bands as
shown. Exponential or normal tailing (D) represents
a slight departure of the band from a symmetrical
curve, and is present in every chromatographic system.
However, it is barely noticeable in most LC separations, and has little practical significance for well-
designed systems. Often sample bands in LC exhibit
less tailing than in gas chromatography. Occasionally,
when the recovery of highly purified fractions is required, it becomes important to further reduce this
modest exponential tailing (see discussion of 10).
For a more detailed account of band tailing in LC, see
ref. 7.

REFERENCES

1. B. A. Bidlingmeyer and L. B. Rogers, Separation
    Sci., 7, 131 (1972).
2. J. C. Giddings, Dynamics of Chromatography, Marcel
    Dekker, New York, 1965.
3. L. R. Snyder, in Gas Chromatography. 1970, R.
    Stock and S. G. Perry, eds., Institute of Petro-
    leum, London, 1971, p. 81.
4. R. P. W. Scott, J. Chromatog. Sci., 9, 641 (1971).
5. Lab Notes #4, January 1971, Chromatronix, Inc.,
    Berkeley, Cal.
6. J. C. Sternberg Advan. Chromatog., 2, 205 (1966).
7. L. R. Snyder, J. Chromatog. Sci., 10, 200 (1972).
8. L. R. Snyder, J. Chromatog. Sci., 7, 352 (1969).
9. L. R. Snyder, Anal. Chem., 39, 698 (1967).

10. E. P. Horwitz and C. A. A. Bloomquist, J. Chromatog. Sci., in press.

GENERAL BIBLIOGRAPHY

J. C. Giddings, Dynamics of Chromatography, Marcel Dekker, New York, 1965, Chaps. 2, 5-7.
C. Horvath and S. R. Lipsky, J. Chromatog. Sci., 7, 109 (1969).
J. F. K. Huber, Chimia, Supplement, 1970 (5th International Symposium on Separation Methods: Column Chromatography, Lausanne, 1969), p. 24; J. Chromatog. Sci., 7, 85 (1969).
B. L. Karger, in Modern Practice of Liquid Chromatography, J. J. Kirkland, ed., Wiley-Interscience, New York, 1971.
J. H. Knox and co-workers, J. Chromatog. Sci., 7, 614, 745 (1969); 10, 549, 606 (1972).
L. R. Snyder in Gas Chromatography. 1970, R. Stock and S. G. Perry, eds., Institute of Petroleum, London, 1971, p. 81, and prior references.

CHAPTER THREE

CONTROL OF SEPARATION

Chapter 2 reviewed some of the fundamentals of LC separation. In this chapter we will reduce these fundamentals to a general strategy for attacking any separation problem. This strategy or approach to the design of a successful LC system can be broken down into the following six steps.

    1.   Select one of the four LC methods (Chapter 11).

    2.   Select or prepare a suitable column (Chapter 6).

    3.   Select initial experimental conditions (Chapters 7-10).

    4.   Carry out an initial separation.

    5.   Evaluate the initial chromatogram and determine what change (if any) in resolution is required.

    6.   Establish conditions required for the necessary final resolution.

At this point we will assume we are at step 5--that is, we have a given method and column type, and have made an initial separation. This chapter describes how to carry out steps 5 and 6, by taking advantage of the theory set down in Chapter 2. We will also derive some additional relationships as we proceed. However, our final strategy for achieving steps 5 and 6 is essentially a "table-lookup" operation, which requires only a few seconds and avoids all mathematical calculation.

3.1   ESTIMATING R_S

Our next step following the initial separation (step 4) is to evaluate the chromatogram in quantitative terms. This means we must estimate R_S for two or more adjacent bands. For the moment, assume that only two sample bands are involved. Shortly we will consider how to extend the R_S concept to include a complex chromatogram which contains several bands.

Our approach to estimating R_S is based on comparison with a standard set of resolution curves: Figures 3.1 through 3.6. In each case (e.g., Figure 3.1), examples are given of different values of R_S, beginning with $R_S = 0.4$ (almost complete overlap) through $R_S = 1.25$ (almost complete separation). In Figure 3.1 the two bands in each example are of equal height (1/1 ratio). In subsequent figures (3.2-3.6) the band-height ratio changes to 2/1, 4/1, 8/1, ..., 128/1. Figures 3.1-3.6 allow us to estimate R_S for an actual pair of bands from an initial separation, by visually comparing these bands with one of the examples in these figures. Some examples are given below to show how this works. Since the essence of this approach to estimating resolution is speed and convenience, it is helpful to have enlarged copies of Figures 3.1-3.6 (see, e.g., ref. 1) mounted on the wall beside each LC unit. Estimates of R_S can then be made on the spot, by glancing from the chromatogram on the recorder to the R_S curves on the wall.

Consider next the six hypothetical chromatograms in Figure 3.7. We will now estimate R_S for each of these examples. In (a) of Figure 3.7 the heights of

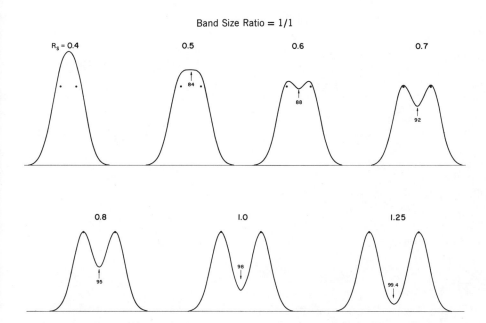

Band Size Ratio = 1/1

Figure 3.1   Standard resolution curves for a band-
size ratio of 1/1 and R_S values of 0.4-1.25 (1).
Reprinted by permission.

the two bands are in an approximate ratio of 2/1,
which directs us to Figure 3.2. There we see a good
match with the example for R_S = 0.7. Therefore, R_S
= 0.7 for (a) of Figure 3.7. Similarly in (b) of
Figure 3.7, the band height ratio is about 8/1, and
in Figure 3.4 there is a good match for R_S = 1.0.
Examples (c) and (d) of Figure 3.7 differ in that the
smaller band elutes first, rather than second. In
this case it is necessary to find a match with the
mirror image of the actual chromatogram, that is, mentally reverse the two bands. For (c) we see a band-
height ratio of about 1/4 (or 4/1), and in Figure 3.3
we have a good match with the mirror image of the

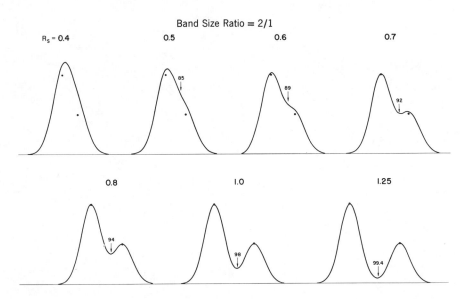

Figure 3.2   Standard resolution curves for a band-size ratio of 2/1 and R_S values of 0.4-1.25 (1). Reprinted by permission.

example for R_S = 0.7.   Similarly, (d) matches the curves for R_S = 0.5 in Figure 3.2 (2/1).   Example (e) of Figure 3.7 involves two bands of widely different heights; so much so that one of the two bands is off-scale.   Matches for this case can be found in Figure 3.6; either R_S = 1.25 for a relative band-height ratio of 128/1, or R_S = 1.0 for a peak-height ratio of 32/1. The accuracy of estimates of R_S is seen to be somewhat less when one of the two bands is off-scale.

Example (f) of Figure 3.7 shows a more complex chromatogram.   The resolution of such a chromatogram can be defined in one of two ways, depending on which bands are of interest.   Where every band in the chromatogram may be important, then the two adjacent bands

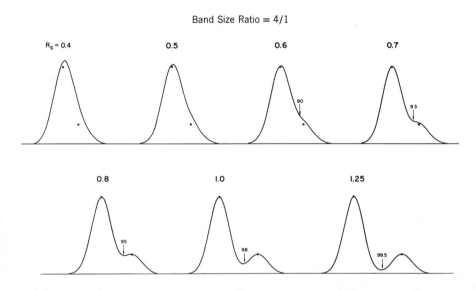

Figure 3.3   Standard resolution curves for a band-
size ratio of 4/1 and R_S values of 0.4-1.25 (1).
Reprinted by permission.

with the poorest resolution determine R_S for the entire separation.   In (f) either bands 1-2, or 3-4
have the lowest value of R_S.   Comparison of either of
these band pairs with Figures 3.1 or 3.2 suggests R_S
= 0.8, and this is the value of R_S for the chromatogram.

Another possibility is that not every band in a
chromatogram is of interest.   For example, assume
that we are concerned only with band No. 5 in (f).   In
this case R_S for the chromatogram is determined by the
pair of bands, 4-5 or 5-6, which is most poorly separated.   For band 5 in (f), resolution is poorest for
bands 4 and 5, and their value of R_S determines R_S for
the chromatogram: R_S = 1.0, by comparison with Figure
3.1.

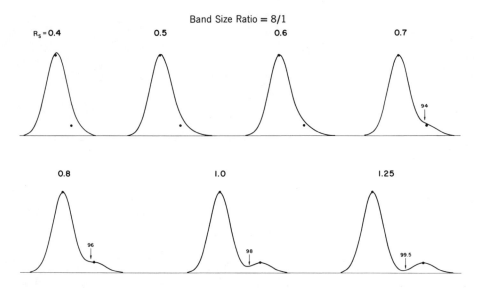

Figure 3.4   Standard resolution curves for a band-size ratio of 8/1 and R_S values of 0.4-1.25 (1).   Reprinted by permission.

For a further discussion of this approach to estimating resolution see ref. 1.

3.2   HOW LARGE SHOULD R_S BE?

Having estimated R_S for a given pair of bands or the entire chromatogram, the next step is to determine how large R_S must be to achieve our immediate separation goal.   This in turn depends upon what that goal is. Several possibilities exist, as we will now see.

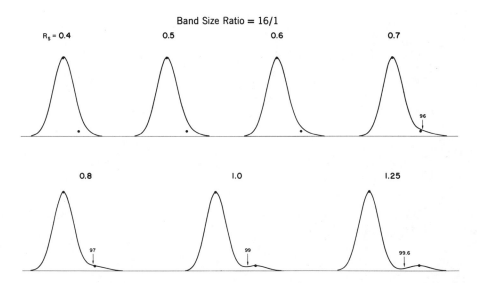

Figure 3.5 Standard resolution curves for a band-size ratio of 16/1 and R$_S$ values of 0.4-1.25 (1). Reprinted by permission.

How Many Compounds Are in the Sample?

We may be interested only in knowing how many compounds are present in a sample--for example, in examining a compound or sample for purity. Or we may wonder if a particular band in the chromatogram is really two severely overlapping bands, rather than a single compound. In either case we are faced with a band, or pair of bands, that we suspect (but do not know) is really two severely overlapping bands.

The nature of the problem is illustrated by the example for R$_S$ = 0.5 in Figure 3.3. A noticeable tail appears to the right of the band center, but it is conceivable that this might represent band tailing (as in Figure 2.12a) rather than two overlapping

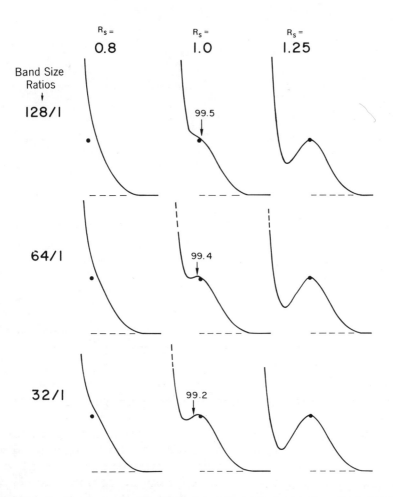

Figure 3.6   Standard resolution curves for band-size ratios of 32/1, 64/1, and 128/1, and R_S values of 0.8-1.25 (1).   Reprinted by permission.

bands.   To solve this problem, R_S should be increased to 0.7 (see Figure 3.3).   If the band in question is really two overlapping bands, it will now be quite apparent.   An obvious shoulder as in R_S = 0.7 will

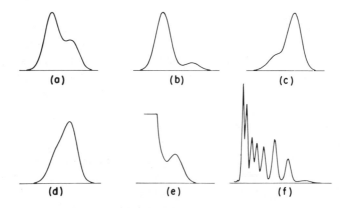

Figure 3.7 Examples of separations with different R$_S$ values (see text).

appear. If, on the other hand, band tailing rather than two overlapping bands is involved, there will be little change in band shape as R$_S$ is increased from 0.5 to 0.7.

We might likewise ask whether example (d) in Figure 3.7 represents one distorted (reverse-tailing) band or two overlapping bands. Here the presence of two bands appears more likely, but we must keep in mind that peculiar band shapes are occasionally observed in LC. To be sure in this case, R$_S$ should be increased from 0.5 (Figure 3.2) to about 0.7.

What Compounds are Present?

More often we are interested in the identification or qualitative analysis of the various components of a sample. The simplest way of identifying an unknown sample band is to compare its t_R value with that of some known compound suspected to be present. When the retention times of the unknown and known compounds

match, this is a good indication that the two compounds are the same. However, the resolution of two
bands affects the accuracy with which their t_R values
can be measured. This is illustrated in Figure 3.8a.
Here two overlapping bands are shown (dashed curves),
along with the composite curve (solid line) that would
be seen in an actual chromatogram. The arrows above
the first band in (a) indicate (from left to right)
the apparent and true positions of the band center, and
therefore the apparent and true values of t_R for this
band. From this we see that poor resolution can result in the displacement of the apparent band center
toward the overlapping band, with a resulting error in
the measurement of t_R. The standard resolution curves
(Figures 3.1-3.6) provide information on whether this
is a problem in a given case. For each R_S example two
black dots are superimposed on top of the two overlapping bands. These dots indicate the true positions
of the two band centers. By comparing the horizontal

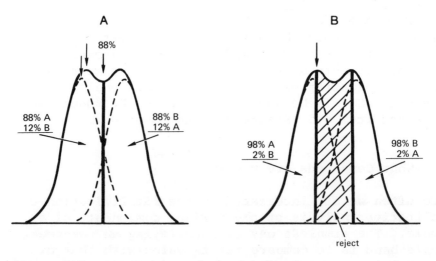

Figure 3.8   Increasing the purity of recovered fractions by rejecting a center fraction.

position of the dot with the apparent t_R value of the overlying band, we can immediately see how close the apparent band center is to the true band center, and whether R_S needs to be increased for an accurate measurement of t_R. As an example, consider example (c) of Figure 3.7. Comparing this example with Figure 3.3, we see that R_S must be increased from 0.7 to 0.8, for an accurate t_R measurement.

In some cases an accurate retention time measurement does not identify the unknown band. Either no known compound has the same t_R value, or more than one compound has a retention time close to that of the unknown. In this case we must isolate the unknown band in sufficient purity to allow its subsequent identification by the usual analytical techniques (e.g., infrared, mass spectrometry, etc.). Isolation and characterization are also recommended for the positive confirmation of bands whose preliminary identification has been made by comparisons of t_R values.

The standard resolution curves (Figures 3.1-3.6) can be used as guides in obtaining fractions of an unknown compound in some desired purity. On these curves (for higher values of R_S) are seen numbered arrows between the two band centers; for example, 92 for $R_S = 0.7$, in Figure 3.2. The arrows in each case indicate the cutpoint that divides the two bands into fractions of equal purity, with the number giving the percent purity of each of the resulting two fractions. In the latter example, a cutpoint at the arrow would produce two fractions of 92% purity. Or, in Figure 3.8a, the equal-purity cutpoint produces a fraction that is 88% A and 12% B, as well as a corresponding fraction that is 88% B and 12% A. Thus, the standard resolution curves immediately tell us whether a given separation can yield fractions of the required purity, and if not, by how much R_S must be increased. For example, assume we require 98% pure fractions for infrared characterization, and have example (a) of Figure

3.7 in our initial separation. Comparison with Figure
3.2 indicates R_s = 0.7, and the equal-purity cutpoint
will give fractions of 92% purity. However, Figure
3.2 also shows that an increase of R_s to 1.0 will provide fractions of the necessary purity (98%).
     The estimation as above of fraction purity and
equal-purity cutpoints assumes that the detector response or sensitivity is equal for each of the two
overlapping compounds; that is, equal quantities of
each compound would give equal areas (and approximately
equal band heights) in the final chromatogram. In LC,
unlike gas chromatography, this is seldom the case.
However, the detector responses for two overlapping
bands are often similar, so that the error introduced
by the use of Figures 3.1-3.6 is usually minor. In
any case, in the absence of detailed information on
the detector response of the two compounds, no better
estimates of purity and cutpoint can be made. We
should also note that the purity of recovered fractions
can be increased <u>without</u> increasing R_s, if we are will-
ing to sacrifice yield. This is illustrated in Figure
3.8b, where it is seen that rejection of an intermediate fraction (cross-hatched area) increases the purity
of the outside fractions from 88% to 98%.

     Quantitative Analysis

Generally we are interested in how much of one or more
compounds is present in the sample. We can determine
sample concentrations on the basis of band height,
band area, or band recovery plus weighing. Band height
or area methods usually assume a proportionality between height or area and the concentration of a given
sample component. This proportionality constant is
determined by injecting known concentrations of the
sample of interest and constructing a calibration
curve. Figure 3.9 illustrates the measurement of band

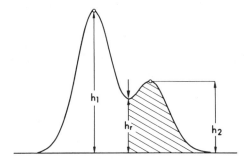

Figure 3.9   Quantitation by band height or area.

heights (h_1, h_2) and band areas (cross-hatched area
for band 2). For overlapping bands, as in this example, it is customary to divide the areas of the two
bands by means of the drop line from the valley (the
line h_r in Figure 3.9).
        Which method we choose--band height or area--de-
pends on several factors, as discussed in Section
13.1. For the moment we want to focus on the effect
of resolution on each type of quantitation. That is,
what value of R_S is required for accurate quantitation
by band height or area? The effect of resolution on
the accuracy of band-height measurements is shown in
the standard resolution curves. Since the dark dots
in each case indicate the true positions of the underlying band centers, a vertical displacement of a dot
from the apparent band center is a measure of the
error introduced by poor resolution. Thus, in Figure
3.3 (4/1) we see that a resolution of 1.0 is required
for an accurate band-height measurement; lesser values of R_S would lead to apparent band heights that are
too large. As a rough guide to the overall effect of
resolution on the accuracy of band height quantitation,
we can say the following: if R_S = 1, relative band
heights can be varied from 128/1 to 1/128, with an
accuracy of better than $\pm 3\%$; that is, the relative

concentrations of the two bands can change by an overall factor of about 15,000.

Consider next how poor resolution will affect the
accuracy of band area measurements. In Figure 3.10
are shown the measured (cross-hatched) areas of the
minor band, for $R_S = 1$ and several different ratios
of the heights of the two bands. The measured area of
the minor band is expressed here as a percentage of
its _true_ area (i.e., as in the absence of band overlap). As the relative concentration of the minor band
decreases, we see a decreasing accuracy in the measured area of the minor band (100% = no error). The
error in the area of the major band is always positive, and is never larger than 1%.

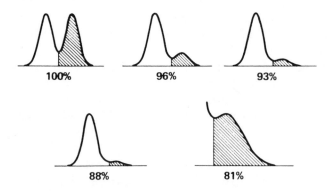

Figure 3.10   Error in quantitation by band area (100%
equals no error in minor band).

It can be seen that if the peak height ratio of
two overlapping bands is x (x > 1), the errors in the
measured band areas are related as

(error in major band) = $(\frac{-1}{x})$ (error in minor band).

Thus, in Figure 3.10, the error in the major band for
the 16/1 example would be +(12/16)%, or +0.8%.

It is seen from Figure 3.10 and the preceding discussion that poor resolution has a much greater effect on band area than on band height. Thus, for $R_S = 1.0$ and an accuracy of at least $\pm 3\%$, it is possible to vary band heights from 128/1 to 1/128. Band area, on the other hand, cannot be varied by more than 3/1 to 1/3 (Figure 3.10). This means that resolution must generally be better for quantitating by band area, than for quantitating by band height.

The error introduced by poor resolution into quantitation by band area can be estimated from the height of the valley h_r relative to the height of the minor band h_2, as in Figure 3.10:

h_r/h_2	Relative Error in Apparent Area (Minor Band)
0.25	−1%
0.4	−2%
0.6	−5%
0.75	−10%

Thus, in Figure 3.3, we can estimate the following errors in the measured areas of the minor bands: greater than 10% for $R_S = 0.8$, about 3% for $R_S = 1.0$, and less than 1% for $R_S = 1.25$.

When the concentration of a sample component (band) is large enough, quantitation can also be achieved by simply collecting the band, removing the solvent (by evaporation), and weighing. This avoids the need for calibration, and it is particularly attractive when dealing with the one-time analysis of unknown mixtures (e.g., a competitor's product). It is obviously less attractive for the repetitive, quantitative analysis of similar samples, because it is

more tedious than the band-height or -area methods.
The selection of cutpoints for band collection and the
error introduced by poor resolution parallel the use
of band area quantitation. Most of what has been said
concerning band-area quantitation, therefore, applies
equally to quantitation by recovery and weighing.

Preparative Separation

In preparative separation we are interested in purity
and recovery. Estimating the purity of recovered fractions was discussed above in connection with qualitative analysis. The fractional recovery that can be
expected for a given band can be estimated from the
standard resolution curves plus Figure 3.11. The procedure is as follows, illustrated with example (a) of
Figure 3.7. First determine the value of R_s and of
the band-height ratios: for (a) R_s = 0.7, and the
band height ratios are 2/1 for the major band and 1/2
for the minor band. From Figure 3.2 the purity of
each fraction (for an equal-purity cutpoint) is 92%.
For this same (equal-purity) cutpoint, the recovery of
each band can now be estimated from Figure 3.11. Thus,
the 2/1 curve, for a purity of 92%, shows a recovery
of 96%. This is the recovery of the major band (2/1).
The recovery of the minor band is similarly seen to
be about 82% (1/2 curve and 92% purity, Figure 3.11).
This example also illustrates a general principle:
For equal-purity fractions, the recovery of the minor
band in its fraction will be less than the corresponding recovery of the major band.
      In determining the resolution needed for a given
separation, for whatever purpose, a margin of safety
is generally desirable. For example, if we decide
that R_s should be equal to about 1.0, we might aim for
an actual value of $R_s \geq 1.1$. This recognizes several
contingencies: (1) Our value for the required change

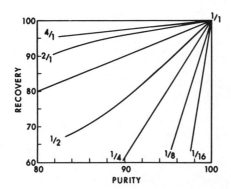

Figure 3.11   Estimating the recovery of each band (compound), using the equal-purity cutpoint.

in R_s may be slightly in error (because of the _estimation_ rather than _measurement_ of R_s), (2) our estimate (Section 3.5) of the experimental conditions required for the new value of R_s will be approximate, and (3) the normal variability of column performance over a period of time can lead to a corresponding small variation in R_s.

3.3   CONTROLLING RESOLUTION: GENERAL

Equation 2.11 is the key to controlling resolution in LC:

$$R_s = (\tfrac{1}{4})(\alpha-1)\sqrt{N}\left[\frac{1}{1+k'}\right] \qquad (2.11)$$

$$\text{(i)} \quad \text{(ii)} \quad \text{(iii)}$$

Terms (i)-(iii) of Eq. 2.11 can each be varied to improve resolution, as illustrated in Figure 3.12. First, an increase in the separation factor α results in a displacement of one band center, relative to the

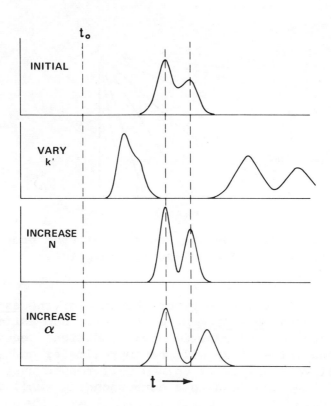

Figure 3.12 The effect on sample resolution of changes in k', N, and α.

other, and a rapid increase in R_s. The time of sep-aration and the heights of the two bands are not much changed for moderate changes in α. Second, an in-crease in the plate number N results in a narrowing of the two bands and an increase in band height; again separation time is not directly affected.* Last, a

*This assumes that the ratio of sample/weight of col-umn packing remains constant. Also, an increase in N often requires an increase in separation time (see Section 3.5).

change in k' can have a dramatic effect on separation,
as shown in Figure 3.12.   If k' for the initial separation falls within the range 0.5-2, a decrease in k'
leads to a rapid worsening of separation, and an increase in k' can provide a significant increase in
resolution.   However, as k' is increased, band heights
rapidly decrease and separation time increases.

Sections 3.4-3.6 will examine the effect on resolution of terms (i)-(iii) in greater detail.   It will
also be seen how each of these terms can be varied for
predictable change in R_S.   First, however, let us consider how each of these three terms relates to an overall strategy for improving separation.   Three different examples of poor resolution are illustrated in
Figure 3.13.   When R_S must be increased and k' for the
initial separation is small (Figure 3.13a), k' should
first be increased into the optimum range $1 \leq k' \leq 10$

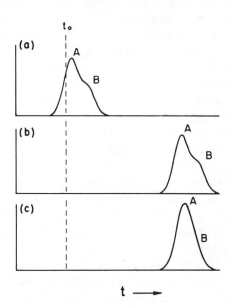

Figure 3.13   Different resolution problems, requiring
different separation strategies.

(see Section 3.4 for further details). No other
change in separation conditions will give as large an
increase in R_s for as little effort.

When k' is already within the optimum range of
values, and resolution is still marginal (Figure 3.13b)
the best solution is usually an increase in N. Normally this means an increase in separation time (see
below and Section 3.5). However, the necessary change
in experimental conditions is easily predicted, and
little effort will be spent in achieving the required
increase in N and R_s. A possible exception is the development of a routine separation that will be used
for the analysis of hundreds of samples. In this case,
it may be more attractive to attempt an increase in α,
rather than N, because the separation time per sample
(and the related cost and effort) can be reduced.

The final chromatogram, (c) of Figure 3.13, shows
two bands in the optimum k' range, but with R_s quite
small. Here the necessary increase in N will probably
require a very long separation time, and it might even
be impossible to achieve (e.g., when $\alpha = 1$). In this
case what is needed is an increase in α.

Predicting the right conditions for the necessary
change in α is seldom a straightforward procedure, and
it often involves much effort. Thus, an increase in
α can provide the shortest possible separation times,
but the effort required to discover the right experimental conditions may represent a greater investment
than we care to make. So, in the example of Figure
3.13b, a change in α is unattractive when we have a
limited number of separations of a given type to carry
out. However, an increase in α may well be preferable
when a large number of such separations is involved.
Other possibilities in addition to separations shown
in Figure 3.13 include poor separation with k' > 10,
and $R_s = 0$ for $k' \approx 0$. In each case the first change
that should be made is to adjust k' into the range
$1 \leq k' \leq 10$.

The above discussion stresses the importance of separation time, which in general should be as short as possible. Of course, we also want to minimize the effort involved in a given separation. These two factors--separation time and total effort--must each be considered when we are faced with a given separation problem (cf. discussion of Section 3.5). Apart from the question of time and effort, the maximum attainable resolution is also limited by the column pressure P which our LC unit can provide. The larger the value of P, the greater is the potential resolution. The importance of P in this respect is illustrated by some specific examples in Section 3.5.

For the moment let us consider a general expression for R_s as a function of column pressure P and separation time t.* First, however, it is useful to define the so-called effective plate number N_{eff}.

$$N_{eff} = N \left[\frac{k'}{1 + k'} \right]^2 . \qquad (3.1)$$

This can be substituted into Eq. 2.11 to give

$$R_s = \frac{1}{4} (\alpha-1) \sqrt{N_{eff}} . \qquad (3.1a)$$

Thus N_{eff} represents the combined contribution of N and k' to resolution. Since α does not depend upon column pressure and separation time, we can examine the effect of these parameters on R_s through their effect on N_{eff}. By combining Eqs. 2.11 and 2.8 we arrive at a general expression for N_{eff} (see Appendix III for derivation):

$$N_{eff} \quad \underbrace{\left[\frac{K^{(1-n)/2}}{D} \right]}_{(i)} \quad \underbrace{P^{(1-n)/2}}_{(ii)} \quad \underbrace{t^{(1+n)/2}}_{(iii)} \quad \underbrace{\left[\frac{k'^2}{1+k'^{(5+n)/2}} \right]}_{(iv)} \qquad (3.2)$$

---

*The reader may want to skip to Section 3.4, since the balance of this section is somewhat advanced and not essential to the following discussion.

Equation 3.2 assumes that column length L is adjusted
to give the necessary value of t, and t is in turn
determined by the required value of N_{eff} (see discussion of Section 3.5). The coefficient n of Eq. 2.8
is typically close to a value of 0.4 in modern LC.
Therefore, N_{eff} is seen to be proportional to about
$P^{0.3}$ [term (ii)]; a tenfold increase in P will yield
about a twofold increase in N_{eff}, if t is held constant. Similarly N_{eff} is approximately proportional
to $t^{0.7}$ [term (iii)].

The importance of high-pressure operation in modern LC is apparent in Eq. 3.2. Thus, classical LC
typically involved gravity flow with column pressure
of the order of 0.5 psi. Modern LC is often carried
out at pressures up to 5000 psi, which represents a
10^4 increase in P. From Eq. 3.2 this means an increase in N_{eff} by about a factor of 16, or a decrease
in separation time t (same value of N_{eff}) by a factor
of roughly 100. However, the advantage in modern LC
of further increases in operating pressure (above
5000 psi) may be marginal. An increase in P to
10,000-20,000 psi would provide a further increase in
N_{eff} of only 25-50%, while operating pressures of this
magnitude will require much more expensive equipment.
There are also fundamental complications associated
with the use of such pressures: generation of heat
within the column, variation of k' values with P, and
so on.

The significance of terms (i) and (iv) of Eq. 3.2
are examined in Appendix VI and Section 3.4. For a
detailed discussion of the further significance and
limitations of Eq. 3.2, see ref. 2.

3.4   RESOLUTION VERSUS k'

The effect of k' on resolution is given in term (iii)
of Eq. 2.11. Note that k'/(1+k') is the fraction of

sample molecules X in the stationary phase (see Eq.
2.2). Since differential migration depends upon the
preferential retention of sample molecules in the stationary phase, R_s is proportional to this quantity,
which has the following values as a function of k':

k'	$[k'/(1 + k')] \propto R_s$
0	0
1	0.5
2	0.67
5	0.83
10	0.91
∞	1.00

When k' is initially small (≤1), R_s increases rapidly
with increase in k'. For values of k' greater than 5,
however, R_s increases very little with further increase in k'. At the same time it should be realized
that separations which involve k' values greater than
10 result in long separation times, and excessive band
broadening (Eq. 2.6), to the point where detection becomes difficult. Therefore, there is an optimum range
of values of k', in terms of resolution, separation
time, and band detection. We see from the above tabu-
lation of values that this optimum k' range is about
1 ≤ k' ≤ 10. In a moment we will refine this estimate
by means of more quantitative reasoning.
     We have observed earlier that k' values in LC are
controlled by means of solvent strength. When k' values must be increased, a so-called <u>weaker solvent</u> is
used. <u>Stronger solvents</u> are used to reduce k' values.
This is illustrated in Figure 3.14, for the separation
of a mixture of anthraquinones on a bonded stationary
phase column (reverse-phase liquid-liquid chromatography). In this series of separations, solvent

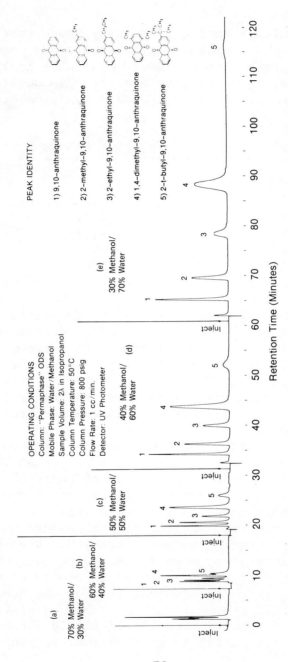

Figure 3.14 The effect of solvent strength on resolution (6). Reprinted with permission.

70

strength increases from pure water as solvent to pure
methanol. Intermediate mixtures of these two solvents
have intermediate solvent strengths. In Figure 3.14a
we can see that 70% methanol/water as solvent is much
too strong, resulting in low k' values and poor separation. In the separation of Figure 3.14e, 30% meth-
anol/water as solvent is much too weak; that is, separation time is too long (60 min), and band No. 5 has
widened to the point where it is barely detectable.
50% methanol/water as mobile phase (Figure 3.14c) is
seen to be a good compromise; resolution of all five
components is adequate, band heights are reasonable,
and the separation time is less than 10 min. Solvent
composition has no effect on band migration in gel chromatography (see Chapter 10), where the stationary
phase must be varied to achieve changes in relative
band migration.

The optimum value of k' can also be approached
from the standpoint of maximum resolution per unit
time.* For a given value of α, R_s is proportional to
N_{eff}, and

$$N_{eff} = \frac{N\, k'^2}{(1 + k')^2}$$

$$= \left(\frac{L}{H}\right) \frac{k'^2}{(1 + k')^2} \qquad (3.3)$$

Assume first that solvent velocity u is held constant,
which means that H is also constant. Separation time
t is related to t_o and k' as $t = t_o(1 + k')$ (Eq. 2.12),
and $t_o = L/u$, so that

$$L = \frac{t\, u}{1 + k'} \qquad (3.3a)$$

---

*The reader may want to skip the following paragraph,
since this is somewhat advanced and not essential to
the following discussion.

Inserting Eq. 3.3a into 3.3 gives

$$N_{eff} = (\frac{t\ u}{H}) \frac{k'^2}{(1 + k')^3} \qquad (3.3b)$$

For t, u, and H constant, this expression has a maximum value for $k' = 2$. That is, with separation time held constant by adjusting L for each value of k', and solvent velocity u unchanged, maximum resolution is achieved for $k' = 2$. If we allow solvent velocity to vary, as well as L, a similar analysis in terms of Eq. 3.2 (see ref. 2) gives maximum R_s per unit time equal to $4/(n + 1)$. For typical values of $n \approx 0.4$, k' should then be about 3. Similarly, if L is held constant and both u and k' are allowed to vary, the optimum value of k' is about 5 (see ref. 3). Thus, depending upon circumstances, the optimum value of k' can vary between about 2 and 5. The exact (optimum) value of k' is in any case not critical, since term (iv) of Eq. 3.2 changes only slowly with k' in the region $2 \le k' \le 5$; thus in practice we strive for $1 \le k' \le 10$.

As a final example of the importance of resolution and k', consider the so-called general elution problem illustrated in Figure 3.15a. This shows the separation of a multicomponent mixture by normal elution from silica; that is, without changing conditions during the separation. Here the various components of the sample have a wide range in k' values, and as a result there is poor resolution of the bands in the early part of the chromatogram (bands 1-5 in Figure 3.15a), and broadening of later bands (bands 10, 11) to the point of difficult detection. The problem is that k' values for the first bands are too small, while the k' values of the last bands are too large. No change of solvent, as in Figure 3.14, can solve this problem, because the k' values of individual bands differ widely (by about 50-fold) and can never simultaneously fall into the optimum range $1 \le k' \le 10$.

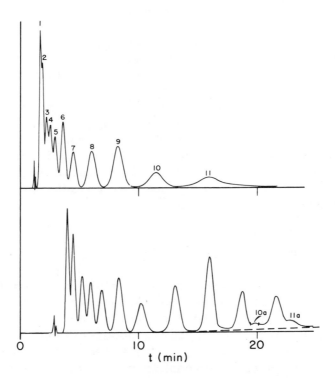

Figure 3.15   The general elution problem and the use of gradient elution (4).   (a) Normal (isocratic) elution from silica.   (b) Gradient elution from silica.

    One solution to the general elution problem is to change conditions during separation, so as to optimize k' for individual bands as they move through the column. This can be achieved in a number of ways, as will be discussed in Section 13.3.   The most common technique for doing this is gradient elution, as illustrated in Figure 3.15b.   Here the same sample is eluted as in Figure 3.15a, but with the composition of the solvent varied <u>during</u> the separation.   Elution is begun with a weak solvent, which elutes early bands (1-5) with k'

values near their optimum values. Solvent composition
is changed during separation to successively stronger
solvent, so that later bands elute with k' values that
are reduced into the optimum range. Normally gradient
elution involves a change in the composition of a binary solvent, for example, methanol/water, as in Figure 3.14. In this case gradient elution might involve
initial elution with 30% methanol/water, with subsequent increase in the % methanol as separation proceeds.
Note that the various bands in Figure 3.15b are of
roughly equal width. This is characteristic of separations by gradient elution, unlike the case of isocratic
separation (e.g., Figure 3.15a).

3.5   RESOLUTION VERSUS N

The effect of N on resolution is given in term (ii) of
Eq. 2.11. The quantity N has also been related to experimental conditions in Section 2.3 and Appendix I.
More will be said in Chapter 6 about optimizing N for
a given column. However, in this section we want to
consider the interrelationship of N, separation time,
column length L, and column pressure P. This treatment
assumes that short lengths (e.g., 1 m each) of a given
column type are available, and that these column
lengths can be combined in series to give any desired
overall length L (e.g., 2, 3, or 4 m). In practice,
this is the way that column length is normally adjusted
in modern LC.
   We will begin with some examples of how N can be
varied in different ways and go on to discuss which of
these techniques is appropriate for a given separation
problem. Then we will show the basic relationships between the necessary experimental conditions for some
required increase in R_s, that is, from R_1 in the initial separation, to R_2 in the final separation. Finally, this treatment will be reduced to a simple, rapid

scheme for predicting the necessary change in separation conditions. It should be emphasized that the latter scheme requires no calculations, nor--for that matter--any understanding of the underlying theory.

Assume that we start with short column lengths whose performance is given by the H versus u plot of Figure 2.6. Now consider the examples in Table 3.1 of calculated plate numbers N as a function of different experimental conditions. The first example might correspond to an initial separation, for which R_S and N must be increased by some known factor. Here an initial column length of 100 cm is assumed, along with certain values of column pressure P, separation time t, and solvent velocity u. With u = 2 cm/sec, H is given by Figure 2.6 as 0.156 cm, and therefore N = L/H = 100/0.156 = 640 plates.

TABLE 3.1   INCREASING N IN VARIOUS WAYS (LC)

Procedure	L (cm)	P (psi)	t (sec)	u (cm/ sec)	H (cm)	N
(Initial separation)	100	10^3	150	2.0	0.156	640
Decrease P, L constant	100	180	840	0.36	0.078	1280
Increase L, P constant	164	10^3	405	1.22	0.128	1280
t constant, increase L, P	320	10^4	150	6.4	0.247	1280

Now assume that N for the above example must be doubled to a final value of 1280 plates. The second

example of Table 3.1 assumes that we increase N by simply decreasing column pressure P (holding L constant).
Separation time t is proportional to 1/P (Section 2.5),
so that t is increased by the same factor (5.6-fold)
by which P was reduced. Solvent velocity u is proportional to P (Eq. 2.12), and therefore is reduced by
5.6-fold. The new value of H from Figure 2.6 is 0.078,
and N = 100/0.078 = 1280 plates. Thus, N has been
doubled relative to our initial separation, simply by
reducing P. The price we have had to pay in this approach is a 5.6-fold increase in separation time.

In the third example of Table 3.1 we propose to
increase N by increasing L, while holding P constant.
Here a 1.64-fold increase in L means an increase in t
(Eq. 2.13) by the factor $(1.64)^2$ = 2.7, and a decrease
in u by the factor 1/1.64 (Eq. 2.12). The value of H
for u = 1.22 cm/sec is 0.128 (Figure 2.6), and N =
164/0.128 = 1280 plates. Again we have doubled the
initial value of N. Note, however, that the time required in this case is less than in the preceding example, where L was held constant: 405 versus 840 sec.
The reason is that now we are operating at a higher
column pressure, (1000 psi) which means, other factors
being equal, that column efficiency per unit time will
be greater (Eq. 3.2).

In the last example of Table 3.1 we increase N by
holding t constant, while increasing L and P. With L
increased by a factor of 3.2, P must be increased by
$(3.2)^2$ = tenfold (Eq. 2.13). As a result, u is increased by a factor of 3.2 (Eq. 2.12), and H = 0.247
(extrapolation of Figure 2.6). Finally, N = 320/0.247
= 1280 plates. Again we have doubled N, but now at the
price of increased column pressure P. Figure 3.16 provides actual examples of the increase in N and resolution by each of these various ways: decreasing P with
L constant (b, c versus a); increasing P and L, with
t constant (d versus b, c); increasing L with P constant (e versus d).

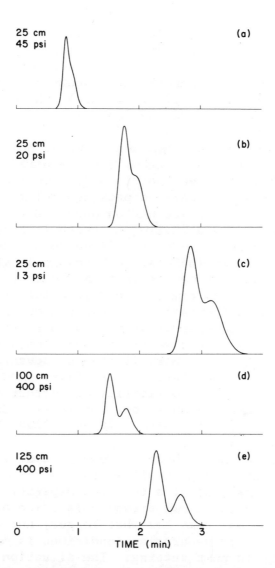

Figure 3.16   Separation of 1-nitronaphthalene (k' =
2.4) and methyl 1-naphthoate (k' = 2.8) on 10% water-
silica (Waters Porasil® A), using 0.2 vol% acetonitrile-
pentane (UV detection) (5).   Reprinted with permission.

What are the practical implications of the data in Table 3.1? The easiest way in which we can increase N and resolution is by a change in P (L constant). This involves only a change in the pump setting (either decrease in pressure or in flow rate). However, the resulting increase in separation time may be considerable.

If we increase column length, while holding P constant, some additional work will be required. Thus, we will have to connect up additional column lengths, and in some cases we may have to pack new columns--if additional column lengths are not on hand. But our extra effort is repaid by a smaller increase in separation time, compared to the use of a fixed column length.

Finally, if we increase both column length and pressure, resolution can be increased without any increase in separation time. Apart from the extra effort involved in increasing L, however, our LC unit may not be able to operate at much higher pressures. Because resolution is greater at higher column pressures (Eq. 3.2), it is normally to our advantage to operate LC pumps near their upper limit. And if this is the case, then the possibility of further large increases in column pressure does not exist. So increasing N while holding t constant is an option that is not always available to us.

Now that we have these various ways of increasing N and R_s, which method should we choose for a given problem? In the case of a one-time separation, a simple decrease in column pressure is quite attractive. The separation may take somewhat longer, but the effort required to change separation conditions is minimal-- just a change in pump setting. The situation is somewhat different when it comes to developing an LC method for carrying out a large number of identical separations or analyses--as in some routine assay procedure. Here the additional work required to increase column length (with P held constant) is quite worthwhile, because

the resulting saving in time per analysis (compared to
a separation with L held constant) is multiplied by
the large number of individual separations. A final
possibility, that of increasing both column pressure
and length, will occur less often--for the reasons
discussed above. In some cases we may develop an initial separation on a lower pressure LC unit, because
of greater convenience. A pressure increase by a factor of 10 (from 500 to 5000 psi) is then possible by
changing to another LC unit. Or, we may simply have
operated our LC unit initially at a pressure lower than
its maximum (e.g., Figure 3.16a-3.16c), and again we
can increase P while holding t constant.

The various relationships that determine what separation conditions are required for a given change in
R_s (from an initial value R_1 to a final value R_2) are
derived in Appendix IV and are summarized below, for
each of the three ways of increasing N.

1.   Decrease P, Hold L Constant

The necessary pressure in the final separation (P_2) is
related to the initial pressure P_1 and the necessary
change in resolution as

$$\frac{P_2}{P_1} = \left[\frac{R_2}{R_1}\right]^{-2/n} . \qquad (3.4)$$

The column parameter n, as defined by Eq. 2.8, can be
determined from plots of log H versus log u for the
column in question. The corresponding increase in separation time--from an initial time t_1 to a final time
t_2-- is, then,

$$\frac{t_2}{t_1} = \frac{P_1}{P_2} . \qquad (3.5)$$

2.   Increase L, Hold P Constant

The necessary increase in column length from an initial value L_1 to the final value L_2 is

$$\frac{L_2}{L_1} = \left[\frac{R_2}{R_1}\right]^{2/(1+n)} . \qquad (3.6)$$

The corresponding increase in separation time is

$$\frac{t_2}{t_1} = \left[\frac{L_2}{L_1}\right] . \qquad (3.7)$$

3.   Increase L and P, Hold t Constant

The increase in L is

$$\frac{L_2}{L_1} = \left[\frac{R_2}{R_1}\right]^{2/(1-n)} \qquad (3.8)$$

The increase in column pressure is

$$\frac{P_2}{P_1} = \left[\frac{L_2}{L_1}\right]^2 . \qquad (3.9)$$

The form of the preceding relationships (Eqs. 3.4-3.9) allows us to use a table-lookup method for determining the required change in separation conditions, given values of R_1 and R_2 as determined in Sections 3.1 and 3.2. The necessary tables are shown as Tables 3.2-3.5, each table corresponding to some value n for the particular type of column we are using. It is advantageous (but not essential) to know n for every column that is either purchased or prepared. In the event n is not known for a given column, a value of n = 0.4 (Table 3.3) can be assumed as a first approximation.

The use of Tables 3.2-3.5 for estimating required separation conditions will now be illustrated by a few examples. The procedure is identical, regardless of the value of n, and we will asume n = 0.4 (Table 3.3)

TABLE 3.2   VARIATION OF RESOLUTION WITH TIME, PRESSURE, AND COLUMN LENGTH
Column parameter n = 0.3

(a) Varying Pressure and Time (Column Length Constant)
Relative Change in Time or Reciprocal Pressure

R_s** Needed / R_s →	0.4	0.5	0.6	0.8	1.0	1.25
0.4	--	4.4	15	100	450	2000
0.5	0.2	--	3.4	23	100	450
0.6	0.07	0.3	--	6.7	30	130
0.8	0.01	0.04	0.15	--	4.4	19
1.0	0.002	0.01	0.03	0.2	--	4.4
1.25	0.0005	0.002	0.008	0.05	0.2	--

(b) Varying Column Length and Time (Pressure Constant)
Relative Change in Column Length (Time)

R_s Needed / R_s →	0.5	0.6	0.8	1.0	1.25
0.4	(1.4,2.0)	(1.9,3.6)	(3,9)	(4.5,18)	(6,36)
0.5	--	(1.3,1.8)	(2.0,4.3)	(3.2,10)	(4.7,20)
0.6	(0.8,0.6)	--	(1.5,2.4)	(2.2,4.8)	(3.3,10)
0.8	(0.5,0.2)	(0.7,0.4)	--	(1.4,2.0)	(2.1,4.0)
1.0	(0.3,0.1)	(0.5,0.2)	(0.7,0.5)	--	(1.5,2.0)
1.25	(0.2,0.05)	(0.3,0.1)	(0.5,0.2)	(0.7,0.5)	--

(c) Varying Column Length and Pressure (Time Constant)
Relative Change in Column Length (Pressure)

R_s Needed / R_s →	0.5	0.6	0.8	1.0	1.25
0.4	(1.9,3.6)	(3.2,10)	(8.2,50)	(13,170)	(25,620)
0.5	--	(1.7,2.8)	(3.8,14)	(7.2,50)	(13,170)
0.6	(0.6,0.4)	--	(2.3,5)	(4.3,1.8)	(8.1,65)
0.8	(0.3,0.07)	(0.4,0.2)	--	(1.9,3.6)	(3.6,13)
1.0	(0.14,0.02)	(0.2,0.6)	(0.5,0.3)	--	(1.9,3.6)
1.25	(0.08,0.006)	(0.12,0.015)	(0.3,0.08)	(0.5,0.3)	--

*Gauge pressure between pump and column inlet; alternatively, pressure can be replaced by pump flow rate.
**Initial R_s, equal to R_l.

TABLE 3.3 VARIATION OF RESOLUTION WITH TIME, PRESSURE, AND COLUMN LENGTH

Column parameter n = 0.4

(a) Varying Pressure and Time (Column Length Constant)

R_s Needed	Relative Change in Time or Reciprocal Pressure					
$R_s \longrightarrow$	0.4	0.5	0.6	0.8	1.0	1.25
0.4	--	3.1	8	32	100	300
0.5	0.3	--	2.4	10	32	100
0.6	0.1	0.4	--	4.2	13	39
0.8	0.03	0.1	0.2	--	3.1	9
1.0	0.01	0.03	0.08	0.3	--	3.1
1.25	0.003	0.01	0.03	0.1	0.3	--

(b) Varying Column Length and Time (Pressure Constant)

R_s Needed	Relative Change in Column Length (Time)				
$R_s \longrightarrow$	0.5	0.6	0.8	1.0	1.25
0.4	(1.4,1.9)	(1.8,3.2)	(2.7,7.3)	(3.7,14)	(5.1,26)
0.5	--	(1.3,1.7)	(1.9,3.6)	(2.7,7.3)	(3.7,14)
0.6	(0.8,0.6)	--	(1.5,2.3)	(2.1,4.4)	(2.8,8.0)
0.8	(0.5,0.3)	(0.7,0.4)	--	(1.4,1.9)	(1.9,3.6)
1.0	(0.4,0.2)	(0.5,0.3)	(0.7,0.5)	--	(1.4,1.9)
1.25	(0.3,.07)	(0.4,0.1)	(0.5,0.3)	(0.7,0.5)	--

(c) Varying Column Length and Pressure (Time Constant)

R_s Needed	Relative Change in Column Length (Pressure)				
$R_s \longrightarrow$	0.5	0.6	0.8	1.0	1.25
0.4	(2.1,4.4)	(3.9,15)	(10,100)	(21,440)	(44,1900)
0.5	--	(1.8,3.7)	(4.8,23)	(10,100)	(22,480)
0.6	(0.5,0.3)	--	(2.6,6.8)	(5.5,30)	(11,120)
0.8	(0.2,.04)	(0.4,0.1)	--	(2.1,4.4)	(4.4,19)
1.0	(0.1,.01)	(0.2,.03)	(0.5,0.2)	--	(2.1,4.4)
1.25	(0.05,.002)	(0.1,.01)	(0.2,.05)	(0.5,0.2)	--

TABLE 3.4. VARIATION OF RESOLUTION WITH TIME, PRESSURE, AND COLUMN LENGTH
Column parameter n = 0.5

(a) Varying Pressure and Time (Column Length Constant)

Needed Rs →	Relative Change in Time or Reciprocal Pressure					
Rs	0.4	0.5	0.6	0.8	1.0	1.25
0.4	--	2.5	5.0	25	63	150
0.5	0.4	--	2.1	10	2.5	63
0.6	0.2	0.05	--	5.0	12	30
0.8	0.04	0.1	0.2	--	2.5	6.3
1.0	0.016	0.04	0.08	0.4	--	2.5
1.25	0.007	0.016	0.03	0.16	0.4	--

(b) Varying Column Length and Time (Pressure Constant)

Needed Rs →	Relative Change in Column Length (Time)				
Rs	0.5	0.6	0.8	1.0	1.25
0.4	(1.3,1.8)	(1.7,2.5)	(2.5,5.3)	(3.4,10)	(4.6,18)
0.5	--	(1.3,1.6)	(1.9,3.4)	(2.5,6)	(3.2,11)
0.6	(0.8,0.6)	--	(1.5,2.1)	(1.9,3.8)	(2.6,7)
0.8	(0.5,0.3)	(0.7,0.5)	--	(1.3,1.8)	(1.7,3.2)
1.0	(0.4,0.17)	(0.5,0.3)	(0.8,0.6)	--	(1.3,1.8)
1.25	(0.3,0.09)	(0.4,0.14)	(0.6,0.3)	(0.8,0.6)	--

(c) Varying Column Length and Pressure (Time Constant)

Needed Rs →	Relative Change in Column Length (pressure)				
Rs	0.5	0.6	0.8	1.0	1.25
0.4	(2.4,5.8)	(5.0,25)	(16,250)	(39,1500)	(95,9000)
0.5	--	(2.1,4.4)	(6.5,42)	(16,250)	(39,1500)
0.6	(0.5,0.2)	--	(3.1,10)	(7.6,58)	(17,290)
0.8	(0.15,0.02)	(0.3,0.1)	--	(2.4,6.0)	(6.0,36)
1.0	(0.06,.004)	(0.13,0.017)	(0.4,0.17)	--	(2.4,6.0)
1.25	(.03,.0006)	(0.06,0.003)	(0.17,0.03)	(0.4,0.17)	--

TABLE 3.5. VARIATION OF RESOLUTION WITH TIME, PRESSURE, AND COLUMN LENGTH
Column parameter n = 0.6

(a) Varying Pressure and Time (Column Length Constant)

Relative Change in Time or Reciprocal Pressure

R_s Needed $R_s \longrightarrow$	0.4	0.5	0.6	0.8	1.0	1.25
0.4	--	2.1	3.9	10	21	44
0.5	0.5	--	1.8	4.8	10	21
0.6	0.3	0.6	--	2.6	5.5	11
0.8	0.1	0.2	0.4	--	2.1	4.4
1.0	0.05	0.1	0.2	0.5	--	2.1
1.25	0.02	0.05	0.1	0.2	0.5	--

(b) Varying Column Length and Time (Pressure Constant)

Relative Change in Column Length (Time)

R_s Needed $R_s \longrightarrow$	0.5	0.6	0.8	1.0	1.25
0.4	(1.3,1.7)	(1.7,2.9)	(2.4,5.8)	(3.2,10)	(4.2,18)
0.5	--	(1.3,1.7)	(1.8,3.2)	(2.4,5.8)	(3.2,10)
0.6	(0.8,0.6)	--	(1.4,2.0)	(1.9,3.6)	(2.5,6.3)
0.8	(0.6,0.3)	(0.7,0.5)	--	(1.3,1.7)	(1.7,2.9)
1.0	(0.4,0.2)	(0.5,0.3)	(0.8,0.6)	--	(1.3,1.7)
1.25	(0.3,0.1)	(0.4,0.2)	(0.6,0.3)	(0.8,0.6)	--

(c) Varying Column Length and Pressure (Time Constant)

Impractical for n ≥ 0.6.

in these examples. First, assume that we have a one-
time separation where R_1 = 0.6 and a final resolution
R_2 = 1.0 is required. This problem suggests that we
decrease pressure, while holding L constant. In Table
3.3 it is seen that pressure must be reduced by a factor of 1/13, which means that separation time will be
increased by 13-fold.

In a second example, assume we are developing a
routine LC assay. An initial value of R_s = 0.6 is
found, and the final value required is 1.0. In this
case an increase in L, holding P constant, is indicated. From Table 3.3b we see that L must be increased
by a factor of 2.1, and separation time will be 4.4
times as great as initially. As expected, this is a
smaller increase in t than in the preceding example
(4.4 X versus 13 X).

If it is important to hold t constant, we can
refer to Table 3.3c for the right values of L and P.
For the same example of R_1 = 0.6 and R_2 = 1.0, column
length must be increased fivefold and column pressure
30-fold, if t is not to be increased.

For additional discussion of the use of Tables
3.2-3.5, and for further illustrative problems and examples of their application, see ref. 5.

3.6 RESOLUTION VERSUS α

The effect of α on R_s is given by term (i) of Eq. 2.11.
We desire that α be as large as possible, other factors
being equal. An increase in α is a powerful technique
for improving resolution, because R_s can be <u>increased</u>
while t is <u>decreased</u>. However, determining the necessary conditions (i.e., change in compositions of moving or stationary phases, or varying separation temperature) for a change in α is often difficult, usually
requiring a trial-and-error approach. While in most
cases adequate guidelines for predicting changes in α

are absent, a few rules are offered in later chapters
on the individual LC methods and in Section 13.2.

When is a change in α the most likely solution to
a separation problem? We have already seen in Section
3.3 that no separation is possible if $\alpha = 1$. Therefore,
when $(\alpha - 1)$ is less than about 0.1, a change in α
should be our first goal. Because large values of α
can yield quite rapid separations (e.g., requiring less
than a minute), a change of α may be necessary when developing a routine assay for a large number of samples;
as, for example, in process control applications, clin-
ical analyses, and so on. The separation of complex
samples in routine fashion may also require that we
adjust <u>different</u> α values for the various adjacent
bands, so as to give chromatograms where the individual
bands are more or less regularly spaced (as in Figure
3.17b versus 3.17a); that is, similar α values for all
adjacent band pairs. A particularly nice example of
this is seen in the amino acid analysis of Figure 9.8.

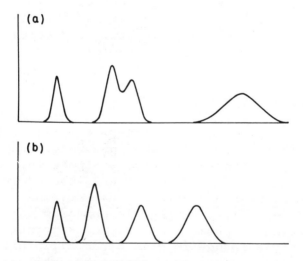

Figure 3.17   The improvement of routine LC separations
by altering the α values of all adjacent bands.

Finally, large α values should be sought in preparative separations, because then sample size can be greatly increased without seriously impairing resolution. This is illustrated in Figure 3.18.

Apart from the need to vary the composition of liquid or stationary phase, or separation temperature, what is our general approach to increasing α for improved resolution? The general goal is to select conditions which give k' values in the optimum range (1-10), then to vary the latter parameters within this range. This is illustrated below for a liquid-solid chromatography separation of two compounds on silica (7):

| | k' | |
Sample Compounds	Solvent A	Solvent B
Acetonaphthalene	5.5	2.3
Dinitronaphthalene	5.8	5.4
α	1.05	2.04

(A is 23% methylene chloride/pentane; B is 5% pyridine/pentane).

With solvent A, k' values are in the right range, but α is small and separation is difficult. However, by changing to another solvent B of about the same strength (e.g., similar k' values for dinitronaphthalene), α is changed significantly with improved resolution. Further examples of this approach are discussed in Section 8.

In other cases we may change one variable, for example, pH in ion-exchange chromatography, while holding other conditions constant. From such a study it can be found that there is an optimum pH in terms of α values. This is illustrated in Figure 3.19 for the separation of the four major mononucleotides by ion-exchange chromatography. Here, a pH of 3 is seen to

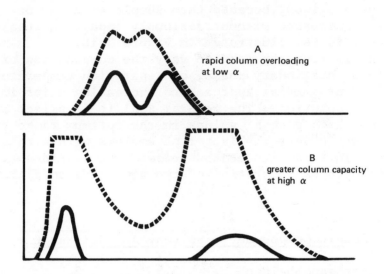

Figure 3.18   The advantage of large α values in preparative separation by LC.

give approximately equal peak separations in terms of
t_R values.   If the resulting k' values had been non-
optimum for pH = 3, then some other separation parameter would have required adjustment (e.g., ionic
strength of the solvent) so as to vary k' into the
optimum range, while retaining satisfactory α values.

At the beginning of Section 3.6, we noted that a
change in α permits us to increase R_S while decreasing
separation time t.   How is this achieved in actual
practice?   A hypothetical example will be used by way
of illustration.   Assume that R_S = 0.5 initially, a
final value of R_S = 1.0 is desired, and α is initially
equal to 1.05.   Further assume we are able to vary conditions (e.g., composition of mobile phase) in such a
way that k' for the second band is held constant (so
that t remains constant), while α is increased to

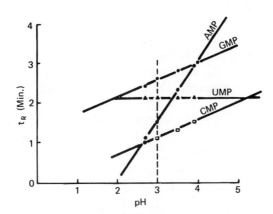

Figure 3.19 Anion-exchange separation of different mononucleotides as a function of pH (8). Reprinted by permission of _Applied Science_.

1.20   This will also result in a change of R_s from 0.5 to 2.0 (Eq. 2.11). Now we propose to reduce R_s from 2.0 to 1.0, by reducing L and holding P constant. Referring to Eq. 3.6, we see that the necessary change in L is determined by the _ratio_ R_2/R_1, which means that a reduction in R_s from 2.0 to 1.0 is equivalent to a change from 1.0 to 0.5. Assuming n = 0.4, we see in Table 3.3b that a change in R_s from 1.0 to 0.5 requires a reduction in L, by the factor 0.4. Separation time is thereby reduced to one-fifth the original value. Thus the final separation, with increased α and decreased L, shows a doubling of R_s in a much shorter separation time.

REFERENCES

1. L. R. Snyder, J. Chromatog. Sci., _10_, 200 (1972).
2. L. R. Snyder, in _Gas Chromatography. 1970_, R.

Stock and S. G. Perry, eds., Institute of Petroleum (Elsevier), London (Amsterdam), 1971, p. 81.
3. L. R. Snyder, J. Chromatog. Sci., $\underline{7}$, 352 (1969).
4. L. R. Snyder, J. Chromatog. Sci., $\underline{8}$, 692 (1970).
5. L. R. Snyder, J. Chromatog. Sci., $\underline{10}$, 369 (1972).
6. Du Pont application sheet, Instrument Products Division.
7. L. R. Snyder, J. Chromatog., $\underline{63}$, 15 (1971).
8. Gas-Chrom Newsletter, Vol. 13, No. 6, Nov./Dec., 1972, Applied Science Labs. Inc., State College, Pa.

CHAPTER FOUR

EQUIPMENT

4.1 INTRODUCTION

The apparatus needed to carry out modern liquid chromatography is very different from the relatively simple
and unsophisticated equipment used for classical LC
separations. While useful and sometimes elegant separations can be carried out with modest apparatus,
relatively sophisticated equipment is needed for separating complex mixtures or for producing highly precise quantitative analyses. Equipment for modern LC
may be assembled from component parts which are available from commercial suppliers, or completely assembled
units may be purchased from various manufacturers.
Appendix V is a partial list of LC manufacturers, plus
various types of component parts and complete instruments that now are commercially available. It should
be recognized, however, that developments in LC apparatus are moving rapidly; therefore, the latest information on currently available apparatus should be secured
from the manufacturers before purchases are attempted.
    The question often arises, what LC equipment is
best? Actually, there is no one "best" equipment,
since the requirements for a particular apparatus depend on the specific problems which are to be solved.

Guidelines will be presented in this chapter to assist the reader in making the proper selection for his own needs. Since detectors are so important in LC, information on these devices is given separately in Chapter 5.

To obtain high-performance operation, the equipment for modern LC must be produced to careful specifications. An excellent high-efficiency column placed in a poorly designed LC apparatus will produce disappointing results. Consequently, it is important that the LC apparatus be designed and fabricated with the same care and skill which is required to produce high-efficiency LC columns. Table 4.1 summarizes the requirements of an apparatus for high-performance liquid chromatography.

A general schematic of apparatus used for modern LC is shown in Figure 4.1. The components in this system comprise a relatively simple apparatus. As will be described below, other accessory components may be needed for conducting relatively sophisticated separations. This is illustrated in the schematic of a commercial LC instrument shown in Figure 4.2. In the remainder of this chapter, the components in LC equipment will be described in the order shown in Figure 4.1. It should first be pointed out, however, that it is virtually impossible to design a single apparatus that will be simultaneously ideal for high-speed analytical chromatography, gel permeation chromatography, and large-scale preparative chromatography. Consequently, the purpose of this chapter is to acquaint the reader with the various components of an apparatus which are available, and to provide him with insights into the advantages and disadvantages of the various designs, so that he may make a proper choice of equipment that best suits his needs.

TABLE 4.1. CRITERIA FOR MODERN LC EQUIPMENT

Performance Requirements	System Characteristics	Equipment Needs
Versatility	Useful with many chemical types of samples	Chemically resistant materials Variety of detectors Unique column packings
Speed	Selective, highly effi-cient columns High carrier velocity High data output	High-pressure pump Low dead-volume fittings and detectors Fast recorder, automatic data handling
Reproducibility	Control of operational parameters	Precise control of: column and detector temp. carrier composition (gradient) flowrate detector response
Sensitivity	High detector response Sharp peaks	Careful detector design for good signal/noise ratio Efficient columns

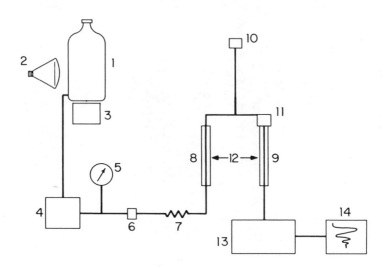

Figure 4.1   General schematic of LC equipment.   1,
reservoir; 2, heating element; 3, magnetic stirrer;
4, high-pressure pump; 5, pressure gauge; 6, line filter; 7, restrictor for pulse dampening; 8, pre-column
(optional); 9, analytical column; 10, pressure sensor/
pump protection device (optional); 11, sample introduction system; 12, thermostatted oven (optional); 13, detector; 14, recorder or data handling system.

4.2   MOBILE PHASE RESERVOIRS

For analytical liquid chromatography, solvent reservoirs should hold about 1 liter of the mobile phase.
In preparative applications, much larger volumes will
be needed.   Figure 4.3 is a schematic of a reservoir
which is used in one of the commercial instruments.
This reservoir is made of stainless steel, which is
inert to most mobile phases and is not subject to
breakage.   Many different forms of reservoirs have been
used, and simple units may be constructed from glass

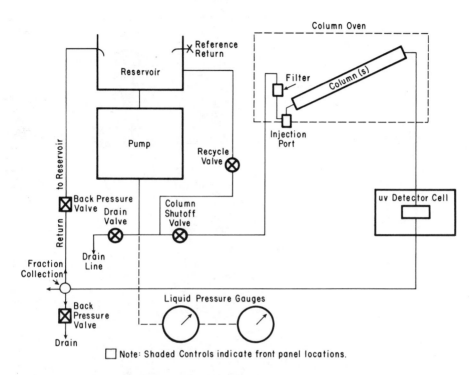

Figure 4.2   Schematic of commercial liquid chromatograph.   Reprinted by permission of Du Pont Instrument
Products Division.

flasks or bottles of appropriate size.
     Some reservoirs are designed so that the mobile
phase may be degassed in situ. Degassing is required
to eliminate dissolved gases, particularly oxygen,
which may react with the mobile or stationary phases
within the column (1). In addition, degassing the
carrier reduces the possibility of bubbles forming in
the detector during the chromatographic separation.
To permit in situ degassing, reservoirs sometimes are
equipped with a heater, a stirring mechanism, and in-
lets for applying a vacuum and a nitrogen purge. Such

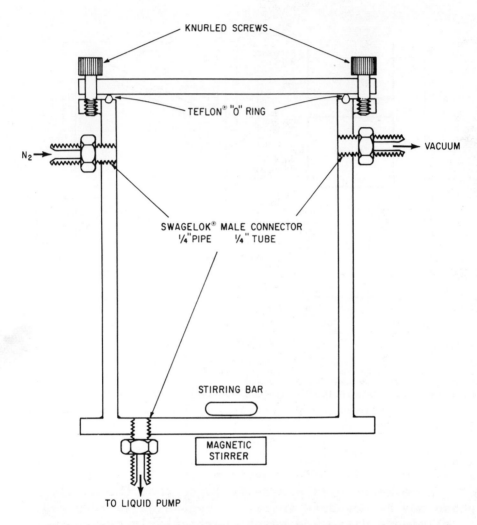

Figure 4.3 Mobile phase reservoir. R. A. Henry, in
__Modern Practice of Liquid Chromatography__, J. J.
Kirkland, ed., Wiley-Interscience, New York, 1971.
Reprinted by permission of publisher.

facilities are a part of the reservoir shown in Figure

4.3. This reservoir is equipped with a magnetic stirrer to facilitate the vacuum degassing. Not shown is a band heater which surrounds the vessel so that its contents may be heated to facilitate the degassing process. Degassing is particularly important with water and other polar solvents. To prevent the formation of bubbles, degassing can be carried out by applying a vacuum to the filled reservoir for a few minutes while stirring the mobile phase vigorously. When possible, warming the carrier facilitates degassing. Vacuum degassing must be carried out carefully so as not to disturb the equilibrium of solvents presaturated with water or organic stationary phases.

Degradation of oxidizable samples, solvents, or column stationary phases can be essentially eliminated by removing the last traces of oxygen from the mobile phase. A nitrogen blanket on the reservoir prevents oxygen from redissolving in the carrier after degassing, and also acts as a safety precaution to reduce the possibility of ignition of the vapors from flammable solvents.

The reservoir can function as more than just a container for the solvent. As indicated below in the discussion of pumps, certain types of reservoirs can serve as a pump for the mobile phase.

4.3 PUMPING SYSTEMS

One of the most important parts of a modern LC instrument is the pumping system. In modern LC, the resistance to flow of the long, narrow columns packed with small particles is relatively high, and high pressures are required. Pumps of this type may be grouped into two major categories: mechanical pumps which deliver the mobile phase essentially at a constant flow rate, and pneumatic pumps which deliver the mobile phase with a constant pressure. Each of these systems has its unique advantages and disadvantages; hence, there

is no one pumping system that is best for all types of
LC operations. The criteria for a suitable modern LC
pumping system are listed in Table 4.2.

TABLE 4.2.   REQUIREMENTS FOR MODERN LC ANALYTICAL
             PUMPING SYSTEM

Made of materials chemically resistant to carrier

Capable of outputs of at least 500 psi, preferably
  4000-6000 psi

Pulse-free, or have a pulse-dampener

Have a flow delivery of at least 3 ml/min for analysis

Deliver carrier with constancy of at least 1-2%

Desirable: small holdup volume for rapid solvent
  changes, gradient elution with external device,
  and/or recycle

     As indicated in Table 4.2, it is important that
the seals in the pumping system be made of materials
which will resist the solvents which are to be utilized. Table 4.3 lists recommended materials for use
with various solvents. Since virgin Teflon® and various forms of filled Teflon® will tolerate all solvents,
seals of these materials are recommended and most often
used. It should be noted that the gaskets, "O"-rings,
and so on, employed in the rest of the chromatographic
system should be given the same consideration.

TABLE 4.3.   RECOMMENDED MATERIALS FOR PUMP SEALS AND
GASKETS

Solvent Type	Recommended Material
Aromatic hydrocarbons, phenols	Viton®-A
Alcohols, dimethylformamide	Buna-N
Saturated hydrocarbons	Buna-N or Viton®-A
Ketones, ethers	EPR*
Aqueous systems	EPR*, Viton®-A, Buna-N
All solvent systems	Teflon®, Rulon®, and other polytetra-fluoroethylenes

*Ethylene/propylene rubber

Mechanical Pumps

The screw-driven, syringe-type (constant displacement)
pump has been incorporated into several commercial instruments. Figure 4.4 is a schematic of a single
stroke, displacement-type pump. The plunger is actuated by a screw feed which is driven through a gear
box by a stepping motor. Flow rate is electrically
controlled by varying the voltage on the motor, which
changes the rate of delivery of the fluid by the piston. The main advantage of this type of pump is that
it can deliver a high-pressure (3000-7000 psi), pulse-
free supply of liquid (250-500 ml). This type of pump
is considered to deliver a constant flow independent
of the operating pressure. However, minor compressi-
bility of liquids (e.g., ~ 3% at 6000 psi) and pump
seals does result in some flow change as the pressure
is increased. A disadvantage of the positive-displacement pump is that it has a limited solvent capacity

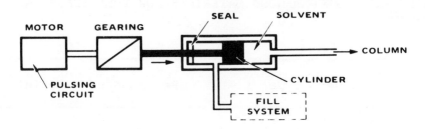

Figure 4.4   Constant displacement pump.   N. Hadden
et al., **Basic Liquid Chromatography**, Varian Aerograph,
Walnut Creek, Calif., 1971.   Reprinted by permission
of publisher.

and must be stopped for refilling. The changing of
solvents in this pump can be somewhat awkward; therefore, some models are equipped with devices to facilitate this operation. This type of pump is expensive,
but convenient to operate, since the output is easily
controlled. With two pumps in tandem, multishaped for-
ward and reverse gradients can be produced.

Another widely used mechanical pump is the reciprocating type. Figure 4.5 shows a schematic of a
piston-type, constant volumetric-flow pump. Each stroke
of the pump plunger pushes a small volume of mobile
phase into the chromatographic system, and on the reverse stroke, the piston cavity refills from the reservoir as a result of the ball-check valve system. The
rate of flow output from the pump is easily controlled
by adjusting the length of the stroke of the plunger.
Diaphragm reciprocating pumps work similarly to piston
pumps, except that a membrane of stainless steel or
fluorocarbon polymer is in contact with the mobile
phase. This diaphragm is actuated by a piston working
on a cavity of oil, and each stroke of the piston
causes the oil to actuate the diaphragm, which results
in a pulsating supply of liquid from the pump.

The specific advantage of the reciprocating pumps

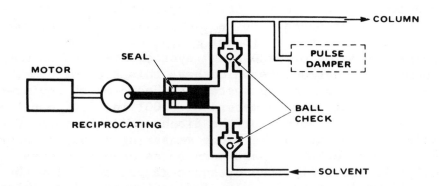

Figure 4.5 Reciprocating pump. N. Hadden et al.,
Basic Liquid Chromatography, Varian Aerograph, Walnut
Creek, Calif., 1971. Reprinted by permission of pub-
lisher.

is that their internal volume can be small, and their
delivery is continuous. Consequently, there is no re-
striction on the size of the reservoir which can be
used. This type of pump is particularly useful for recycle chromatography (2) (see also Section 10.5). An
advantage of the reciprocating pumps is that they deliver a fairly constant flow rate regardless of the back
pressure of the LC column. However, a measurable deviation from constancy is observed when the pressure
is increased, mainly due to leakage in the check valves
in the pump. Recently, more expensive reciprocating
pumps have appeared on the market which can partially
compensate for these flow changes. Multihead, sini-
soidal drive pumps have also been developed to minimize the output pulsations. When using mechanical
pumps, it is desirable to use a pressure relief device.
This protects the pump seals if a blockage occurs in
the high-pressure side of the chromatographic system.
A pulsating mobile phase can cause disturbances
to certain detectors (e.g., differential refractometer)

and limit detection sensitivity by the amount of baseline noise produced. To reduce pulsations in the carrier flow from the action of reciprocating pumps,
pulse-dampening or "snubbing" systems are used in most
commercial instruments which have this type of pump.
An effective dampening system for "home-made" equipment
is a combination of 5 meters of 0.001 in. i.d. capillary
tubing and the volume of the diaphragm or Bourdon-tube
which is present in the gauge measuring column pressure.
Arrangement of these parts can be as seen in Figure
4.1; alternatively, the pressure gauge can follow the
capillary for more accurate indication of column inlet
pressure. The capillary tubing acts as a flow restrictor, and the diaphragm or Bourdon-tube as a capacitor;
the combination of these two components usually reduces the pressure pulsations to manageable levels. Pulsations in the mobile phase may affect detector outputs, but apparently have no noticeable effect on the
plate height of the column (3).

     It should be noted that pulse-dampening devices
increase the volume of the system between the pump and
the chromatographic column. This additional volume increases the inconvenience of changing from one carrier
system to another, since for satisfactory operation
all the dead volume between the pump and the column
must be completely replaced with the new carrier. The
increased volume of the pulse-dampening system can have
a particularly deleterious effect on gradient elution,
where the added volume acts as an unwanted mixing vessel which tends to degrade the desired characteristics
of the solvent gradient.

     Pneumatic Pumps

The second type of pump used widely in modern LC equipment is the pneumatic device. Figure 4.6 shows a schematic of a simple direct pressure pump. In the

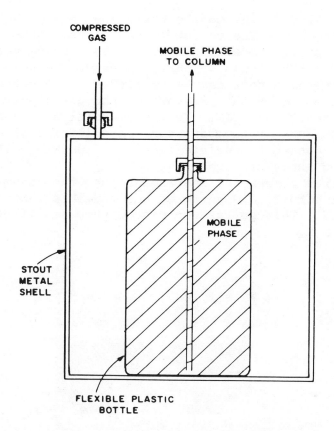

Figure 4.6   Simple direct pressure pump.   S. G. Perry,
R. Amos, and P. I. Brewer, Practical Liquid Chromatography, Plenum Press, New York, 1972.   Reprinted by
permission of publisher.

simplest form of this type, gas pressure works on a
collapsible container filled with the mobile phase.
Alternatively, the mobile phase is contained in a
special reservoir such as a coil of tubing or some
other appropriate pressure vessel.   These systems are
arranged to minimize problems with the high-pressure

gas dissolving in the mobile phase and forming bubbles in the detector. The output pressure from the simple, direct-pressure pumps is controlled by varying the gas pressure on the liquid or on the pressure vessel. The maximum pressure that can be delivered by this type of pump is limited by the pressure of the gas supply, which is about 3000 psi. Most commercial direct-pressure pumps have maximum pressure capabilities of about 1500 psi, due to the safety aspects of working with compressed gases at higher pressures.

Several commercial instruments use the pneumatic amplifier pumps. A diagram of one model is shown in Figure 4.7. This gas-driven piston pump amplifies the

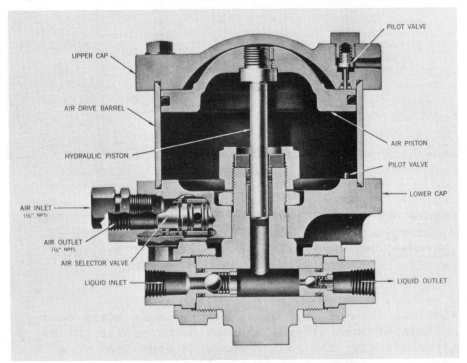

Figure 4.7 Pneumatic amplifier pump (courtesy of Haskel Engineering and Supply Co.).

gas pressure by using a large-area pneumatic piston
and a small-area liquid piston; it acts much like an
air-actuated syringe. The pneumatic amplifier pump
may be equipped with a power-return stroke which permits refilling of the carrier from the solvent reservoir in about 1 sec, so that the detector baseline is
interrupted only momentarily during this refilling
operation. The advantages of the pneumatic amplifier
pumps are that high pressures can be achieved rather
economically and the output of mobile phase is pulse-
free and relatively constant. This type of pump is
preferred by many workers for trace component and high-
precision quantitative analyses because of the low
level of noise which is generated in detectors by the
pumping action. Another advantage of the pneumatic
amplifier pumps is that the flow can be programmed
easily by varying the gas supply. The disadvantages
are (a) they can deliver only limited volume of mobile
phase before being refilled (although the refilling is
quite rapid), and (b) the flow rate from the pump is
constant only as long as the pressure drop across the
chromatographic system remains constant.

Comparison of Pumping Systems

The advantages and disadvantages of the various types
of pumping systems which have been used in modern LC
are summarized in Table 4.4. It should be noted that
each of these pumping systems has some unique advantage, for example, cost, convenience, or performance.
However, none of the pumps have all of the desired
characteristics; therefore selection of the pumping
system must be based on individual need.

TABLE 4.4. PROPERTIES OF PUMPS FOR MODERN LC

I. Mechanical Pumps

  A. Positive-Displacement Pumps (Syringe Type)

Advantages	Disadvantages
Pulse-free	High cost, particularly for two-pump gradient elution
Flow rate essentially independent of carrier viscosity and column permeability	Limited solvent capacity
Rapid pressure buildup	Inconvenient in changing solvents
Convenient electronic flow control	

  B. Reciprocating and Diaphragm Pumps

Advantages	Disadvantages
Essentially constant carrier flow rate regardless of solvent viscosity and column permeability	Pulsating output
Small internal volume	Limited range of volume output
Modest cost with some models	Slow buildup of pressure
Simple mechanical flow control	

106

II. Pneumatic Pumps

    A. Simple Direct-Pressure Pumps

Advantages	Disadvantages
Lowest cost	Limited carrier capacity
Simple operation	Limited pressure output
Pulse-free	Flow rate dependent on carrier
Rapid pressure buildup	viscosity and column
	permeability

    B. Pneumatic Amplifier Pumps

Advantages	Disadvantages
High pressure at low cost	Flow rate dependent on carrier
Pulse-free	viscosity and column
Convenient flow control, easily	permeability
programmed	
Capable of high carrier delivery rates	
Easy carrier change	
Rapid pressure buildup	
Excellent reliability	

4.4   OTHER UNITS IN THE HIGH-PRESSURE MOBILE PHASE DELIVERY SYSTEM

Filter

A filter is normally placed in the line following the pump to remove fine particles of particulate material which can clog the inlet of the column. Generally, a 2-μ sintered stainless steel filter is adequate for this purpose, and most commercial instruments contain such a device. When using columns containing particles in the < 10-μ range, some workers have found it necessary to use a 1/2-μ filter in the line. These filters must be constructed of materials to withstand the pressures of the system. When permitted by the pumping system, a low pressure-drop filter sometimes is placed in the inlet to the pump. This device protects the pump against dust and other foreign material which may affect check valves and seals, resulting in decreased pump life and pumping accuracy.

Pressure Monitoring Devices

A device for monitoring the column input pressure should be inserted in the line between the pump and the chromatographic column. This pressure-monitoring device indicates if there has been a plugging of the column or a failure of the pumping system. Knowledge of the pressure of the system assists in the optimization of separation parameters and also permits the calculation of column permeability (see Section 2.5 and Appendix II).

Two types of pressure monitoring devices are in current use. The first is the pressure transducer or strain gauge. These electronic devices are somewhat expensive, but are capable of high precision and can be equipped with both high- and low-pressure alarm or

cutoff systems to protect the pumping system. Bourdon-
tube or diaphragm gauges are more widely available and
are less expensive. However, gauges are less precise
and offer no protection to the pumping system. When
these gauge-type devices are used in conjunction with
constant-volume pumps, it is desirable to have a pres-
sure-release system on the outlet of the pump so that
it is protected against high-pressure overload. Table
4.5 lists the characteristics of the two systems for
measuring pressure in LC instruments.

TABLE 4.5.   CHARACTERISTICS OF PRESSURE-MEASURING
             DEVICES

Diaphragm or Bourdon Tube Types	Pressure Transducer (Strain Gauge)
Less costly Simpler design More robust	More precise Lower internal "dead" volume Can be obtained with alarm or cutoff circuit

The dead volume in a pressure-indicating device
is troublesome when changing solvents. This is more
critical with the diaphragm or Bourdon-tube pressure
indicators, since the internal volume of these units
is usually larger than that in the strain gauges. Air
left in the dead volume of a Bourdon gauge will dissolve in the mobile phase. This dissolved air can
cause bubbles in the detector, even though the mobile
phase was initially degassed. The dead volume within
these pressure sensing systems is particularly a

problem when changing solvents and using the refractive
index detector. Residues of solvents left in the pressure gauge can cause severe baseline drift which per-
sists for long periods. Flow-through pressure sensing
devices have been used to minimize this problem; however, the best approach is to completely dry out the
lines between the pump and the column (plus pressure
sensor) by evacuation after disconnecting them from
the system.

4.5   EQUIPMENT FOR GRADIENT ELUTION

Gradient elution is often used to solve the general
elution problem (Section 3.4). Such a situation exists
when a complex mixture of solutes having widely varying k' values must be separated. Gradient elution in
liquid chromatography is somewhat analogous to pro-
grammed-temperature gas chromatography, except that the
change in the k' of the solutes during the chromatographic run is accomplished by changing the polarity,
pH, or ionic strength of the carrier, rather than the
temperature. For controlling the retention of compounds during the separation of complex mixtures, gradient elution is more powerful than either temperature
or flow programming (4) (see Section 13.3 for details).
     Devices for conducting gradient elution can be
divided into two types: those that mix solvents at
atmospheric pressure and are then pumped at high pressure into the column and those in which the solvents
forming the gradient are pumped at high pressure into
a mixing chamber preceding the column. Changes in the
mobile phase composition during gradient elution can
be stepwise or continuous. Changes in solvent composition can be programmed on a linear basis, but often
an exponential change is preferable for optimum operation. Gradient elution is difficult to use with certain types of detectors, and this aspect is discussed
more fully in Chapter 5.

Low-Pressure Gradient Formers

Gradient systems in which the solvents are mixed at
atmospheric pressure and then pressurized by the high-
pressure pump have an advantage in that they can be
simple to assemble, relatively inexpensive, and permit
very large changes in the capacity factors to be made
during a chromatographic run. In this approach, a
single high-pressure pump is required to transport the
gradient solvent from the mixing vessel to the chromatographic column. The pump used in this arrangement
should be of the reciprocating type with a small internal volume, so that mixing effects within the pump
itself are minimized.

Figure 4.8 is a schematic of one system which uses
the low-pressure gradient mixing method. In this equipment, a series of vessels containing mobile phases of
increasing solvent strength is arranged from right to
left. The separation is started with the weakest solvent contained in the vessel to the extreme right, and
the gradient is produced by sequentially opening the
valves connecting the reservoirs containing solvents
of increasing strength, going progressively to the
left. Thus, the gradient is generated by mixing solvents at atmospheric pressure, and this mixed mobile
phase is fed to the column with a single high-pressure
pump. Useful gradients may be generated with systems
consisting of only two reservoirs containing solvents
of different strength. However, by using a large number of vessels containing solvents of different
strengths, very sophisticated gradients may be generated which are capable of resolving very complex mixtures containing compounds of very wide chemical differences (5,6).

Another form of the low-pressure gradient mixing
system is shown by the schematic in Figure 4.9. In
this equipment, the starting carrier in reservoir 2 is
allowed to feed the high-pressure pump 2. To generate

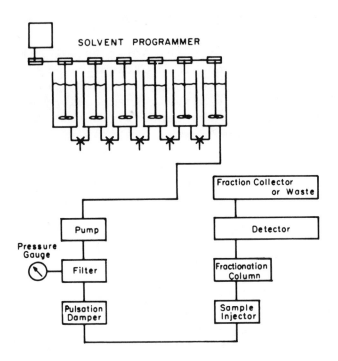

Figure 4.8   Gradient LC apparatus with low-pressure mixing.

the gradient, the low-pressure metering pump 1 then feeds the higher strength solvent contained in reservoir 1 into the gradient chamber leading to the high-pressure pump. Low-volume reciprocating pumps are also required for this approach, and the gradient is usually limited to a solvent strength range which can be produced with the use of only two solvents. However, it is possible to add a third solvent reservoir to this gradient system by using suitable switching valves. Some commercially available modular low-pressure gradient mixing devices are listed in Appendix V.

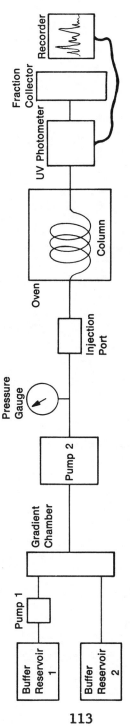

Figure 4.9 Two-pump system for low-pressure mixing of gradient. Reprinted by permission of Varian Aerograph.

113

High-Pressure Gradient Formers

This type of gradient device involves high-pressure
mixing of the mobile phase solvents. The gradient is
generated by programming the delivery of the high-
pressure pumping system. Since the output of pumps
may be controlled electrically, it is possible to generate almost any type of gradient. Figure 4.10 is a
schematic of one form of high-pressure mixing apparatus for gradient elution. In this approach, the output from two high-pressure pumps is programmed into a
mixing cell before flowing into the column. This is a
very convenient system and used extensively in commercial instruments. However, this arrangement is quite
expensive, since two high-pressure pumps are required.
The type of gradient which can be produced by this
system is also limited by the availability of only two
solvents to produce the gradient.

An interesting variation of the high-pressure mixing approach which requires the use of only one high-
pressure pump is shown in Figure 4.11. The weakest
solvent is contained in reservoir A, and this base solvent directly feeds the high-pressure pump. Valves in
the flow system are arranged so that at the beginning
of the gradient run, this base solvent flows from the
pump through valve A, into the mixing chamber and then
into the column. At this time, valves B and F are
closed. The higher strength solvent in reservoir B
has previously been allowed to flow through valve B
and the holding coil C and out valve E, so that the
holding coil system is filled with the higher strength
solvent. To start the gradient, valves B and F are
momentarily opened electrically, so that the flow of
the solvent from reservoir A pressurizes the solvent
in the holding coil. This forces the higher strength
solvent B from this holding coil through valve D to
the mixing chamber and the column. By electronically
proportionating the opening and closing of the valves,

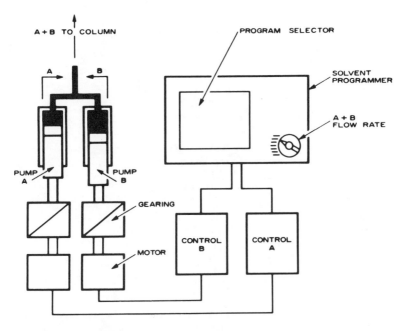

Figure 4.10   High-pressure mixing of gradient with programmed pumps.   N. Hadden et al., Basic Liquid Chromatography, Varian Aerograph, Walnut Creek, Calif., 1971.   Reprinted by permission of publisher.

the solvent from the two reservoirs may be mixed in a wide variety of concentration profiles. This system still has the limitation of producing a gradient from only two solvents.

High-pressure mixing systems have a unique advantage for LC when used in the isocratic or constant-solvent-composition mode. The relative concentration of the two mobile phases can be easily selected by setting a dial on the electronic system controlling the pump outlets. Therefore, solvent strength for

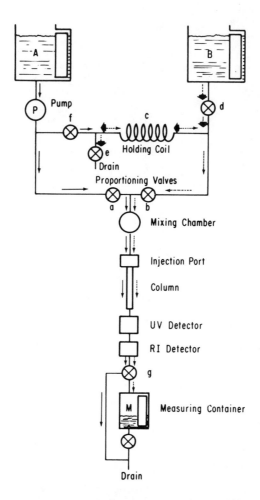

Figure 4.11   High-pressure gradient system with one pump.   Reprinted by permission of Du Pont Instrument Products Division.

isocratic separations can be optimized rapidly and conveniently.   In the high-pressure mixing gradient system, design of the mixing cell is critical.   The slow diffusion of liquids, coupled with the relatively low fluid velocities, makes turbulent mixing difficult.

The characteristics of the low- and high-pressure mixing systems for gradient elution LC are compared in Table 4.6. It should be noted that there is the usual compromise in performance, convenience, and cost, so that a choice between the two approaches should be based on individual need.

TABLE 4.6. COMPARISON OF GRADIENT SYSTEMS

Low-Pressure Mixing	High-Pressure Mixing
More versatile	More convenient
Simpler equipment	Easier solvent change
Lower cost	Automation easier
Greater change in elution power possible	

Figure 4.12 shows an isocratic separation (chromatogram B), compared to a linear gradient separation of the same mixture (chromatogram A). In the gradient run, improved separation of the early-eluting peaks is obtained, and elution of the more strongly retained components has been significantly speeded up. The peaks for the more strongly retained compounds have been greatly sharpened, and as a result, the sensitivity for these components has been increased. In addition, two minor components in the gradient run are evident that were not apparent in the isocratic separation.

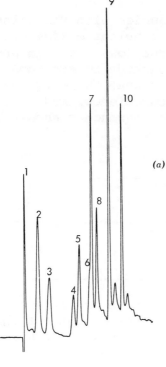

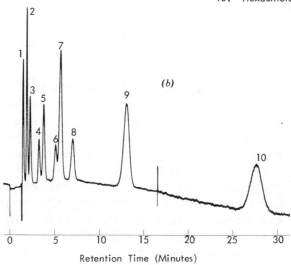

(a)

(b)

PEAK IDENTITY

1. Benzene
2. Monochlorobenzene
3. Orthodichlorobenzene
4. 1,2,3-trichlorobenzene
5. 1,3,5-trichlorobenzene
6. 1,2,4-trichlorobenzene
7. 1,2,3,4-tetrachlorobenzene
8. 1,2,4,5-tetrachlorobenzene
9. Pentachlorobenzene
10. Hexachlorobenzene

Retention Time (Minutes)

118

Figure 4.12 Comparison of gradient elution and isocratic separations. Comparison of gradient elution
and constant composition for the separation of chlorinated benzenes. Chromatogram A: linear gradient 40/60
methanol-water to methanol at 8%/min. Chromatogram B:
constant-composition 50/50 methanol-water. Column: 1
m x 2.1 mm i.d.; precision-bore stainless steel; packing: Permaphase® ODS; sample: 5 µl of chlorinated benzenes in isopropanol; detector: UV photometer at 254
nm; temperature: 60°C; column inlet pressure: 1200 psig.
R. A. Henry, in Modern Practice of Liquid Chromatography, J. J. Kirkland, ed., Wiley-Interscience, New
York, 1971. Reprinted by permission of publisher.

4.6 SAMPLE INTRODUCTION DEVICES

How a sample is introduced into an LC column is a very
important factor in obtaining high column performance.
Sampling is more critical in LC than in gas chromatography, although sampling techniques are somewhat similar for these two chromatographic methods. Sampling
is equally important for all forms of LC, whether GPC
or the other LC methods. The more diffused is the plug
of sample in the mobile phase introduced into the column, the wider is the separated component bands at the
end of the column. Therefore, ideally the sample
should be introduced as an infinitely narrow band onto
the chromatographic bed (see p. 33-35).
    The performance of very high-performance columns
containing very small particles is particularly influenced by sampling. Many columns containing particles
of <10 µ appear to be of the "infinite-diameter" type
(see Chapter 13 for discussion of this special type).
With an "infinite-diameter" column, the solute band,
as a result of radial diffusion during the development
of the chromatogram, never reaches the wall area containing most of the packing inhomogeneity. Thus,

particularly with columns of very small particles, careful injection of the sample in a very small spot in the
center of the column bed inlet is ideal. Other conditions of sampling can result in serious extra-column
band broadening, especially for solutes with small k'
(7).

Syringe Injections

One of the most widely used and simplest forms of sample introduction is the injection port. Figure 4.13
is an exploded view of one such device. With this
approach, the sample is injected with a microsyringe

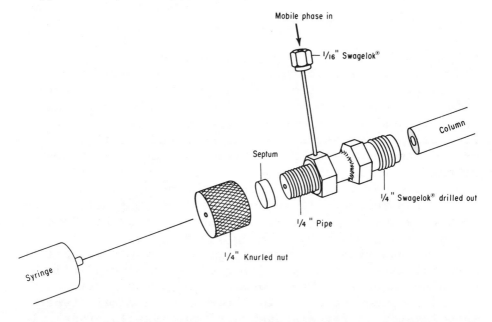

Figure 4.13 On-column injection port. R. A. Henry,
in Modern Practice of Liquid Chromatography, J. J.
Kirkland, ed., Wiley-Interscience, New York, 1971.
Reprinted by permission of publisher.

through a septum contained in a low-volume inlet system.
On-column injection places the sample directly into the
packing at the inlet of the column. This form of sample
introduction produces the highest separation efficiency,
since the sample is placed as a very sharp pulse at the
center of the column packing. Injection ports should
be designed so that the inlet system has a minimum volume and is cleanly swept by the incoming mobile phase.

Injection is made with microsyringes designed to
withstand pressures up to at least 1500 psi. Septum
materials are generally made from elastomeric silicone
or Neoprene®. However, fluoroelastomeric septa, silicone, and Neoprene® septa with Teflon® on one face, as
well as nylon- and fiberglass-reinforced septa, are
also commercially available. These special septa can
be used with carrier liquids (e.g., chloroform, dichloromethane) which tend to attack the ordinary elastomers.

While on-column injection yields the highest column efficiencies, this sampling technique does create
some practical problems. Since the injection is made
directly into the column packing, small particles of
packing often plug the needle. Small pieces of the
elastomeric septum material are also deposited on the
top of the column, which disturbs the packing and can
cause serious band broadening. To eliminate these difficulties, many commercial instruments use a swept-
port injection device, which permits a modification of
the on-column technique. In this approach, the sample
is injected through a low-volume inlet port directly
onto the top of a rigid column plug. This technique
is particularly convenient, since manufacturers often
fabricate columns with rigid end plugs for superior
transport and handling. The swept-port approach is a
useful sampling alternative, but does result in a
slight lowering of column efficiency for chromatographic peaks with small k' values.

It is generally not feasible to make syringe injections above about 1500 psi through sampling ports

containing elastomeric septums. Therefore, at high pressure, a <u>stop-flow injection</u> technique normally is used with syringes. The pump is first turned off until the column inlet pressure is essentially atmospheric. The sample is then injected in the usual fashion and the pump turned on. Pneumatic pumps are particularly useful in stop-flow sampling, since the pressure from these devices can be almost instantly resumed by actuating an on-off valve before the injection port. Since this approach can cause damage to syringe-type and reciprocating pumps, a three-way valve is normally used to divert the flow for stop-flow injection with these systems. Stop-flow injection can be used without affecting the efficiency of the LC process, since diffusion in liquids is very slow. With properly designed inlet systems, unwanted mixing of the sample with the mobile phase is not a problem.

Microsampling Valve

Samples can also be introduced into columns with microsampling valves, such as that shown in Figure 4.14. In this unit, the sample is held in an external loop in the valve; in other valves the sample is contained in an internal cavity. Microsampling valves are available with sample capacities from 0.5 μl up to several milliliters. Valves with external loops permit the introduction of larger sample aliquots. Some valves can be operated at pressures up to 5000 psi without leakage.

The question often arises, which type of sampling is best: syringe or sampling valves? Table 4.7 lists the major advantages of these two sampling modes. Generally, when scouting chromatographic conditions and defining the conditions for separation, the syringe injection technique is more convenient. However, use of a sampling valve is recommended once a routine procedure is established.

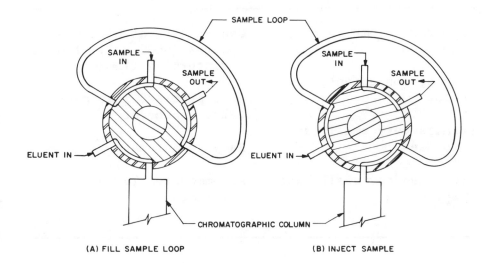

Figure 4.14 Six-port microsampling valve for LC.
C. D. Scott, in Modern Practice of Liquid Chromatography, J. J. Kirkland, ed., Wiley-Interscience, New
York, 1971. Reprinted by permission of publisher.

4.7 COLUMNS AND COLUMN HARDWARE

Column Materials

The unpacked column and the associated hardware both
contribute significantly to good column performance.
These units must be constructed of materials that will
withstand both the pressures to be used and the chemical action of the mobile phase. Most columns are made
of stainless steel tubing. However, heavy-wall glass

TABLE 4.7.   MAJOR ADVANTAGES OF SAMPLING MODES

Syringe	Sampling Valve
Injected volume readily changed	Can be used at high pressures (5000 psi) without disturbing flow
Simple, convenient, low cost	More precise, less operator dependent
Permits very small sample volumes	Easy to automate
Produces very small band spreading	Accommodates large sample volumes

columns are sometimes used; columns that will withstand pressures up to about 600 psi are commercially available. For operation at high pressures, glass-lined metal columns also can be used (see Appendix V).
    The type of tubing used for the columns can have an important effect on column efficiency. Columns made with mirror-finish, precision-bore stainless steel or Trubore® glass show threefold lower H values than columns made with ordinary seamless stainless steel tubing and prepared with the same packing (8). The very smooth internal walls of the precision-bore stainless steel and the glass columns apparently are important in obtaining a homogeneous packed bed. However, such differences are seen only with high-performance packings. Less efficient columns do not show as significant an effect with the different types of column blanks (9). The importance of columns with highly

polished walls also has been indicated in work with
particles of less than 10 μ (7).

Column Fittings and Plugs

Column end fittings should be designed with minimum
dead volume and no dead corners or pockets; these can
act as miniature mixing vessels and contribute to
extra-column band broadening (Section 2.3). Compres-
sion fittings with minimum dead volumes and small-
diameter passages now are used by most LC instrument
manufacturers.
    Porous plugs are used in the ends of columns to
retain the packing. These plugs must be homogeneous
to ensure uniform flow, with a minimum of unwanted
spreading of sample bands. Porous nickel, stainless
steel, or porous Teflon® plugs (or frits) have been
used for this purpose. Porous Teflon® generally should
be used only in the column outlet, since less-rigid
forms of this material can compress in the column inlet at high pressures. When on-column injection is to
be carried out, a plug of Teflon® wool can be advanta-
geously used in the column inlet. All of these plugs
should be about 1/8 in. long, and the porosity of the
plugs should be substantially smaller than the size of
the packing particles.

Column Dimensions and Configurations

Straight sections of LC columns in lengths of 25-150
cm are normally preferred. Some columns may also be
bent into a "U"-shape with little loss in efficiency,
providing the diameter of the bend is larger than about
3 in. Coiled columns are sometimes used, but are often
less efficient than columns prepared in straight sections. However, coiled columns of satisfactory

efficiency apparently can be produced providing the
ratio of the coil radius to the column radius is more
than 130 (10). This configuration minimizes the "race-
track" flow effect that occurs within coiled columns;
that is, where the distance that a solute must travel
in a coiled column is less near the inner wall than
it is near the outer wall. Coiled columns are generally packed in straight lengths and then bent into the
desired configuration, since columns that have already
been coiled are difficult to pack homogeneously. The
bending of heavy-wall tubing should be avoided, but
satisfactory results have been reported with the coil-
ing of certain soft tubing, such as copper and aluminum.

Long columns may be assembled from shorter lengths
of straight sections, using low-volume capillary con-
nectors. The data from one study (5) showed only a 3%
decrease in efficiency (per column length) when connecting short lengths to make longer columns. However,
it is important that matching columns be used to assemble longer lengths. If a column of good efficiency is
connected to a column of poor efficiency, a column of
average efficiency is not obtained; instead, a total
column of poor efficiency is likely. Similarly, columns of different diameter or containing different
stationary phases should not be connected (11).

Pre-columns (Figure 4.1) generally are desirable
and are required in liquid-liquid chromatography with
mechanically held stationary phases. The pre-column
ensures that the mobile phase is completely saturated
with the stationary phase before it passes into the
carefully prepared analytical column. The pre-column
need not be carefully packed, and can consist of larger
diameter tubing (e.g., 4.8 mm i.d., 50 cm long). For
LLC, 20-30% of the stationary liquid phase on a 120-
140 mesh diatomaceous earth support is convenient.
Pre-columns should be changed frequently to ensure
true equilibration between the mobile and stationary

phases in the chromatographic column.

"Guard" columns are sometimes used to protect carefully made analytical columns from extraneous materials in samples (e.g., water, highly retained components) which could cause changes in the performance characteristics of the analytical column. Guard columns usually consist of short lengths (5-15 cm) of the analytical column. These units are installed between the sampling device and the analytical column, and are frequently replaced if required.

The internal diameter also has a significant effect on the efficiency of LC columns. For analytical studies, columns 1-4 mm i.d. are normally used, with 2-3 mm i.d. appearing to be the best compromise between efficiency and convenience. Larger i.d. columns are common in analytical gel permeation chromatography. Columns of <1 mm i.d. are difficult to pack and impose very rigid requirements on the sample and detector systems. As the internal diameter and volume of the column is decreased, the volume of the inlet and detector becomes relatively larger. Consequently, band broadening due to extra-column effects in the inlet and detector are more likely. Columns of larger internal diameter are used for preparative work, with columns up to about 1 in. i.d. being useful for isolating unknowns for identification by other techniques. A more detailed discussion of large-diameter columns for preparative LC is given in Chapter 12.

4.8   COLUMN THERMOSTATS

It is important to control the column temperature in liquid-liquid and ion-exchange chromatography. Temperature variations within the column should be maintained within $\pm 0.2°C$; larger changes in column temperature can result in significant variations in retention time. In the case of liquid-liquid chromatographic systems with mechanically held liquid stationary phases,

close temperature control is even more important because changes in temperature can result in loss of
stationary phase from the analytical column (due to
changes in solubility). It is important to be able to
work at higher temperatures in gel permeation chromatography, but precise control of temperature is not
important.

Air Baths

Air baths are more convenient and generally preferred
for maintaining the temperature of liquid chromatographic columns. With air baths, ±1°C is easily maintained around the column which results in the temperature of the packing varying by no more than about
±0.2°C. Column air baths in LC are very similar to
those used in gas chromatographic instruments. These
involve high-velocity air blowers and electronically
controlled thermostatic systems.

Jacketed Columns

Alternatively, LC columns may be jacketed and the temperature controlled by circulating fluid through the
system from a constant-temperature bath. Circulating
water baths are commercially available which can maintain the temperature of the liquid to better than
±0.01°C, with resulting control of the column to at
least ±0.1°C. These are generally less convenient.

4.9   FRACTION COLLECTORS

Fraction collectors are normally not needed in high-
performance LC, since separation is accomplished in
minutes and manual collection is very convenient.
Most commercial instruments have a built-in sample

collection port on the outlet of the detector.   In conventional gel permeation chromatography with larger diameter columns, fraction collectors are often used, because larger volumes of carrier are employed and the separations are usually much slower.   Manufacturers of suitable fraction collectors for this operation are found in the laboratory guides listed in the bibliography for Appendix V.

4.10   FLOW RATE MEASUREMENT

A knowledge of the flow rate of the mobile phase, and how constant it is during the study, is required to obtain precise data on both retention times and peak height or peak areas for quantitation.   In gel permeation chromatography the flow rate must be quite precise, since retention volume is used to correlate data on molecular weight.   Variations in carrier flow rate can occur as a result of the failure of the pumping system, or with constant-pressure pumps if there is a change in the back pressure of the system resulting from partial pluggage of the column, temperature changes, and the like.

Volumetric Methods

Volumetric methods of measuring flow rate are most commonly used.   The technique is to collect the mobile phase for a known period of time in a calibrated vessel.   This technique is very widely used in gel permeation chromatography, but is somewhat time-consuming with small-diameter analytical columns.

Flowmeters

A convenient but relatively inaccurate type of flow measurement is the flow meter. The lack of reproducibility and the need for calibrating the flow meter for each mobile phase (because of differences in solvent density), make this method useful only as a semiquantitative technique.

Gravimetric Methods

The most accurate method for determining the flow is by gravimetry, in which the carrier is collected for a known period of time and then weighed. This is an inconvenient technique, but it is particularly useful for calibrating pumps because of its high precision.

Flow-Tube Method

One of the most convenient techniques is the use of the flow tube. An air bubble is introduced into the column eluent stream which is passing through transparent, volume-calibrated tubes connected to the detector outlet. The air bubble is then timed while traveling between two points representing a certain volume. With this device, flow rates can be quickly measured with the precision of < 1%.

4.11   DATA HANDLING SYSTEMS

A high-speed recorder should be used with LC equipment, since in many instances separation is completed very rapidly. Table 4.8 lists the characteristics of the desirable recorder for high-speed LC.

TABLE 4.8.   DESIRABLE RECORDER CHARACTERISTICS FOR
HIGH-SPEED LC

Pen response of 1 sec full scale or less

High input impedence

Good AC noise rejection

Floating input

Variable chart speeds (e.g., 4 in./min to 4 in./hr)

As discussed in Section 13.1, digital electronic
integrators and computerized systems are of considerable value for precise quantitative LC analysis and
are particularly used for routine quantitative work.

4.12   SAFETY ASPECTS IN LC

Because high pressures are normally used in modern LC
systems, neophytes often worry about the safety of
such systems. However, it should be remembered that
liquids are difficult to compress (e.g., only about
4% decrease in the volume of methanol from 15 psi to
7500 psi). Therefore, very little energy is stored
up at the pressures normally used in modern LC, and
these pressures pose no particular hazard. Should a
break or leak occur within a high-pressure system
supplied by a constant-volume pump, the pressure immediately goes essentially to atmospheric, with only
the leakage of a small amount of solvent. It should
be noted that even with a leak, pneumatic pumps will
maintain the pressure of the system as long as mobile
phase is available.
    The main safety consideration comes in the

handling of the flammable or toxic solvents which may be used. This potential hazard is eliminated by using the ordinary precautions of working in a well-ventilated laboratory, using safety glasses, and so on. Placing the LC instrument in an area with an overhead exhaust often is not adequate, since the vapors from some solvents are heavier than air. To reduce the possibility of fire, some commercial instruments offer the option of purging the reservoir and column compartment continuously with nitrogen. Pressure-overload devices also protect some equipment from possible hazards due to solvent spills resulting from leaks.

REFERENCES

1.   R. E. Leitch, J. Chromatog. Sci., 9, 531 (1971).
2.   K. J. Bombaugh and R. F. Levangie, J. Chromatog. Sci., 8, 560 (1970).
3.   C. G. Horvath, B. A. Preiss, and S. R. Lipsky, Anal. Chem., 39, 422 (1967).
4.   L. R. Snyder, J. Chromatog. Sci., 8, 692 (1970).
5.   L. R. Snyder and D. L. Saunders, J. Chromatog. Sci., 7, 195 (1969).
6.   R. P. W. Scott and P. Kucera, J. Chromatog. Sci., 11, 83 (1973).
7.   J. J. Kirkland, Gas Chromatography, 1972, S. G. Perry, ed., Applied Science Publishers, Essex, England, 1973, p. 39.
8.   J. J. Kirkland, J. Chromatog. Sci., 7, 361 (1969).
9.   B. L. Karger and H. Barth, Anal. Lett., 4, 595 (1971).
10. H. Barth, E. Dallmeier, and B. L. Karger, Anal. Chem., 44, (1972).
11. J. Kwok, L. R. Snyder, and J. C. Sternberg, Anal. Chem., 90, 118 (1968).

BIBLIOGRAPHY

L. V. Berry and B. L. Karger, Anal. Chem., <u>45</u>, 819A, (1973).

R. M. Cassidy and R. W. Frei, Anal. Chem., <u>44</u>, 2250 (1972). (Design of injection port and pressure relief device for home-made equipment.)

D. Chilcote and C. D. Scott, Chem. Instrum., <u>3</u>, 113-24 (1971). (Use of a small dedicated computer with a high-resolution liquid chromatograph.)

T. H. Gouw and R. E. Jentoft, in <u>Guide to Modern Methods of Instrumental Analysis</u>, T. H. Gouw, ed., Wiley-Interscience, Chapter II, pp. 43-82, 1972. (Contains section on instrumentation for modern LC.)

R. A. Henry, Liquid Chromatography Technical Bulletin, No. 73-1, E. I. du Pont de Nemours & Co., Instrument Products Division, Wilmington, Del., 19898.

R. A. Henry, in <u>Modern Practice of Liquid Chromatography</u>, J. J. Kirkland, ed., Wiley-Interscience, New York, 1971. (Discussion of all components in an LC system, except detectors.)

Instrument and Control Systems, April 1971, p. 103-10. (Survey of metering pumps.)

1973/74 International Chromatography Guide, J. Chromatog. Sci., January, 1973. (Directory of manufacturers, instruments, accessories, etc., for chromatography.)

B. L. Karger and L. V. Berry, Anal. Chem., <u>44</u>, 93 (1972). (Inexpensive pneumatic pumping system for continuous operation.)

J. J. Kirkland, Anal. Chem., <u>40</u>, 391 (1968). (Description of high-performance UV detector and home-made apparatus for LC.)

1973/74 Laboratory Guide, Anal. Chem., August, 1973. (Contains guide to available equipment and other products for LC.)

S. G. Perry, R. Amos, and P. I. Brewer, <u>Practical Liquid Chromatography</u>, Chapter 7, Plenum Press, New York, 1972. (General discussion of LC equipment.)

C. D. Scott, W. F. Johnson, and V. E. Walker, Anal. Biochem., <u>32</u>, 182-84 (1969). (Sampling value for high-pressure LC.)

J. Q. Walker, M. T. Jackson, Jr., and J. B. Maynard, <u>Chromatographic Systems; Maintenance and Trouble-shooting</u>, Academic Press, New York, 1972. (Contains information on LC components and equipment.)

N. Hadden, et al, <u>Basic Liquid Chromatography</u>, Varian Aerograph, Walnut Creek, Calif., 1971.

CHAPTER FIVE

DETECTORS FOR LIQUID CHROMATOGRAPHY

5.1 INTRODUCTION

Sensitive detectors that can continuously monitor the column effluent are a major requirement in modern LC. Unfortunately, the physical properties of both mobile phase and sample components are often quite similar, which makes the detection of eluted LC bands difficult. Various approaches to this problem have been pursued in the development of modern LC detectors:

- Elimination of the mobile phase before detection.
- Differential measurement of a bulk property possessed by both sample and solvent.
- Selection of a measurable sample property which is not possessed by the mobile phase.

This has led to the appearance of a variety of special detectors for LC. Although no one of these detectors is universally applicable to all LC problems, there is usually a satisfactory detector for a particular application.

The need for detector sensitivity must be stressed. In high-performance LC, samples as small as 10 μg can be required. If we wish to detect individual components at a level 1/1000 of this amount (i.e., compound concentrations of 0.1%, we will need a system which can detect 10 ng or less of a solute eluting in about 1 ml of carrier. Therefore, whether or not a particular detector has sufficient sensitivity for a desired application is of major importance.

Table 5.1 lists the characteristics of an ideal LC detector. Unfortunately, no presently available detector possesses all of these properties, nor is it likely that any such detector will be developed.

TABLE 5.1.   CHARACTERISTICS OF IDEAL LC DETECTOR

The Ideal Detector Should:

Have high sensitivity and predictable response.

Respond to all solutes, or else have predictable specificity.

Be unaffected by changes in temperature and carrier flow.

Respond independently of the mobile phase.

Not contribute to extra-column band broadening.

Be reliable and convenient to use.

Have a response which increases linearly with the amount of solute.

Be nondestructive of the solute.

Provide qualitative information on the detected peak.

None of the present LC detectors are as versatile or as universal as might be desired, and some workers regard the detector as the weak link of modern LC technology. There is no LC equivalent to the thermal conductivity or flame ionization detectors of gas chromatography. However, in spite of these limitations, presently available detectors allow a wide range of applications, and in most cases, LC applications are not limited by the detector. Moreover, development of new detectors is continuing, and significant improvements can be expected.

Two types of LC detectors are in use: the bulk property or general detectors, and solute property or selective detectors. Bulk property detectors measure a change in some overall physical property of the mobile phase plus solute. The solute property detectors are sensitive only to the solute.

To select a detector for a particular application, some information is needed on the specifications of these devices. Unfortunately, these specifications are seldom given in common terms by the various suppliers. Therefore, considerable confusion exists regarding the abilities of some detectors. To complicate matters further, for most samples of interest there are no published reference values for many of the properties measured by different detectors. Therefore, to gain the needed information on detector performance, a similar application in the literature has to be found. Alternatively, information may be obtained from one of the applications laboratories of the various instrument manufacturers. Most often, the sample is simply tested in the laboratory with the detector under consideration.

There are several important detector parameters which should be considered when evaluating these devices. First, the noise of the detector should be known. Detector noise can arise from instrument electronics, temperature fluctuations, line voltage surges,

and other sources resulting in a change of detector
output signal which is not directly attributed to the
solute. Variation in some detector signals also can
occur as result of flow changes, pulses from the pump,
and the like.

Also needed is information on detector drift,
which is a steady movement of the baseline either up-
or down-scale. In Figure 5.1, we see a representation
of both high-frequency noise, which appears as a "fuzz"
and widens the baseline, and short-term noise, which
is a variation of the recorder tracing appearing as
random peaks and valleys on the baseline. Also shown
is a representation of detector drift. Drift tends to
camouflage noise, and a determination of noise cannot
be made accurately unless the drift is small in rela-
tion to the magnitude of the noise. Detector noise is
often quoted as a percentage of full-scale response at
specified sensitivity, as a peak-to-peak voltage, or
sometimes as a standard deviation. At this point, it
should be noted that many problems with apparent detector noise and drift in the laboratory are a function
of the LC system (e.g., solvent impurities, column temperature variations), rather than some inherent limitation in the detector itself.

Information is also needed on the absolute sensitivity of the detector. This is the total change in a
physical parameter which is required for a full-scale
deflection of the recorder at maximum sensitivity, with
a defined amount of noise. To make direct comparisons
between detectors, the sensitivities must be quoted at
equal noise magnitudes.

Another useful parameter of detectors is the relative sample sensitivity or detection limit, which is
the minimum concentration of solute that can be detected. This sensitivity often is taken to be the concentration of a solute in the column effluent which
results in a peak that is twice the value of the noise.
It should be noted that relative sample sensitivity is
usually dependent on the chromatographic system. For

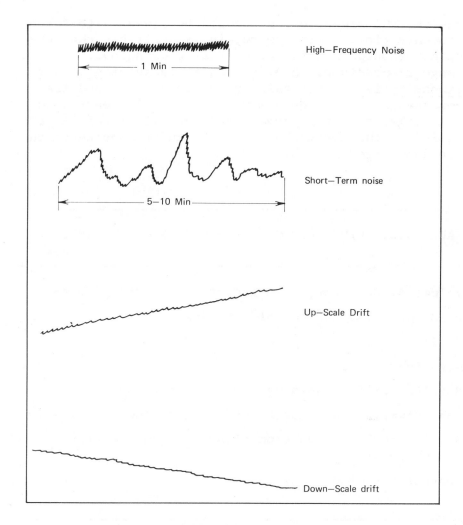

Figure 5.1   Detector Noise and Drift.

instance, the detectability of general detectors such as the differential refractometer varies with the mobile phase for a given sample.

For a detector to function as a useful quantitative

device, the signal output should be linear with solute concentration. Furthermore, it is desirable that the detector output be linear over as wide a concentration range as possible. A wide dynamic range of <u>detector linearity</u> is of particular value when major and trace components must be determined in a single separation.

Other detector parameters generally involving the design of the particular detector under consideration are summarized in Table 5.2.

TABLE 5.2.   OTHER DETECTOR PARAMETERS OF INTEREST

Volume of detector, also geometry and arrangement.

Band spreading due to inlet tubing.

Facilities for back pressure.

Air-tight fittings.

Baseline drift caused by flow changes.

Sensitivity as a function of carrier flow.

Sensitivity as a function of ambient and carrier temperature.

Convenience of operation.

The cell should be designed with the minimum volume which is compatible with the other requirements of the detector. The peak broadening due to the volume of the detector cell is especially significant with early-eluting compounds ($k' < 2$) when using low-volume,

highly efficient columns. To minimize the broadening
of these early-eluting peaks, the effective volume of
the detector should be < 1/5, preferably < 1/10, of the
volume of the peaks of interest. In very high-performance LC columns with < 10-μ particles, peaks of < 50 μl
are sometimes of interest; therefore detectors with
effective volumes of < 5 μl are desirable. Recent
studies have shown that early-eluting peaks from low-
volume, high-efficiency columns of < 10-μ particles are
substantially broadened by a standard 8-μl cell, as
compared to a 1-μl cell (1). With compounds of k'
≥ 2, no significant difference in plate height was observed. Detector cell cavities also should be kept
smooth with no unswept volumes (so-called "dead" volume), which can cause unwanted mixing. When placing
detectors in series, the first unit should have the
smallest cell volume. In addition, connecting tubings
from the column to the detector and between detectors
should have a minimum internal diameter and short coup-
ling distance to minimize extra-column band spreading
(see Section 2.3).

It is desirable that detector cells be able to
operate under moderate pressure (e.g., 50-100 psi).
Gas bubbles, which can form in the detector, can be
eliminated by imposing 15-30 psi of back-pressure on
the detector cell (as well as by prior degassing of
solvent). Fittings on the detector and the rest of
the flow system should be air-tight to prevent diffusion of air into the mobile phase. It is not widely
recognized that air can diffuse into small holes from
which pressurized liquids cannot escape. In this way,
air can leak through insufficiently tightened compres-
sion fittings and dissolve in previously degassed
carrier.

Susceptibility of the detector response to changes
in carrier flow and temperature may have a significant
bearing on the usefulness of a certain detector for a
particular problem. Precise quantitative analysis

requires that a detector be little affected by these factors. Of particular importance in routine applications is how trouble-free a particular detector will be, and how easily it can be operated by relatively unskilled personnel.

5.2   ULTRAVIOLET AND VISIBLE PHOTOMETERS

Detectors based on ultraviolet absorption are probably the most widely used in modern LC. These ultraviolet (UV) devices are relatively insensitive to flow and temperature changes and have a high sensitivity for many solutes. Obviously, the samples must absorb in the UV to be detected. UV photometric detectors with sensitivities of 0.005 absorbance units full-scale with $\pm 1\%$ noise are now commercially available. With this high level of sensitivity, solutes with relatively low absorptivities can be monitored, and it is possible to detect a few nanograms of a solute having moderate UV absorption.

At present, monochromatic photometric detectors are most commonly used in LC. Figure 5.2 shows a schematic of one of the commercially available UV photometric devices. UV radiation (254 nm) from the low-pressure mercury lamp shines on the entrance of the flow cell, b, is transmitted through the mobile phase in the cell and then strikes the analytical photocell, a. Radiation from the back of the lamp shines directly into a neutral-density filter, c, and then to the reference photocell, r. The photocells then generate a current which is proportional to the intensity of the light. As shown by the schematic in Figure 5.3, the current from the phototubes is fed to log amplifiers which use specially matched diodes to generate a log function. To ensure long-term stability and low noise, the diodes are enclosed in a stable-temperature environment which is maintained by a temperature controller. From the log amplifier, the signal is fed

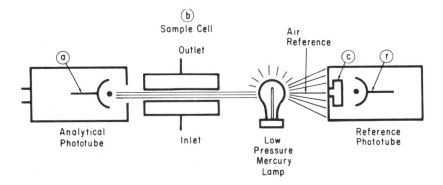

Figure 5.2 Schematic of UV photometric detector.
Reprinted by permission of Du Pont Instrument Products
Division.

to two operational amplifiers where balance and zero
suppression are introduced. The output signal to the
recorder is attenuated with a simple bridge network.

In addition to the widely used UV detectors operating at 254 nm, photometers operating at other wavelengths, for example, 220, 280, 313, 334, and 365 nm,
are available. The multiple wavelength operation of
photometric detectors is obtained by using appropriate sources and filters.

Figure 5.4 shows a cross section of one cell used
in a commercial UV detector. A split-stream or "H"
flow pattern is utilized in this cell to minimize
noise and drift due to possible flow fluctuations.
The mobile phase flows into the center of the optical
path, is split, and goes in opposite directions. At
both ends of this optical path, the mobile phase
sweeps the cell windows and is recombined in an upper
bore. This cell has an optical path with dimensions
of 1 mm i.d. and 10 mm in length and an internal volume of 8 μl. This volume is characteristic of cells
in commercial UV detectors, and is typical of the

Photometer Electronics

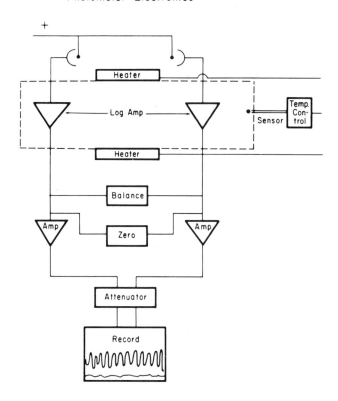

Figure 5.3   Schematic of UV photometer electronics. Reprinted by permission of Du Pont Instrument Products Division.

volumes used in cells for many types of modern LC detectors.

It is not necessary to work at the maximum absorption of a peak when using the UV photometric detector. Although the spectrum of cytidine in acidic solution shown in Figure 5.5 has an absorption maximum at 280 nm, detection can be carried out at 254 nm for nanogram

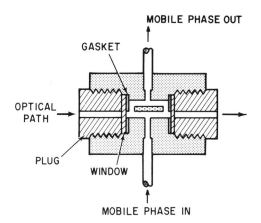

Figure 5.4 Cross section of split-stream flow cell
for UV detector. S. H. Byrne, Jr., in <u>Modern Practice of Liquid Chromatography</u>, J. J. Kirkland, ed.,
Wiley-Interscience, New York, 1971. Reprinted by
permission of publisher.

amounts. For maximum sensitivity, it would be desirable to work at 280 nm with this particular compound.
However, it is also important that the wavelength providing the highest selectivity be used, that is, the
wavelength that gives the largest signal relative to
possible interfering substances. Figure 5.6 illustrates this point. Chromatograms of aflatoxins in
peanut butter were obtained with UV detection at 254
and 365 nm (2). The aflatoxins absorb more strongly
at 365 nm. More importantly, many impurities that
absorb and interfere at 254 nm do not absorb at 365
nm. This enhanced detection selectivity makes the
365 nm wavelength far more desirable for determining
the aflatoxins in sample extracts.

    If a compound has any ultraviolet absorption, it
will usually absorb at 254 nm. Therefore, this wavelength is often satisfactory for compounds having UV

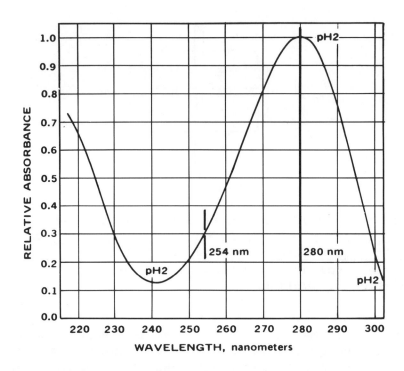

Figure 5.5 UV absorption spectrum for cytidine in
acidic solution. Reprinted by permission of P. L.
Biochemicals.

absorption. The minimum structural requirement for UV
detection at 254 nm is usually a double-bonded chromo-
phore to which is attached some structure with unpaired
electrons which enhances UV absorption. Some possible
minimum functions are indicated in the structure below,
but these are not limiting.

$$X = Y,$$
$$\mid$$
$$W$$

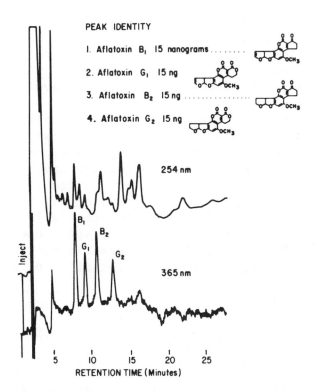

Figure 5.6   Detection of aflatoxins at two different
wavelengths.   Column, Zorbax®-SIL 25 cm x 2.1 mm i.d.;
mobile phase, 60% CH_2Cl_2/40% $CHCl_3$ (both 50% H_2O saturated)/0.1% methanol; pressure, 1500 psi; flow, 0.7
cc/min; temperature, ambient; detector, UV photometer,
254 nm (0.02 As f.s.), 365 nm (0.01 As f.s.).   Reprinted by permission of Du Pont Instrument Products Divn.

where,
      X = C, S, P, etc.,
      Y = O, S, P, N, etc.,
      W = UV-enhancer, such as S, P, etc.

Spectrophotometers are also useful LC detectors. A monochromator plus a continuous-spectrum light source permits a wide selection of wavelengths, both UV and visible. Therefore, such devices have the versatility and convenience of being able to work at the exact absorption maximum of a solute, or at a wavelength which provides maximum selectivity. Recently, several commercial spectrophotometric detectors specially designed for modern LC have been made available (see Appendix V). These devices have high-energy sources, relatively wide band-pass monochromators (narrow wavelengths generally are not required for LC), and stable, low-noise electronics. These factors permit signal-to-noise ratios and linearity of response comparable to the UV photometric detectors. Of course, spectrophotometric detectors cost more than monochromatic photometers. Table 5.3 summarizes the characteristics of UV detectors for modern LC.

TABLE 5.3. CHARACTERISTICS OF UV PHOTOMETRIC DETECTORS

Capable of very high sensitivity (but samples must have UV absorption)
Good range of linearity, ca. 5×10^4-10^5 (for some models)
Can be made with very small cell volumes (small band broadening influence)
Relatively insensitive to carrier flow and temperature changes (except at high sensitivity)
Very reliable
Easy to operate
Nondestructive to sample
Widely varying response for different solutes
Gradient elution capability

UV detectors are ideal for use with gradient elution, and many common UV-transmitting solvents of varying strengths are available as mobile phases in LC.
Therefore, a selection of solvents for use with the UV
detector is not usually a practical limitation. The UV
detector is often very useful for the trace analysis of
UV-absorbing solutes, since the UV-transmitting solvent
and other non-UV-absorbing compounds in the sample are
not detected and do not interfere.

Visible photometers are also used as detectors in
LC. These devices are potentially useful as both general and selective detectors, by utilizing well-known
color reactions which can be performed with the column
eluent and monitored by absorption photometry. An example of this approach is the ninhydrin reaction, which
is routinely used in commercially available instruments
for the determination of amino acids separated by ion-
exchange chromatography. There are hundreds of color
reactions which can produce general, as well as selective colors with a wide range of solutes eluting from
a column. It should be pointed out, however, that this
approach requires small-volume, fast reactions so that
the chromatographic peak is not broadened during the
color-forming step. Nevertheless, the colorimetric detector has excellent potential, and we can expect much
from this approach in the future.

5.3   DIFFERENTIAL REFRACTOMETERS

After the UV detector, the second most widely used LC
detector is the differential refractometer. This device continuously monitors the difference in refractive index (RI) between a reference mobile phase and
the mobile phase containing the sample as it elutes
from the column. Since this is a bulk property or
general detector, it responds to all solutes under the
proper operating conditions.

The commercially available RI detectors operate
on two different principles. The first is the Fresnel
refractometer, which is shown schematically in Figure
5.7. This device is based on Fresnel's law of reflec-
tion, which states that the percentages of light reflected at a glass/liquid interface varies with the angle
of incidence and the refractive indices of the two
phases. To obtain maximum sensitivity, this refractometer is operated using incident light at the liquid/
glass interface which is slightly less than the critical angle. Fluctuations caused by noise and temperature are minimized by using a differential measurement,
where the refractive indices of the sample and reference streams are continuously compared. In Figure 5.7,
light from source lamp SL passes through a source mask,
M1, an infrared blocking filter, F, and an aperature
mask, M2. The energy is collimated by the lens, L1.

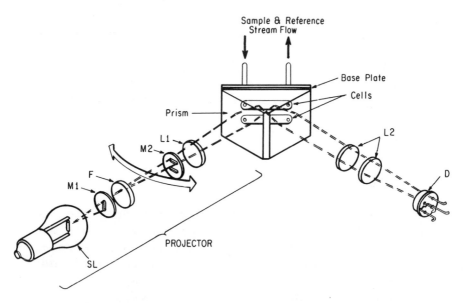

Figure 5.7 Fresnel refractometer. Reprinted by permission of Du Pont Instrument Products Division.

The mask, M2, yields two collimated beams which enter the cell prism and are focused on the glass/liquid interfaces. Sample and reference cell cavities are formed with a Teflon® gasket which is clamped between the cell prism and a stainless steel base plate. All of the optical components in the projector at the left of the schematic are mounted on a separate optical bench which rotates around a pivot to allow a change in the angle of incident light. Light is transmitted through the two interfaces in the cell, passes through the thin liquid film, and impinges on the surface of the cell base plate. This energy is then reflected, and lens L2 focuses this reflected light on the dual-element photodetector, D.

The cells in the Fresnel RI detector are very small (~ 3 μl volume) and are cleanly swept by the carrier, so that many workers prefer to use this device with very high-performance LC columns. However, the cell windows of this detector must be kept very clean for proper operation. This type of refractometer has limited range of linearity and two different prisms must be used to cover the useful refractive index range (about $\eta = 1.33$ to 1.63).

The second type of RI detector is the deflection device, shown schematically in Figure 5.8. Light from source A is limited by mask B, collimated by lens C, and passes through the detector cell D. The cell has separate reference and sample compartments which are separated by a diagonal piece of glass. When the composition of the mobile phase changes in the sample cell, the change in refractive index causes a deflection in the final position of the light beam on the photodetector. As the incident light passes through the cell it is first deflected, then reflected from the mirror E back of the cell and again deflected. The lens C focuses the light on the position-sensitive photodetector F, which produces an electrical signal proportional to the position of the light. The output

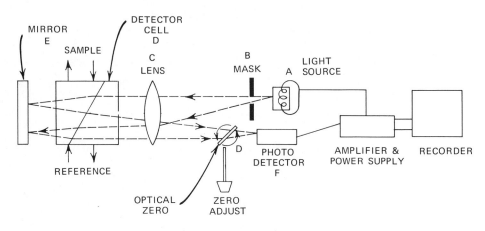

Figure 5.8   Deflection refractometer.   Reprinted by
permission of Waters Associates.

signal is then amplified and relayed to a potentiomet-
ric recorder.   This deflection system has the advantage
of a wide range of linearity.   Also, only one cell is
needed for the entire refractive index range.   The cell
in this detector is generally not as small nor as
cleanly swept as the cells of the Fresnel type, but it
is less sensitive to buildup of contaminants on the
cell windows.
      Table 5.4 summarizes the characteristics of the
refractive index detectors.   These devices have excellent versatility, since they respond to all sample components if the carrier is properly selected.   For maximum RI detector sensitivity, the carrier should have a
refractive index as different from the sample as possible.   A limitation of the RI detectors is that they
have only modest sensitivity, even under optimum conditions.   Because of this sensitivity limitation, this
detector is generally not useful for trace analysis
with narrow-bore, high-performance, analytical columns.
However, in optimum situations, it is possible to

quantitate peaks at the 0.1% concentration level.

TABLE 5.4   CHARACTERISTICS OF REFRACTIVE INDEX DETEC-
TORS

Excellent versatility--any solute can be detected
Moderate sensitivity for solutes
Generally not useful for trace analyses
Efficient heat exchanger required
Relatively insensitive to carrier flow changes, if
  properly thermostatted
Sensitive to temperature changes
Reliable, fairly easy to operate
Difficult to use with gradient elution
Nondestructive

To obtain optimum sensitivity and stability with
refractometer detectors, efficient heat exchangers in
the lines between the outlet of the column and the
detector cell are required. Unfortunately, these heat
exchangers increase the dead volume of the detector,
which can produce significant band broadening with very
high efficiency analytical columns. On the other hand,
if the temperature of the column effluent is closely
maintained by the heat-exchange system, flow programming of the mobile phase is possible (Section 13.4).
The acute sensitivity of refractometers to temperature
change represents one of the limiting features of this
form of detection. In practice, it is difficult to
maintain the temperature of the refractometer cell
with a precision that permits its use at the sensitivity claimed by the manufacturers.
Refractometers are convenient and reliable, although generally not as trouble-free and easy to

operate as the UV photometric detectors. RI detectors
are impractical to use with gradient elution. Rarely
is it possible to use this detector when carrier changes are made during the chromatographic run, since it
is exceedingly difficult to precisely match the refractive indices of the flowing reference and sample
streams. Despite the limitation of sensitivity and
impracticality for use in gradient elution, the differential refractometer is one of the most widely used
LC detectors, particularly in gel chromatography where
sensitivity is not so important.

5.4   SOLUTE TRANSPORT DETECTORS

Another important column monitoring device is the solute transport detector. This device is called the wire
or chain detector by some workers, and is known by
others as the flame ionization detector. With this
detector the volatile mobile phase is evaporated before the detection process is carried out. A portion
of the column eluent is continuously fed onto a transport system, such as a moving chain, wire, belt, disc,
or coil. Here the volatile mobile phase is evaporated,
and the nonvolatile sample residue remaining on the
transport system is then detected by some process such
as flame ionization.
    A major advantage of the solute transport approach
is that detection is essentially independent of the
chromatographic process, since the volatile mobile
phase is continuously removed from the transport system before detection is carried out. Since more volatile compounds will be evaporated with the mobile
phase prior to the detection process, this method of
detection is limited to relatively nonvolatile samples
(e.g., boiling above about $n-C_{18}$).
    Many forms of the transport detector have been
proposed and offered commercially. Some of these
specific models are no longer sold because of inherent

limitations in sensitivity, reliability, and universality.   Figure 5.9 shows a schematic of the latest version of a wire transport detector, which is probably the most satisfactory device of this type yet developed.   Wire from a feed spool, A, is continuously fed through a cleaner oven, B, where organic contaminants on the wire are oxidized.   The cleaned wire passes into block C and is coated with a thin film of the column eluent containing the separated solutes.   The wire then is led through an evaporation oven, D, where the solvent is removed.   The wire with the solvent residue is then carried into an oxidation oven, E, through which air or oxygen is flowing.   There, the solute is burned at about 850 °C and converted to carbon dioxide which is aspirated into a hydrogen stream by means of a molecular entrainer, F.   This gaseous mixture passes over a nickel catalyst bed, G, which is held at about

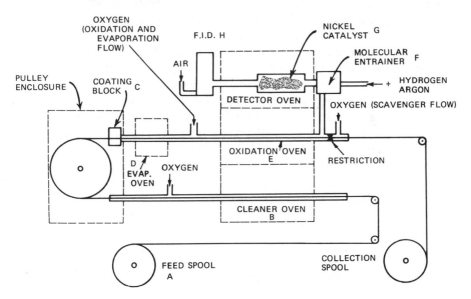

Figure 5.9   Schematic of wire transport detector.   R. P. W. Scott and J. G. Lawrence, J. Chromatog. Sci., 8, 65 (1970).   Reprinted by permission of publisher.

350°C. The catalyst assists in converting the carbon dioxide to methane, which then passes to a flame ionization burner, H, for detection.

This particular transport detector has a predictable response that is proportional to the number of carbon atoms in the solute molecule. It also has a linear response over a wide range of solute concentration. The chief disadvantages of this and other transport detector designs are bulkiness, high cost, relatively poor sensitivity (even though a flame ionization burner is used), and operational inconvenience. The lack of sensitivity of these devices is mainly due to the fact that the transport system picks up only a small part of the column effluent, at the most only a few percent of the total (but see modifications described in ref. 3). Extra-column band broadening effects with this detector (in present forms) can be significant, which degrades the performance of high-efficiency columns. Table 5.5 lists the general characteristics of transport detectors for modern LC.

While the solute transport detectors have not been widely used to date, there has been a recent increase in interest. Some experimenters are reporting improved sensitivity and stability, due mainly to further improvements in the transport system. Other detectors now used in gas chromatography, both general and selective, can probably be utilized with this approach, once the present limitations of the transport systems have been removed (4). There is little doubt that this detector will eventually become one of the most useful of the devices for monitoring modern LC columns.

5.5   RADIOACTIVITY DETECTORS

The radioactivity detector is a specific device for monitoring radio-labeled solutes as they elute from LC columns. For the detection of weak beta emitters

TABLE 5.5   CHARACTERISTICS OF TRANSPORT DETECTORS

Universal for all solutes containing carbon in the reduced form (for flame ionization detector)

Relatively large size, high cost

Insensitive to mobile phase changes; therefore, excellent for gradient elution

Fairly complicated mechanically, less reliable than most other detectors

Moderate sensitivity; generally not useful for trace analysis (present designs)

Insensitive to temperature changes; can be sensitive to carrier flow changes

Destructive to solute utilized

Excellent linearity of response

Moderate extra-column band broadening (present designs)

(i.e., ^{35}S, ^{14}C, and ^3H), scintillators (e.g., anthracene crystals) are placed in contact with the column effluent within a flow cell. As the radioactive species elutes, light pulses are triggered in the scintillant and are detected by a photomultiplier system. This detection device has a wide linear range and is insensitive to solvent changes, making it useful for gradient elution techniques. In optimum systems, radioactive peaks containing ^{14}C have been detected with as little as 100 disintegrations per minute. Chromatographic applications with stronger beta, alpha, and gamma emitters (i.e., ^{131}I, ^{210}Po, ^{125}Sb), using Geiger counting and scintillation systems, have also been reported.

The sensitivity of flow counting cannot compete
with the discontinuous counting of individual samples
from a fraction collector. In continuous measurements,
the signal-to-noise ratio cannot be improved unless the
counting time is increased. This can only be achieved
by a decrease in flow rate, which is contrary with the
requirements for rapid separations. However, for convenience and speed of analysis, the flow counting method is often desired. The flow detector has been found
to be quite useful in drug and pesticide metabolism
studies using radiotracer compounds.

Commercial radioactivity detectors currently lack
low-volume flow cells. While use of low-volume cells
reduces extra-column band broadening, decreased detection sensitivity is obtained; faster passage of radioactive peaks through a low-volume cell results in
shorter counting periods. Nevertheless, while these
commercial devices cannot be used for monitoring narrow-bore, high-performance LC columns, they are satisfactory for less demanding applications (e.g., larger
diameter columns). Since diffusion is slow in liquids,
radioactivity detectors do offer the unique advantage
of stopping the flow of carrier in the cell and counting the radioactive peak for whatever period of time
is required to improve sensitivity. For a recent detailed treatment of the use of the radiometric detector for weak β-emitter measurements, see ref. 5.

5.6 POLAROGRAPHY

Many electro-reducible and electro-oxidizable substances have been detected in column effluents by polar-
ography. This selective electrochemical method measures the current between polarizable and reference
electrodes as a function of applied voltage. Normally,
a constant voltage is imposed between the electrodes,
which results in either oxidation or reduction of the

solute, depending on the voltage. The current that
flows between the electrodes is recorded continuously.
(Some workers call this an amperometric detector, since
the measurement is made at constant voltage.) Figure
5.10 shows a schematic of a polarographic detector that

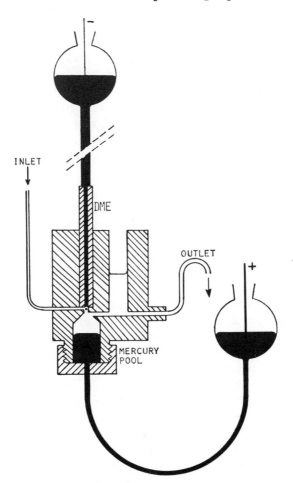

Figure 5.10 Dropping mercury polarographic detector.
J. G. Koen, J. F. K. Huber, H. Poppe, and G. den Boef,
J. Chromatog. Sci., _8_, 192 (1970). Reprinted by per-
mission of publisher.

utilizes a micro dropping-mercury electrode.

At this writing, there are no commercial polarographic detectors available for LC; however, research
forms of this device have demonstrated unique value in
LC. Polarographic detectors are capable of very high
sensitivity, which makes them attractive for use as a
selective detector for trace analysis. For instance,
residues of certain pesticides in naturally occurring
systems have been measured (6), as have very small
amounts of catechol amines in rat brains (7). The
polarographic detector must be used with mobile phases
that have high conductivity, usually an aqueous mixture
with salts, but sometimes can be used with organic solvents needed for the dissolution of nonionic samples.

5.7 INFRARED PHOTOMETRY

Infrared absorption can be used either as a selective
or a general detector in LC. Figure 5.11 shows a schematic of a commercially available infrared detector.
Energy from source A passes through chopper B and then
through the wheel C which contains three gradient interference filters with ranges from 2.5-7, 7-11, and
11-15 μ. The position of this filter wheel can be adjusted so that any wavelength in this range may be
selected for detection. The monochromatic infrared
energy passes through slit D and then through the detector cell, E, which is nominally 3 mm in length.
Energy emerging from the detector cell falls on the
thermoelectric detector, which records the change in
energy from the absorbing sample. The output of the
thermoelectric detector is amplified for presentation
by means of the amplifier and power supply, G.

The sensitivity of this infrared detector is generally not sufficient for use with high-speed narrow-
bore LC columns, but it has been used successfully with
larger diameter gel permeation chromatography systems.

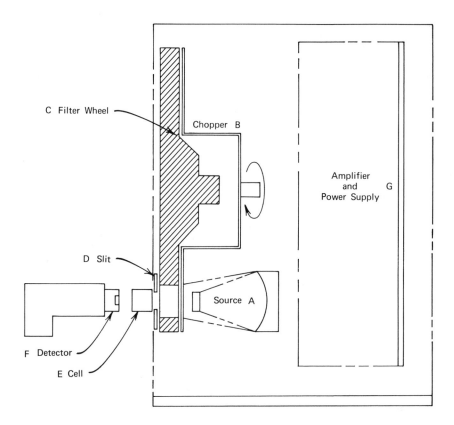

Figure 5.11   Infrared detector for LC.   Reprinted by permission of Wilks Instruments.

By monitoring the C-H stretching frequencies, about 20 µg of a typical organic sample can be detected using this device.   Carbonyl and other strongly absorbing functions provide somewhat lower sensitivities.   The infrared approach is obviously limited to mobile phases which are transparent at the absorption wavelength to be utilized.   However, this detector can also be used for qualitative studies.   The carrier may be interrupted and the wavelengths manually scanned for a qualitative infrared absorption pattern from 2.5-15 μ.

5.8 FLUORIMETERS

The fluorimeter is a very sensitive and selective detector because of its ability to measure fluorescent
energy emitted from certain solutes excited by ultraviolet radiation. In addition, fluorescent derivatives
of many nonfluorescing substances can be prepared (8),
and this approach is particularly attractive for the
selective detection of various classes of compounds
for which sensitive detection methods are lacking.
     Figure 5.12 is a schematic of the right-angle
type fluorimetric detector. Ultraviolet light from

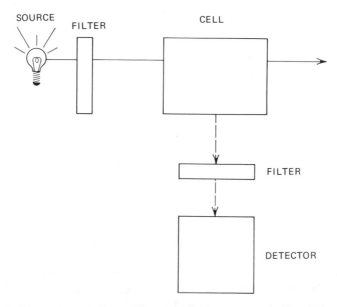

Figure 5.12 Right-angle fluorimeter detector. S. H.
Byrne, Jr., in <u>Modern Practice of Liquid Chromatography</u>,
J. J. Kirkland, ed., Wiley-Interscience, New York,
1971. Reprinted by permission of publisher.

the source passes through a filter which allows only a
certain wavelength to be transmitted to a low-volume
cell containing the column effluent. The fluorescing
sample emits energy at a higher wavelength (sometimes
in the visible region), which then passes through a
filter to remove the incident radiation and is detected
by a photocell. The fluorimetric detector is susceptible to the usual interferences which plague fluorescence measurements, mainly background fluorescence and
quenching effects. However, this device has proved
useful for trace analysis in biological systems. Of
particular current interest is the fluorescent detection of very small amounts of amino acids and peptides
using derivatives of fluorescamine (9).

5.9 CONDUCTIVITY

Conductivity can be used to detect certain ionic solutes in aqueous mobile phases. In principle the conductivity detector should be applicable to certain
nonaqueous systems. So far, however, no use of this
device with high-performance columns in nonaqueous
media has been reported. Conductivity detectors with
low-volume cells suitable for use with high-performance
columns are commercially available. The response of
this detector is temperature-dependent; consequently,
careful control of this parameter must be maintained
during use. Some forms of this detector are susceptible to changes in carrier flow rate. However, response is predictable from conductivity data and electrical conductivity detectors exhibit a linear response
to solutes when properly designed. Unfortunately,
these devices are not generally applicable for use in
totally organic systems.

5.10   OTHER DETECTORS

Many other detectors have been reported for use in LC. These devices have been based on a variety of physical characteristics, for example, dielectric constant, density, vapor pressure, sonic velocity, viscosity, and thermal conductivity. Several years ago, the heat-of-adsorption detector was proposed as a useful device for LC and was marketed commercially. However, due to many inherent limitations, this detector has fallen into disuse and is no longer commercially available. Since all of these detectors are based on bulk proper-ties, they are all limited in sensitivity due to their high susceptibility to temperature changes. Conse-quently, many of these devices do not provide the sensitivity which is needed for use with narrow-bore high-performance LC columns.

5.11 SUMMARY OF DETECTOR CHARACTERISTICS

Table 5.6 is a summary of the specifications for the most popular detectors in modern LC. A comparison of the properties for these detectors will assist in the selection of an appropriate detector for a particular application. While the absolute and relative sensi-tivities of the various detectors are given, it is also useful to know the _practical_ sensitivity of these various detectors. The UV, fluorescence, and polaro-graphic devices can each detect about 1 ng of a favor-able sample. The rest of the detectors listed in this table will detect about 1 μg, except for the infrared detector, which can detect about 20 μg in a favorable situation.

A partial list of the currently available commer-cial detectors for modern LC are shown in Appendix V. It should be noted that each device has different specifications and should not be directly compared on

TABLE 5.6. SUMMARY OF TYPICAL SPECIFICATIONS OF PRESENT DETECTORS

Parameter (units)	UV (Absorbance)	RI (RI-Units)	Transport (amp)	Radioactivity	Polarography (μamp)	Infrared (Absorbance)	Fluorimeter	Conductivity (μMho)
	Selective	General	General	Selective	Selective	Selective	Selective	Selective
Useful with gradients	Yes	No	Yes	Yes	N.A.	Yes	Yes	No
Upper limit of linear dynamic range	2.56	10^{-3}	10^{-8}	N.A.	2×10^{-5}	1.5	N.A.	1000
Linear range	5×10^4	10^4	$\sim10^5$	Large	10^4	10^4	$\sim10^3$	2×10^4
Sensitivity at $\pm1\%$ noise, full-scale	0.005	10^{-5}	10^{-11}	N.A.	2×10^{-6}	0.01	0.005	0.05
Sensitivity to favorable sample	5×10^{-10} g/ml	5×10^{-7} g/ml	10^{-8} g/sec ($\sim5\times10^{-7}$ g/ml)	50 cpm ^{14}C/ml	10^{-10} g/ml	10^{-6} g/ml	10^{-9}-10^{-10} g/ml	10^{-8} g/ml
Inherent flow sensitivity[a]	No	No	Yes	No	Yes	No	No	Yes
Temperature sensitivity	Low	10^{-4}°C	Negligible	Negligible	1.5%/°C	Low	Low	2%/°C

N.A., not available.

[a] Due to sensitivity to temperature changes, some detectors appear to be flow sensitive.

the basis of cost.   Details on each apparatus should
be obtained from the manufacturers before deciding on
the type and model of equipment which is to be acquired.

REFERENCES

1.   J. J. Kirkland, Gas Chromatography. 1972. S. G.
     Perry, ed., Applied Science Publisher, Essex,
     England, 1973,
2.   Du Pont Product Bulletin A-86533, April, 1973.
3.   J. H. van Dijk, J. Chromatog. Sci., 10, 31 (1972).
4.   G. Nota and R. Palombari, J. Chromatog., 62, 153
     (1971).
5.   G. B. Sieswerda and H. L. Polak, Liquid Scintilla-
     tion Counting, Volume 2, Heyden & Son Ltd., New
     York, 1972, Chapter 4.
6.   J. G. Koen and J. F. K. Huber, Anal. Chim. Acta,
     51, 303 (1970).
7.   P. T. Kissinger, C. Refshauge, R. Dreiling, and
     R. N. Adams, Anal. Lett., 6, 965 (1973).
8.   R. W. Frei and J. F. Lawrence, J. Chromatog., 83,
     321 (1973).
9.   S. Udenfriend, S. Stein, P. Böhlen, W. Dairman,
     W. Leimgruber and M. Weigele, Science, 178, 871
     (1972).

BIBLIOGRAPHY

S. R. Bakalyar, Amer. Lab., 3, 29 (1971).   (Use of UV
   detectors in LC.)
S. H. Byrne, Jr., in Modern Practice of Liquid Chroma-
   tography, J. J. Kirkland, ed., Wiley-Interscience,
   New York, 1971, Chapter 3.   (Detailed discussion of
   LC detectors.)
R. M. Cassidy and R. W. Frei, J. Chromatog., 72, 293
   (1972).   (A fluorescence detector for LC.)

R. D. Conlon, Anal. Chem., 41, 107A (1969). (Review of detectors for LC.)

G. Deininger and I. Halasz, J. Chromatog. Sci., 8, 499 (1970). (Modification of a differential RI detector for LC; study of heat exchangers.)

J. F. K. Huber, J. Chromatog. Sci., 7, 172-6 (1969). (Evaluation of five different types of detectors for LC.)

S. Katz and W. W. Pitt, Anal. Letters, 5, 177 (1972). (LC detector based on fluorescence of Cerium III produced from the reaction of Cerium IV with reducible compounds.)

J. J. Kirkland, Anal. Chem., 40, 391 (1968). (High-performance UV photometric detector for LC.)

J. G. Koen, J. F. K. Huber, H. Poppe, and G. den Boef, J. Chromatog. Sci., 8, 192 (1970). (Design and use of polarographic detector for high-efficiency LC.)

M. N. Munk, J. Chromatog. Sci., 8, 491 (1970). (Comparison of several detectors for LC.)

S. G. Perry, K. Amos, and P. I. Brewer, Practical Liquid Chromatography, Plenum Press, New York-London, 1972, Chapter 7. (Discussion of LC detectors.)

J. Polesuk, Amer. Lab., June, 37-40, 42, 44-47 (1970). (Review of detection methods for LC.)

L. Schutte, J. Chromatog., 72, 303 (1972). (Homogeneous and heterogeneous scintillation counting for the continuous detection of radioactive column effluents.)

R. P. W. Scott and J. G. Lawrence, J. Chromatog. Sci., 8, 65-71 (1970). (Improved wire transport detector for LC.)

H. Veening, J. Chem. Ed., 47, A-675-6, A-678, A-680, A-682, A-685-6 (1970). (Review and comparison of commercial detectors for LC.)

CHAPTER SIX

THE COLUMN

6.1 INTRODUCTION

High-performance columns that give a minimum of band
broadening during separation are the heart of a modern
LC system. Therefore, the material put into the column and how well it is packed are of prime concern.
In Figure 2.3, we saw that band broadening occurs by
four processes: eddy diffusion (B), and slow mass
transfer in the mobile phase (C), in the stagnant mobile phase (D), or in the stationary phase (E). What
can we do to prepare columns so as to minimize these
effects which lead to band broadening? Fortunately,
there are several guidelines we can follow. Band
broadening due to eddy-diffusion effects can be minimized by preparing homogeneously packed columns of
small, uniform particles; this is somewhat more easily
accomplished with spherical than with irregular particles. Mobile phase mass transfer can be improved by
using small particles, as the channels between particles are then small. Band broadening due to the
stagnant mobile phase can be minimized by one of two
approaches. First, we can use superficially porous
particles (also porous layer or pellicular beads),
which have solid cores and a thin, porous outer shell
as in Figure 6.1A. These materials minimize the depths

169

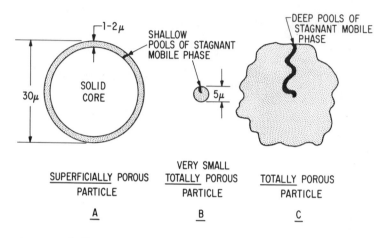

Figure 6.1   Types of particles for modern LC.

of pores within the particle, so that the solute molecules can rapidly move into and out of the shallow
pool of stagnant mobile phase in the porous layer.
The second approach is to use very small particles
(Figure 6.1B). Movement of solute molecules in and
out of very small particles with short diffusion paths
is much faster than in larger totally porous particles
(Figure 6.1C) with deeper pools of stagnant mobile
phase. (Interestingly, the depth of pools of stagnant
mobile phase is roughly comparable for porous layer
beads of ~30 μ and totally porous particles of ~5 μ.)
     As discussed in Section 2.3, broadening of bands
as a result of slow stationary phase mass transfer
(i.e., slow equilibration of solute between the mobile
and stationary phases) occurs less often; eddy diffusion and mobile or stagnant mobile phase effects generally dominate. In modern LC, column packings with
very thin films of stationary phase often are used,
and diffusion of solute molecules into and out of these
thin films is normally not an important factor in

affecting band broadening. However, with ion-exchange
and bonded-phase packings, stationary phase mass transfer can be significant. In such cases, we can reduce
peak broadening by using thin films of the stationary
phase in conjunction with a porous layer bead-type support. These types of ion exchangers are discussed more
fully in Chapter 9.

Column selectivity is an important factor in obtaining a desired separation, as discussed in Chapter
3. Selective columns, that is columns with large α
values, produce the best separations, all other factors being equal. However, in the preparation of columns we also strive for maximum column efficiency and
permeability, that is, a maximum number of theoretical
plates with a minimum to for given column length and
pressure. Other factors which might be considered in
the packing of columns are reproducibility, ease of
preparation, capacity of the column, and cost. Since
the modern liquid chromatographic column is often re-
used for many hundreds of separations, the cost of the
column is usually of minor consequence in analysis.

In this chapter the general aspects of column
preparation will be discussed. Later chapters will
expand on those aspects of column preparation that
are peculiar to the individual LC methods. Before
continuing this general discussion, however, it should
be noted that packed columns can be purchased from the
manufacturers of specific commercial instruments. Not
all types of packed columns are available from each
specific manufacturer, but it is often possible to
adapt commercially available columns to particular instruments. Some packed columns are also available
from independent chromatographic specialty houses (see
Appendix V). Column prices vary depending on type and
length, but are generally in the $100-300 range. Be-
cause of the skills and special equipment which are
needed to pack certain types of columns, particularly
those with particles of < 20 μ, many workers will prefer

to purchase ready-made columns in these cases.

6.2   CHARACTERISTICS OF COLUMN PACKINGS

In this section, we will discuss the general characteristics of commercial packings for modern LC.   Information on specific packings for the various LC methods
is given in the chapters on the individual methods.   A
description of the unpacked column and associated column hardware has already been given in Chapter 4.
   Packings for modern LC can be classified according to the following categories:

- Rigid solids, hard gels, or soft gels.
- Spherical or irregular.
- Porous or superficially porous.

The type of material to be packed into a column is determined to a large extent by the LC method which will
be used (see Chapter 11 for a discussion of the selection of LC methods).   Rigid solid packings are used in
liquid-liquid and liquid-solid chromatography, while
hard gels are often used for separations by ion exchange.   All three types of packings--rigid solid,
hard, and soft gels-- are utilized in exclusion or gel
chromatography.
   The rigid packings (e.g., superficially porous
beads or porous-layer beads) exhibit the best column
efficiency of any of the available packing materials.
Rigid packings can also be used at very high column
inlet pressures, and this favors large N values per
unit time (Section 3.3).   The soft gels (e.g., agarose)
can be used only at low pressures.   These soft particles collapse at higher inlet pressures, and some
varieties can be used only at pressures of a few inches
of water.   Soft gel packings are employed when others
are not applicable, and are used mainly for separating
high-molecular-weight, water-soluble substances.   The

larger-size totally porous particles (e.g., silica gel, diatomaceous earths) are relatively cheap, widely available, and have a high sample capacity as a result of their relatively large surface area. Larger samples can be used with columns of totally porous supports, so that greater detection sensitivity is possible (i. e., larger sample enables the minimum detectable quantity to be a smaller fraction of the total). The main disadvantage of porous particles of >10 μ is lower efficiency and speed; this limitation is eliminated with the use of totally porous particles of ≤ 10 μ.

Other factors being equal, <u>spherical</u> particles are preferred over <u>irregular</u> packings, because columns are more easily produced with somewhat higher efficiency and generally with better reproducibility.

Rigid solid packing materials are widely used in all LC methods. Two different types of particles are available, porous and superficially porous (also <u>porous layer</u> and <u>pellicular</u> beads). These two types of particles have different characteristics, and a comparison of porous versus superficially porous rigid packings is summarized in Table 6.1.

Irregularly shaped particles and packings with wide particle size ranges create difficulties in the packing of homogeneous beds. Inhomogeneous beds favor gross transcolumn variations in carrier velocity, resulting in increased band spreading. Constancy of carrier velocities across the column cross section is especially critical in LC, because of the slow diffusion of sample molecules in liquids. This slow molecular diffusion restricts solute molecules from quickly moving between areas of faster and slower moving mobile phase to achieve a uniform carrier velocity across the column. Therefore, it is important in high-speed LC that these transcolumn carrier velocity variations be minimized.

To eliminate some of the drawbacks of the larger, totally porous particles, several types of chromatographic packings specifically designed for high-

TABLE 6.1 COMPARISON OF POROUS VERSUS SUPERFICIALLY POROUS PACKING PARTICLES

| Property | Totally Porous Packings | | | Superficially Porous Packings |
	Irregular >20 μ	Spherical >20 μ	Spherical or Irregular <20 μ	Spherical,>20 μ
Efficiency	Low to moderate	Moderate	High	High
Speed	Low to moderate	Moderate	High	High
Packing characteristics, reproducibility	Fair	Good	Fair	Excellent
Capacity	High	High	High	Low
Availability	Good	Good	Fair	Good
Cost	Low	Moderate	High	High

174

performance LC have been developed.   The superficially
porous packings (also porous-layer beads and pellicular beads) comprise one example.   Another type consists
of closely sized ion-exchange resins of <10 μ which
have been used for several years and are available commercially.   More recently, very small (< 20 μ) totally
porous particles have become available for liquid-
solid chromatography, and it is anticipated that very
small particles for the other LC methods will shortly
be obtainable.   The available packings for modern LC
are listed in the chapters describing the individual
methods.

As indicated in the left-hand schematic in Figure
6.1A, the superficially porous or porous-layer bead
particle has a solid core and a thin porous shell containing the chromatographically active sorbent.   The
thin, porous surface layer promotes efficient interaction of the solute with the stationary phase.   The porous shell typically is about 1/30 of the thickness of
the total particle, or about 1-2 μ thick for a 30-50-μ
particle.   Solutes rapidly move into this thin porous
shell from the mobile phase, interact with the chromatographic sorbent, and rapidly return to the moving
phase.   Thus, these superficially porous particles provide for higher column efficiency, and this advantage
is most apparent at higher carrier velocities (e.g.,
see Figure 6.9).   The porous crust can be used as a
solid-phase for liquid-solid chromatography, or can
be uniformly coated with a stationary liquid or ion-
exchange resin for partition or ion-exchange chromatography.

A comparison of the structures of porous and superficially porous particles is shown in Figure 6.2.
This figure shows a scanning electron micrograph of
Chromosorb®-W, a diatomaceous earth support.   Note the
irregularity of the particles and the relatively large
pores in the characteristic, honeycomb-like structure
of the diatoms which make up this material.   Figure

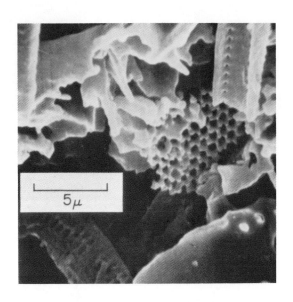

Figure 6.2   Scanning electron micrograph of Diatomaceous earth. W. A. Aue, C. R. Hastings, J. M. Augl,
M. K. Knorr, and J. V. Larson, J. Chromatog., $\underline{56}$, 295
(1971).   Reprinted by permission of editor.

6.3 is a cross-section electron micrograph of the porous surface of a Zipax® particle, one of the superficially porous supports. The surface of Zipax® is made
up of layers of 200 mμ microspheres of silica which
are sintered to produce a mechanically stable porous
shell of known dimensions. The pores in this particle
are relatively large (about 1000 $\overset{\circ}{A}$), which produces a
relatively high surface area, yet allows ready access
of the solutes for rapid chromatographic interaction.
      Because of their spherical shape and relatively
high density, superficially porous supports are more
easily and reproducibly packed into efficient columns. The influence of shape is illustrated in Figure
6.4, which shows that relatively large channels are

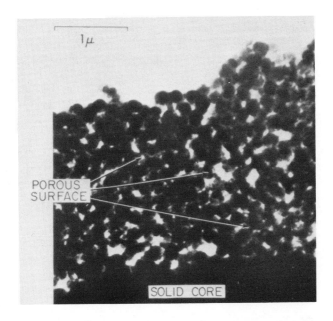

Figure 6.3 Cross section scanning electron micrograph
of the surface of a superficially porous particle
(Zipax®).   J. J. Kirkland, J. Chromatog. Sci., $\underline{7}$, 361
(1969).   Reprinted by permission of editor.

formed between irregularly shaped totally porous particles, because of the difficulty in closely packing
these materials.   On the other hand, the bed of spherical superficially porous beads shown on the left has
(on the average) smaller channels.   These particles
can be formed into a more closely packed structure because of their dense, spherical character.   In this
more closely packed structure, the average distance
that a solute molecule must travel between particles
for a chromatographic interaction is less; therefore,
mobile mass transfer effects and column efficiency are

   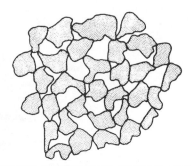

POROUS LAYER BEADS   IRREGULAR, TOTALLY POROUS SUPPORT

Figure 6.4 Comparison of the type of supports in
packed beds.

improved.
     Superficially porous particles have a relatively
low surface area compared to totally porous particles;
thus, sample capacity is three- to fivefold less
than totally porous particles. As a result, smaller
samples and more sensitive detectors sometimes must be
used with columns of these materials. The isolation
of sufficient quantities of sample for subsequent identification by other techniques also can be a problem
when using these low-capacity column packing materials.
Since they are produced in small quantities by special
techniques, superficially porous supports are about
five to twenty times more expensive than porous particles of the same size.
     Particle size is an important factor in the prep-
aration of efficient LC columns. More efficient columns can be prepared with small diameter particles--
both plate height and the permeability of columns decrease with smaller particle size. As suggested in
Figure 6.5, the average distance that a solute molecule must travel between packing particles to undergo

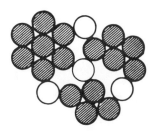

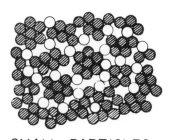

LARGE PARTICLES   SMALL PARTICLES

Figure 6.5   Effects of particle size in packed beds.

a chromatographic interaction is less in a bed of small
particles than in one with larger particles.   Since the
mobile phase mass transfer process within a column is
at least partially diffusion controlled (and diffusion
of solutes in liquids is very slow), the distance that
a solute molecule must travel between chromatographic
interactions should be minimized.   This situation is
optimized by using small particles.   Figure 6.5 shows
the effect of particle size on the H versus u plots
for LSC columns made from silica gel particles.

On the other hand, by using smaller particles to
gain higher column efficiency, we also sacrifice permeability; that is, we must use higher inlet pressure.
However, on balance, columns of smaller particles produce the highest N values for comparable values of P
and t.   This is discussed more fully in Appendix VI.
The disadvantage of small-particle columns is that
they require much smaller dead volumes in the LC system (N is larger, and L is smaller, and therefore V_m is
smaller), and they are more difficult to pack.   With
very small particles (e.g.,<10 μ) special procedures
must be used, as discussed in Section 6.4.   Consequently, in modern LC a compromise in particle size is
often made.   Particles most generally used are those
that are small enough to produce good column efficiency, but large enough to be packed conveniently into

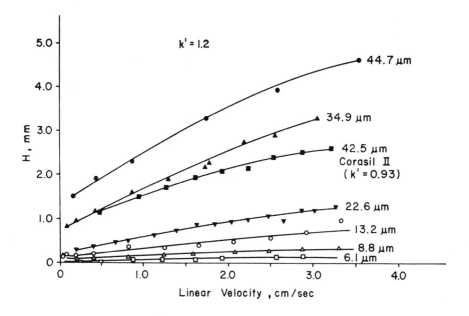

Figure 6.6   Effect of particle size on plate height.
Mobile phase; 90; 9.9; 0.125 parts by volume of hexane-
methylene chloride-isopropanol; sample, 1 µl of ug/ml
of N,N-diethyl-p-aminoazobenzene. R. E. Majors, J.
Chromatog. Sci., 11, 88 (1973). Reprinted by permission of editor.

homogeneous beds. Many workers are currently using
5-20-µ particles for ion exchange, and 20-50-µ particles for the other LC methods.
   Figure 6.7 shows the effect of particle size on
the theoretical plates of an LC column and the pressure required for its operation at a certain mobile
phase velocity. This figure is useful for predicting
the performance of a column containing particles of a
particular size. In addition, the pressure drop plot

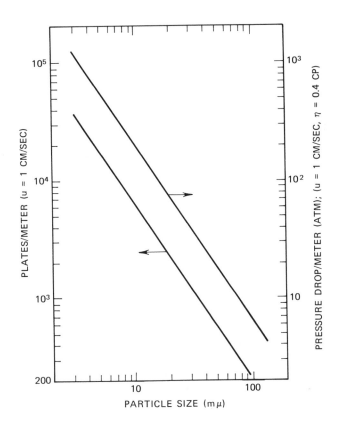

Figure 6.7 Effect of particle size on theoretical
plates and pressure drop of column. Reprinted by
permission of D. L. Saunders.

indicates whether or not the proper density of a column packing has been obtained. Improperly packed columns often show a lower pressure drop than predicted
in this figure. Any unsuspected partial blockages
which have occurred in the column or elsewhere in the
high-pressure system can also be detected.

    For best results, the range in particle size for

a given packing should be relatively narrow, e.g., +10-20% for porous ion-exchange beads, and a total size spread of no more than a factor of 1.5-2 for other packings. A narrow range of particle sizes facilitates the packing of homogeneous columns, because the sizing of such particles across the column is less detrimental to column performance (1). A wide particle size range leads to lower N values and column permeabilities, as compared to narrow-range packings of similar average size. Narrow particle size fractions can be obtained by wet or dry sieving, by solvent elutriation (2), or by air classification (cyclone separation) (3). However, the preparation of very narrow particle fractions (+10-20%), particularly for irregular particles, is generally difficult. Therefore, wider range packings are used for the most part.

Recently, the practicality of working with porous particles of < 10 µ has been demonstrated, and several workers have confirmed the advantages of particles in this size range for very high-performance LC separations (1,4-7). Figure 6.8 shows scanning electron micrographs of some of the particles which have been developed for high-performance LC, including some particles of <10 µ (g,h). It is anticipated that very small particles of a variety of types will now become available for all the LC methods.

A word of caution about all packings for modern liquid chromatography--often there are significant differences in chromatographic characteristics between batches of the various column packings. Manufacturers are endeavoring to minimize these variances, but we should take this problem into account, and evaluate each batch of packings before its use. If there is a critical application for use over a long period of time, it is desirable to stock-pile a relatively large lot of the particular packing to be used. In this way, constant column performance can be ensured, without having to investigate a change in separation parameters

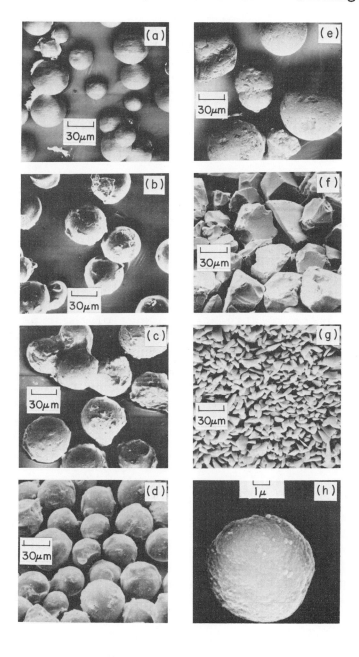

Figure 6.8. Some commercially available packings for
modern LC. (a) Zipax®, (b) Corasil®-I, (c) Corasil®-
II, (d) Perisorb®-A, (e) Porasil®-A, (f) Sil®-X, (g)
LiChrosorb®, (h) Zorbax®-SIL; (a)-(d), superficially
porous; (e)-(h), totally porous. See Appendix V.
Plates a-g, reprinted by permission of R. E. Majors,
Varian Aerograph.

to maintain separation.

6.3 PERFORMANCE OF COLUMN PACKINGS

A comparison of plate height versus carrier velocity
plots for some of the particles used for modern LC is
shown in Figure 6.9 (see also Appendix VI). The H
values and the slope of the plots are indicative of
the approximate relative performance of these packings,
although the values for each support depend somewhat on
the system being studied. Note that the plot for the
diatomaceous earth support used in LLC is much steeper
than the other curves, reflecting the poorer solute
mass transfer within this bed of totally porous particles as the velocity of the carrier is increased.
Totally porous beads for LSC (e.g., Porasil® or Sphero-
sil®) of the same particle size give H versus u plots
of similar slope. The data for a porous-layer bead
support, Corasil®-II, produces a less steep curve, and
the advantages over the diatomaceous earth support are
particularly apparent at the higher carrier velocity.
Data on Zipax®, a superficially porous support, show
a still smaller increase in H as the carrier velocity
is increased. The flattest curves are exhibited by
the columns of very small porous particles (< 10 μ).
The variation in H values for the packings in this figure also is an indication of the effect of particle
size (see Figure 6.6).
    Table 6.2 presents a different type of comparison
for some of the very high-efficiency packings for LC.

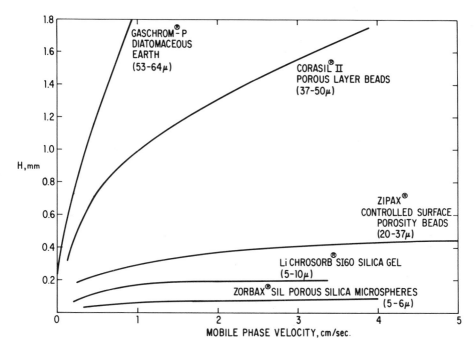

Figure 6.9   Comparative Efficiency of Column Packings.

These data are the best yet reported for these particular materials and may approach a practical limit of
efficiency obtainable with these packings by present
techniques.   Classical LC separations were carried out
with columns exhibiting about 0.02 effective plates/
sec, but high-efficiency LC columns have now been produced which exhibit more than 36 plates/sec (7).   This
performance exceeds that of most packed gas chromatographic columns, and it is comparable to that of open
tubular or capillary gas chromatographic systems.   All
of the values in Table 6.2 are approximate because of
some minor limitations; however, they are useful as a
general guide to the relative performance of some of
the most efficient packings that have been used.

186

TABLE 6.2 PERFORMANCE OF VERY HIGH-EFFICIENCY LC COLUMNS

Packing Type	LC Mode	Particle size, μ	k'	Effective Plates/sec	H (cm) @ u = 1 cm/sec	Ref.
Superficially porous	LLC	20-37	3.3	16	0.016	8
Diatomaceous earth	LLC	5-10	3.5	--	0.013	5
Silica gel (irregular)	LSC	8.8	11	15	0.014	4
	LSC	6.1	1.2	23	0.0040	9
Porous silica microspheres	LSC	8-9	10	14	0.0076	10
	LLC	5-6	11	23	0.0035	10
	LSC	4.6-5.6	1.3	37	0.0029	7

6.4 COLUMN PACKING PROCEDURES

There are several methods for packing high-efficiency
LC columns, but there is no one "best" method for all
types of packings. The optimum procedure is determined mainly by the nature of the material to be packed
and particle size. As discussed in Section 6.2, the
prime goal in the packing of a column is to obtain a
uniform bed with no cracks or channels, and without
sizing or sorting the particles within the column.
Generally we want to pack rigid solids and hard gels
as densely as possible, but without fracturing the
particles during the packing process.

It is always desirable to measure the efficiency
of a new column, using a test mixture containing at
least two solutes: one with $k' \cong 0$ and one with $k' \cong$
2-5. The H value for the unretained solute at a specific carrier velocity (e.g., 1 cm/sec) is a good indication of how well the column is packed. The H value for the retained solute provides insight into the
performance of the column packing material, apart from
how it was packed. During use of the column, it is
desirable that this test mixture be rerun periodically.
Variations in H and k' values indicate corresponding
changes in the column. It is also useful to measure
the permeability of a column (see Appendix II), or
compare the observed pressure of the column with that
indicated in Figure 6.7. This test ensures that the
anticipated packing density has been reached, and that
there is no obstruction in the end plugs, inlet and
outlet tubes, and the like.

As indicated in Section 2.3, it is also desirable
to determine the n value for each column, since this
parameter is useful in optimizing the use of the column for a certain separation (see Chapter 3). As indicated in Figure 6.10, n is determined by measuring H
values at various mobile phase velocities u and calculating the slope of the log H versus log u plot.

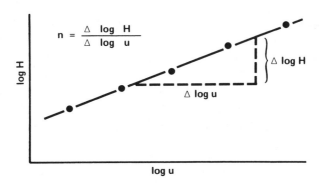

Figure 6.10 Calculating the n value for a column.

Rigid Solids

Empirical studies have resulted in several techniques for the dry-packing of rigid solid materials. A "tap-fill" procedure has been widely used for packing columns of irregular particles of more than about 50 μ, porous spheres of more than about 40 μ, and superficially porous particles of more than about 20 μ. The steps in this procedure (which are not much different than that sometimes used to prepare very high-efficiency gas chromatography columns) are summarized in Table 6.3 (10).

It should be stressed that for the production of high-efficiency columns, mechanical vibration techniques (such as those commonly used in the preparation of gas chromatographic columns) should not be used during the packing process. Lateral vibration techniques tend to size the particles across the column cross section, resulting in an inhomogeneous bed which has the larger particles near the wall and the smaller particles in the center. This sizing process results in increased band spreading, because of faster carrier flow near the walls of the column as compared to the center. The "tap-fill" packing procedure apparently allows a regular consolidation of the packing without

TABLE 6.3 "TAP-FILL" METHOD FOR DRY-PACKING COLUMNS
OF RIGID SOLIDS

---

1. Scrub interior of column tubing with hot detergent
   solution (using long pipe cleaner or a cloth plug
   tied to a string), wash with water, acetone, dry.

2. Fit porous disc in outlet of tubing. Retain porous plug with outlet fitting.

3. Add enough packing to fill 3-5 mm of the column
   (100-200 mg for 2 mm i.d.) via funnel to vertically held tubing.

4. Firmly tap column on floor or bench-top, 2-3 times/
   sec, for 80-100 times, while gently rapping the
   side of the tube at the approximate level of the
   packing.

5. Discontinue rapping the side; very gently tap the
   column vertically for 15-20 sec.

6. Add another portion of packing, continue as above
   until column is filled (requires 30-40 min for 1 m
   x 2.1 mm i.d. column). Rotate column slowly while
   filling.

7. Gently tap column vertically for additional 5 min
   (no rapping on the side).

8. Place rigid plug or Teflon® wool plug in the top
   after carefully removing about 1/8 in. of packing.

9. Place column in instrument and equilibrate with
   carrier until no bubbles are seen in column outlet.

---

undue particle sizing. The precise nature of the "tap-
fill" procedure, coupled with the additional consolidation step, promotes a homogeneous bed structure. Experience indicates that it is desirable to impart sufficient

vertical mechanical energy to the bed to cause particle consolidation but not enough to shake the particles violently so that a sizing process takes place.
The rapping of the column on the side during the packing process is used to keep the packing particles from
hanging up in the column above the bed during the delivery of the packing. This rapping process must be
gentle or it can cause sizing of the particles within
the bed.

Using the "tap-fill" dry-packing procedure, most
workers can reproduce column characteristics within at
least $\pm 10\%$. Column packing machines have been used to
dry-pack columns even more reproducibly and often with
improved efficiency. Packing machines are often used
by suppliers of packed columns, so that a large volume
of columns can be reproducibly packed with good efficiency.

Any packing technique that promotes the rapid es-
tablishment of a dense, stable structure can be used.
For instance, excellent columns of small diatomaceous
earth particles have been prepared by a tamping process (5). In this procedure, small amounts of packing
(3-5 mm in the column) are added to the column and firmly tamped with a Teflon®-tipped rod. Aliquots of packing are tamped in this fashion until the column is
filled. This technique is not generally used for packing Zipax® particles, which tend to creep up around
the tip of the tamping rod when packed by this approach.

Rigid solids can also be packed by high-pressure
slurrying techniques. This approach is particularly
useful for packing columns of < 20-µ particles. In one
form of this technique, the so-called <u>balanced-density
slurry packing</u> method (4,9,11,13), a 15-25% suspension
of the packing is prepared in a liquid of equal density, using solvent mixtures such as tetrabromoethylene/
tetrahydrofuran or methylene iodide/tetrachloroethylene.
(Caution: tetrabromoethylene and methylene iodide are
moderately toxic.) These suspensions are conveniently
prepared by ultrasonic mixing of the packing with the

balanced-density solvent, which simultaneously degasses
the system and creates a stable suspension. The density of the solvent mixture is established by varying
the ratio of the two solvents so that the packing particles remain suspended when mixed in this fashion.
This balanced-density slurry is then rapidly pumped at
high pressure (up to 10,000 psi) into a column blank
fitted with a porous metal screen or plug at the outlet. In this process, the particles are rapidly filtered out of the slurry onto the porous plug at the
bottom of the column, leaving a stable bed which can
then be subsequently conditioned for the desired chromatographic application. Equipment for high-pressure
slurry packing, shown schematically in Figure 6.11, obviously must be constructed from components which will
withstand high pressures. Pneumatic-type pumps are
particularly useful for this operation, since virtually instantaneous pressure can be created for the rapid
slurry-packing process. Efficient columns of < 10-μ
silica gels (LiChrosorb®) have been prepared by this
technique and are now commercially available (4,9).
The technique appears to be most useful for packing
adsorbents and bonded-phase materials, but it cannot
be used for liquid-liquid chromatographic packings with
mechanically held liquids, since the stationary phase
would be solubilized by the balanced-density solvents.
It is also not recommended for particles larger than
about 20 μ.

Another high-pressure slurry-packing technique using stable aqueous suspensions has been used to prepare
efficient columns of <10-μ closely-sized porous silica
microsphere particles (1,6,7). (LSC columns of similar
Zorbax® particles are now commercially available.) The
particles are stabilized in a slurry by ultrasonic mixing in degassed 0.001 M ammonium hydroxide. This ammo-
nia-stabilized slurry is then rapidly pumped into column blanks at high pressure with essentially the same
equipment and technique used for the balanced-density

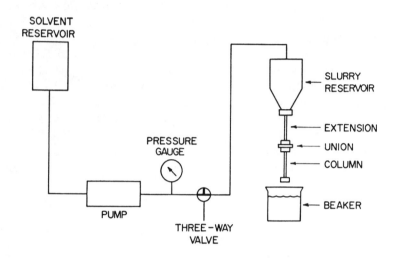

Figure 6.11   Schematic of high-pressure slurry-packing
apparatus. R. E. Majors, Anal. Chem., 44, 1722 (1972).
Reprinted by permission of editor.

slurry-packing approach. Water is removed from the
packing with absolute methanol, and then the column
is equilibrated with a solvent or solvent/stationary
phases that are to be employed for the chromatographic
separation. The stabilized aqueous slurry approach
has the advantage of not using the relatively toxic
halogenated solvents needed for the balanced-density
procedure.

Hard Gels

The balanced-density slurry-packing technique is also
widely used for preparing columns of hard gels. However, the handling of these particles is different.
Hard gel packings must first be allowed to swell in
the solvent which is to be used for the chromatographic
process. Since these organic polymeric gels have a

lower density than the rigid solids, the slurry-packing
technique is carried out with lower density solvents
(e.g., acetone-Perclene® mixtures).

Conventional ion-exchange resins normally are slurry-packed in aqueous systems. The ion exchanger must
first be swollen in the carrier, usually an aqueous
buffer. A "dynamic" packing procedure has been found
to be convenient (12). In this approach, a very thick
aqueous slurry of the swollen ion-exchange resin is
prepared and rapidly pumped with high pressure into a
column blank before particle segregation by sedimentation occurs. An alternative procedure is to prepare a
"cartridge" column of the ion-exchange resin by this
approach. This cartridge column is then connected to
an empty chromatographic column containing the mobile
phase. The ion-exchange resin in the cartridge is then
suddenly extruded by high pressure into the blank column. Several columns may be prepared from a single
cartridge. Equipment for the high-pressure packing of
hard gels is similar to that used for the rigid solids.
However, some hard gels cannot be subjected to very
high pressures.

Alternatively, columns of ion-exchange resins can
be packed by the balanced-density technique. The procedure is to prepare a thick slurry of the swollen ion-
exchange resin in calcium chloride solutions for the
density-matching process. This balanced-density slurry is then forced by high pressure into the column
blank using the technique and equipment described
above. The resulting packed column must be carefully
equilibrated with the chromatographic mobile phase to
ensure complete elimination of the calcium chloride
before use.

Soft Gels

Soft gels cannot be packed dry, nor can they be packed
under high pressure, since most of these materials compress at 100-200 psi or less. Soft gels generally are

packed into columns using a slurry-sedimentation method. First, the gel is solvated in the carrier liquid, and a thick slurry of the material is poured into a column blank which is filled with the carrier. The particles are allowed to collect on a porous plug at the outlet of the column. The bed is formed by filtration as the solvent is allowed to drain from the bottom of the column during the sedimentation process. Since the soft gels normally do not fit the requirements for modern liquid chromatography because of their pressure limitations, more details on the packing procedure for these materials will not be given. Optimum procedures for the preparation of soft gels are described in detail in publications furnished by the manufacturer for the particular packing materials. To obtain the best results with these specialized packings, it is suggested that these directions be followed carefully.

REFERENCES

1. J. J. Kirkland, in Gas Chromatography. 1972. S. G. Perry, ed., Applied Science Publishers, Essex, England, 1973, p. 39
2. P. B. Hamilton, Anal. Chem., 30, 914 (1958).
3. K. Taserik and M. Necasova, J. Chromatog., 75, 1 (1973).
4. R. E. Majors, Anal. Chem., 44, 1722 (1972).
5. J. F. K. Huber, Chimia (Aaran), Suppl. 24-35 (1970).
6. J. J. Kirkland, J. Chromatog. Sci., 10, 593 (1972).
7. J. J. Kirkland, J. Chromatog., 83, 149 (1973).
8. J. J. DeStefano and H. C. Beachell, J. Chromatog. Sci., 8, 434 (1970).
9. R. E. Majors, Anal. Chem., 45, 755 (1973).
10. J. J. Kirkland, J. Chromatog. Sci., 10, 129 (1972).
11. J. J. Kirkland, J. Chromatog. Sci., 9, 206 (1971).
12. C. D. Scott and N. E. Lee, J. Chromatog., 42, 263 (1969).

13. W. Strubert, Chromatographia, <u>6</u>. 50 (1973).

BIBLIOGRAPHY

<u>Gel Chromatography</u>, Bio-Rad Laboratories, 1971. (Pack-
ing and handling of columns of soft gels.)
J. J. Kirkland, J. Chromatog. Sci., <u>7</u>, 7 (1969).
(Techniques of column preparation, effect of column
parameters.)
J. J. Kirkland, Anal. Chem., <u>43</u>, 36A (1971). (Commer-
cial packings for modern LC, column packing proce-
dures.)
R. E. Leitch and J. J. DeStefano, J. Chromatog. Sci.,
<u>11</u>, 105 (1973). (Description and discussion of
modern LC packings.)
D. C. Locke, J. Chromatog. Sci., <u>11</u>, 120 (1973).
(Detailed review of work on chemically bonded phases
for LC.)
R. E. Majors, Am. Laboratory, pp. 27-39 (May, 1972).
(General treatise on modern LC packings.)
F. W. Peaker and C. R. Tweedale, Nature, <u>216</u>, 75
(1967). (Gel preparation and column packing techniques.)
C. D. Scott, Anal. Biochem., <u>24</u>, 292 (1968). (Continuous system for separating ion-exchange resin into
size fractions.)
S. T. Sie and N. van den Hoed, J. Chromatog. Sci., <u>7</u>,
257 (1969). (Preparation of consolidated packings
by a specialized slurry-packing procedure.)

CHAPTER SEVEN

LIQUID-LIQUID CHROMATOGRAPHY

7.1 INTRODUCTION

Liquid-liquid chromatography (LLC), which is sometimes
called liquid-partition chromatography, was developed
in 1941 by Martin and Synge. LLC involves one liquid
as moving phase (sometimes called the carrier), while
another liquid (stationary phase) is dispersed onto a
finely divided inert support. The stationary phase liquid is either immiscible with the moving phase, or it
is chemically bonded to the support. The process of
LLC is similar to simple batch extraction between two
immiscible liquids in a separatory funnel. A successive series of such extractions forms the basis of countercurrent distribution, which is more efficient than
simple one-stage extraction. However, LLC is many times
faster and more efficient than countercurrent extraction. This is the result of the large interface between
moving and stationary phases, with rapid attainment of
an equilibrium distribution of solute between the two
LLC phases. As in extraction or countercurrent distribution, separation of a sample in LLC results from the
differing distribution of the various solutes between
the two liquid phases. Because of the common basis for
both solvent extraction and LLC, extraction data can
be used to predict LLC k' values (e.g., see Section
13.2).

The basic technique of modern LLC is essentially the same as that practiced since its inception. However, recent improvements in packings, column design, and apparatus have permitted a major increase in speed, resolution, and convenience. Much of the information on LLC in the old literature can be adapted to modern practice. Some useful references to classical LLC applications are listed at the end of this chapter, along with more detailed descriptions of the theoretical basis for partition chromatography.

While some general guidelines can be given for the selection of systems for LLC (see Sections 7.3 and 13.2), the selection of practical systems is still largely empirical. Paper chromatography and thin-layer chromatography with cellulose-coated plates can often be used to explore possible systems for LLC, which reduces the time and effort required to determine a useful separation system.

LLC can be applied to a wide variety of sample types, both polar and nonpolar. The basis of this great versatility is the large variety of partitioning phases that are available. Unique chemical interactions can provide separations which are difficult by other chromatographic techniques. LLC generally separates on the basis of the type (and sometimes the number) of substituent groups, and by differences in molecular weight (up to molecular weights of 1000-2000). Therefore, this technique allows the separation of homologs and mixtures of compounds with different functional groups, as illustrated in Table 7.1. While there is no universal partitioning system for all solutes, there is an almost infinite capability for separation, by selecting an appropriate pair of partitioning liquids.

An important practical advantage of LLC is the ability to make columns that can reproducibly effect the desired separations. This is in contrast to some of the other LC methods (e.g., liquid-solid

TABLE 7.1. SEPARATION SELECTIVITY IN LIQUID-LIQUID CHROMATOGRAPHY

Separation By	LLC System (Stationary Phase/Mobile Phase)	Compounds Separated	k'	Ref.
Compound type	ODS-Permaphase®/25% methanol in water (reversed-phase LLC)	1-Naphthylamine	1.1	1
		1-Naphthonitrile	2.3	
		1-Nitronaphthalene	3.5	
		Naphthalene	6.6	
		1-Methoxynaphthalene	9.6	
		1-Methylnaphthalene	15.7	
Homologous series (molecular weight)	Tris-(2-cyanoethoxy)propane/hexane (conventional LLC)	3-Methylpentanal-DNPH[a]	0.2	2
		i-Butanal-DNPH	0.4	
		n-Pentanal-DNPH	0.5	
		n-Butanal-DNPH	0.8	
		Propanal-DNPH	1.3	
		Crotonaldehyde-DNPH	1.6	
		Acrolein-DNPH	2.5	
Number of functional groups (oligomers)	Polyethylene glycol-400/iso-octane-CCl₄, 2:1 (vol/vol) (conventional LLC)	Ethyleneoxide (EO) oligomers (7-14 EO units)	0.4-21	3

a DNPH, 2,4-Dinitrophenylhydrazine derivatives.

chromatography), where the preparation of columns from
different batches of packing material is often less
reproducible than is desired. Liquid-liquid chromatography is also a very gentle technique. Samples are
rarely altered during the course of separation, because
relatively inert support materials are used in conjunction with nonreactive partitioning systems. Due to the
stability of solutes in LLC systems, and the ability to
closely reproduce high-efficiency LLC columns, this
technique is particularly useful for the precise quantitative analysis of mixtures. High-precision assays
in the 90-100% purity range and trace analyses at the
parts-per-million level have been reported.

7.2   COLUMN PACKINGS

In the most widely used form of LLC, the support is
impregnated with a polar stationary phase, while a
relatively nonpolar solvent is used as the mobile
phase. This system is useful for separating polar
compounds, since these are preferably retained by
polar stationary phases. The two LLC phases can also
be interchanged (i.e., less polar stationary phase,
polar moving phase) for the separation of relatively
nonpolar solutes. This is referred to as <u>reversed-
phase</u> LLC.

   With unknown mixtures, it may be difficult to
decide between these two LLC approaches. One answer
is to try the LLC column that is in the instrument (or
readily available), whether it is a conventional or
reversed-phase system. If the sample components elute
with the solvent front after trying mobile phases of
decreasing strength, the liquid phases should then be
reversed for good separation.

   In LLC it is desirable to use supports that are
relatively inert. Such materials should have low surface areas and contain relatively large pores,

which minimize the adsorption characteristics of the support. Although high-surface-area supports are sometimes used for LLC, these materials generally should be avoided. High-surface-area supports used in conjunction with relatively low liquid loadings result in a combination of partition and adsorption. This situation can lead to broad, tailing peaks and irreproducible separations. Columns with high concentrations of stationary phase dispersed on totally porous supports also have been proposed (4-6). Such columns demonstrate higher sample capacities and smaller changes in characteristics due to loss of stationary liquid during use. However, heavily loaded columns with > 20-μ totally porous particles exhibit lower efficiency, particularly at higher mobile phase velocities.

Both totally porous and superficially porous supports can be used for preparing LLC columns. A summary of some of the commercially available supports is given in Table 7.2. Also listed are the newer packings with bonded stationary phases. It should be noted that materials with higher surface areas are generally more useful for liquid-solid chromatography (see Section 8.2). Since developments in packings for modern LC are proceeding rapidly, it is anticipated that additional materials will soon be available. A comparison of plate height versus mobile phase velocity plots of LLC columns for some of these materials was given in Section 6.3, along with other data which characterize the performance of the various materials which are available for LLC.

A comparison of the efficiency of columns made with various types of LLC supports is given in Table 7.3. The values in this table are approximate, but representative. As indicated, the performance of modern LLC columns is now more than 1000 times that of classical LLC columns, which permits analyses formerly requiring hours to be carried out in minutes or seconds.

TABLE 7.2. REPRESENTATIVE COLUMN PACKINGS FOR MODERN LIQUID-LIQUID CHROMATOGRAPHY
(Data from refs. 7 and 8)

Type	Name	Surface Area (m²/g)	Particle Size (μ)	Shape[a]	Approx. Loading for LLC (Wt. %) Min.	Opt.	Max.	Supplier[b]
A. Porous Supports								
Diatomaceous earth	Diachrom®	10	37-44	I	5	10	30	Applied Science, Pierce Chemical (17,28)
	Various others	1-5	>75	I	5	10	30	Applied Science, Supelco (28,29)
Porous silica beads	Porasil®-C[c]	50-100	37-75	S	5[d]	10	15[d]	Waters (11)
	Porasil®-D[c]	25-45	37-75	S				
	Porasil®-E[c]	10-20	37-75	S				
	Porasil®-F[c]	2-6	37-75	S				
Silica gel	LiChrosorb® (Merckosorb®)	200+	10,20,30,40	I	10	20	30	Varian (10)
Porous silica microspheres	Zorbax®-SIL	200+	6-8	S	10	20	30	Du Pont (3)
B. Superficially Porous Pellicular Supports								
"Inactive" silica	Zipax®	1	25-37	S	0.5	1	2	Du Pont (3)
	Liqua-Chrom®	<10	44-53	S	0.6	1	1.8	Applied Science (28)
	Corasil®-I	7	37-50	S	0.5	1	1.5	Waters (11)
	Pellosil®-HS	4	37-44	S	0.5	1	1.5	Reeve-Angel (46)
"Active" silica	Corasil®-II	14	37-50	S	0.5	1	1.5	Waters (11)
	Pellosil®-HC	8	37-44	S	0.5	1	1.5	Reeve-Angel (46)
	Perisorb®-A	10	30-40	S α	0.5	1	1.5	Varian, Merck, EM Labs (10,63,64)
	Vydac® adsorbent	12	30-44	S	1.5	2	3	Applied Science, Separations Group (28,65)

TABLE 7.2. (CONTINUED)

Type	Name and Code	Functionality	Particle Size (μ)	Shape[a]	Supplier[b]
C. Bonded Phase and Coated Supports					
Pellicular, silicate ester	Durapak®- Carbowax® 400/Corasil® ODS/Corasil®	Polyethylene glycol	37–50	S	Waters (11)
Pellicular, bonded silicone monolayer	Bondapak®- C₁₈/Corasil® Phenyl/Corasil®	Octadecyl Phenyl	37–50	S	Waters (11)
Pellicular, bonded silicone monolayer	Vydac®- Reverse phase Polar bonded phase	Octadecyl Nitrile	30–44	S	Applied Science, Separations Group (28,65)
Pellicular, bonded silicone, polymolecular	Permaphase®- ODS ETH	Octadecyl Aliphatic ether	20–37	S	Du Pont (3)
Porous, silicate ester	Durapak®- OPN/Porasil®-C Carbowax® 400/Porasil®-C n-Octane/Porasil®-C	Oxydipropionitrile Polyethylene glycol n-Octane	36–75 75–125	S	Waters (11)
Porous, bonded silicone monolayer	Bondapak®- C₁₈/Porasil®-B	Octadecyl	37–75	S	Waters (11)
Porous, bonded silicone monolayer	Micropak®- CH CN NH₂	Octadecyl Alkylnitrile Primary alkylamine	~10 ~10 ~10	I I I	Varian (10)

TABLE 7.2. (CONTINUED)

Type	Name and Code	Functionality	Particle Size (μ)	Shape[a]	Supplier[b]
Polymer-coated	Zipax®_HCP	Ethylene-propylene copolymer	20–37	S	Du Pont (3)
	Polyamide	Polyamide	20–37	S	Du Pont (3)
	Pellidon®	Polyamide	55–65	S	Reeve-Angel (46)
Polymer, bonded film	Sepcote®_ PM40	Polystryene	~40	S	Separation Technology (70)

[a]S, spherical; I, irregular.

[b]See Appendix V for full name and addresses, listed numerically (e.g., Waters (11)).

[c]Also available under the name of Spherosil® (from Supplier (29), Appendix V).

[d]For Porasil®-C or Spherosil®-XOB 075. For others, loadings roughly proportional to the surface area.

TABLE 7.3   COMPARISON OF COLUMN EFFICIENCIES IN LLC

Support Type	Particle Size (μ)	Max. N_{eff}/sec
Porous (classical)	150	~ 0.02
Porous layer beads (Corasil®-I)	44	1-2
Diatomaceous earth (Kieselguhr)	5-15	1-5
Porous layer beads (Zipax®)	27	10-16
Porous silica microspheres (Zorbax®)	5-6	25

Although LLC with liquid stationary phases provides a unique and reproducible mechanism of separation, it can be one of the more difficult of the LC
methods to maintain over a long period of time. Short
column life and costly column replacement can be the
result of improperly designed LLC systems. Column performance in LLC can be retained for many separations if
several potential difficulties are anticipated and controlled. These problems are discussed in detail in
Section 7.3.
    To reduce or eliminate some of the disadvantages
of LLC systems with mechanically held stationary phases,
packings with surface-reacted or bonded stationary phases have been developed. The advantage of these materials is that pre-columns and/or presaturation of the
two phases is not required, as is usually the case in
conventional LLC with mechanically held stationary
phases (see Section 7.3). In addition, packings with
bonded stationary phases are quite stable, because
there is no opportunity for the chemically bound stationary phase to be eluted during use. A disadvantage
of bonded-phase packings is a lack of systemic information regarding the mode of retention for solutes;

selection of mobile phase to obtain a desired selectivity often is a trial and error process. Some bonded-phase packings also exhibit lower efficiency than conventional liquid-liquid systems. This situation apparently is due to limitations in mass transfer imposed by the highly viscous organic polymer stationary phase (9).

There are two types of surface-reacted or bonded stationary phases that are now commercially available. An esterified siliceous material is shown in the simplified schematic in Figure 7.1. In one form (e.g., Durapak®), the structure consists of a totally porous silica bead esterified with various alcohols to form the corresponding silicate esters. The surface of the esterified particles may be pictured as a forest of organic groupings standing on end--thus the term brushes, which is sometimes used to describe these materials. Esterified porous-layer bead-type supports are also available, with the usual advantage of speed and efficiency and the disadvantage of lower sample capacity.

The esterified siliceous packings have one significant disadvantage: These materials are not

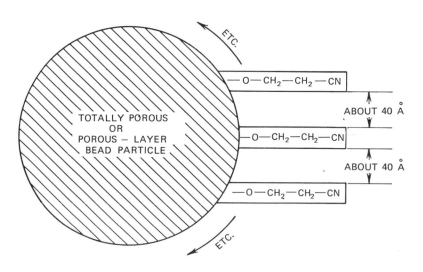

Figure 7.1 Particle with esterified organic coating.

hydrolytically or thermally stable. The organic surface will also exchange with lower alcohols. Consequently, water and the lower alcohols should be excluded from systems when working with these packings.
A second type of surface-reacted packing uses
chemically bonded silicone polymers as stationary phases. These bonded phases cannot be removed by organic
solvents and are thermally and hydrolytically stable.
The organic coating can be essentially a monomolecular
layer (e.g., Bondapak®, Vydac®) or it can be many layers
thick (e.g., Permaphase®). Figure 7.2 is a schematic
representation of Permaphase®, a multilayered chemically bonded silicone packing. There is strong evidence

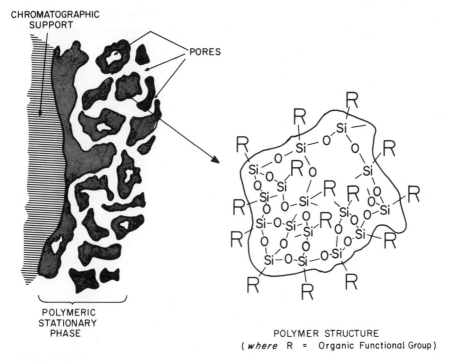

Figure 7.2 Schematic cross section of support with a
chemically bonded polymolecular organic stationary
phase.

that the cross-linked structure of this bonded polymeric material is in the form of a porous, three-dimension-
al gel network when solvated. The solvated gel network
appears to function primarily as a partitioning medium,
although adsorption effects may also take place in some
systems (10).

Columns of bonded-phase packings have been used
continuously for many months without changes in chromatographic characteristics. These bonded-phase materials are available with several types of functional
groups and provide for separations of many types of
solutes. Packings with the chemically bonded silicone
polymers may be used with a wide variety of mobile phases without fear of degradation.

A large number of stationary phases for LLC has
been reported in the literature, and bonded-phase packings are now being used in a wide variety of applications. Some illustrative examples of modern LLC with
both mechanically held and bonded stationary phases are
given in Table 7.4. Figure 7.3 compares the selectivity of a number of commonly used (mechanically held)
stationary phases for the separation of a mixture of
aromatic alcohols, using the same mobile phase.

While a large variety of stationary phases are
available for use in LLC, the number commonly needed
is not very large. Large changes in the k' and α values can be obtained by modifying the mobile phase with
appropriate solvents. (The manner in which selectivity may be systematically changed is discussed in Sections 7.3 and 13.2.) Regardless of the LLC system
used, the optimum capacity factor range (k' = 1-10)
should always be sought, whenever possible, by adjusting the strength of the mobile phase.

There are four different techniques now in use
for preparing liquid-coated packings for LLC, and each
has its own particular advantage for certain applications. The most commonly used technique for applying
mechanically held stationary phases onto the solid support is the solvent evaporation procedure, which is

TABLE 7.4 ILLUSTRATIVE SYSTEMS FOR MODERN LIQUID-LIQUID CHROMATOGRAPHY

Stationary Phase	Mobile Phase	Systems Separated	Ref.
Ethylene glycol	3% CHCl$_3$/heptane	Steroids	11
β,β'-oxydipropionitrile	7% CHCl$_3$/hexane	Insecticides	12
Carbowax 400	Hexane	Nonionic detergents	13
Squalane	20% H$_2$O/CH$_3$CN	Hydrocarbons	13
Cyanoethylsilicone	Water	Coumarins	13
Polyamide	5% hexane/ethanol	Penicillin esters	13
ETH-Permaphase® (bonded-aliphatic ether)	2.5% Methanol/cyclo-pentane	Phenols	10
Vydac®-reverse-phase (bonded C$_{18}$)	33 1/3% H$_2$O/methanol	Fused-ring aromatics	14
Hydrocarbon polymer	79% Methanol, 21% H$_2$O, 0.1% H$_3$PO$_4$	Fat-soluble vitamins	15
Water/ethanol/2,2,4-trimethylpentane (ternary)	Water-rich phase	Steroids	16

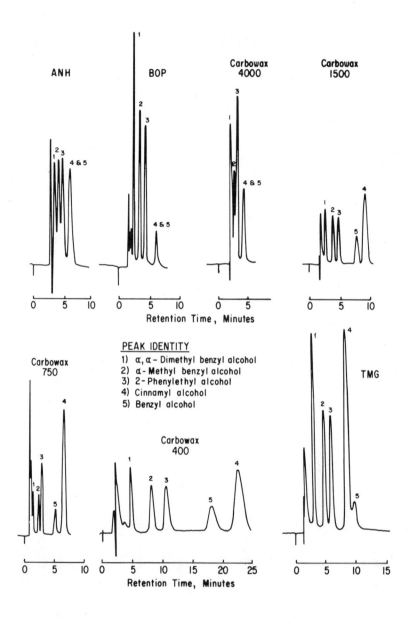

PEAK IDENTITY
1) α,α–Dimethyl benzyl alcohol
2) α–Methyl benzyl alcohol
3) 2-Phenylethyl alcohol
4) Cinnamyl alcohol
5) Benzyl alcohol

Figure 7.3 Comparison of the selectivity of several
stationary phases. Conditions: columns, Zipax® coated
with 1% stationary phase, 1 m x 2.1 mm; mobile phase,
n-hexane; flow rate, 1 ml/min. J. A. Schmit in Modern
Practice of Liquid Chromatography, J. J. Kirkland, ed.,
Wiley-Interscience, New York, 1971. Reprinted by permission of publisher.

also widely used for the preparation of packings for
gas-liquid chromatography. In this technique, a nonvolatile stationary phase is placed on the support by
deposition from a volatile solvent. A useful procedure is to place a known weight of the dry support in
a shallow evaporating dish with an excess volume of a
volatile solvent (e.g., dichloromethane) containing a
known concentration of the stationary phase, so that
the desired weight percent of coating will be obtained
in the final packing. The volatile solvent is slowly
removed from the packing while gently stirring under a
stream of nitrogen (carefully heating with an infrared
lamp if desired), until a dry free-flowing powder is
obtained. It is important that this stirring process
be carried out gently if the fragile diatomaceous earth
supports are used. However, with the mechanically stable superficially porous or totally porous bead supports, solvent may be removed with a rotary evaporator.
    Efficient LLC columns can also be prepared by in
situ coating procedures. In situ coating of supports
for LLC can be carried out by any technique that results in the homogeneous dispersion of the stationary
phase throughout the packing. For example, if the stationary phase has a sufficiently low viscosity, it can
be loaded directly into the porous structure of the
prepacked support. This technique has been effective
in preparing columns with ternary liquid systems composed of water, ethanol, and 2,2,4-trimethylpentane
(16). Two such procedures have been used to coat the
stationary liquid on the support. In the first procedure, the mobile phase (saturated with the stationary

phase) is first pumped through the prepacked columns
to displace the air. Successive small portions of the
stationary phase are then injected into the column with
a syringe, so that the porous structure of the support
is essentially filled with the stationary phase. In
the second method, the stationary phase is pumped
through the dry-packed bed. The column is then treated
with mobile phase equilibrated with the stationary
phase, which displaces the excess stationary liquid
from the inner space between the particles. These two
techniques are effective with stationary liquids of low
viscosity (which also must wet the support), but are
less satisfactory with more viscous stationary liquids.

For stationary liquids with higher viscosity, a
variation in the in situ approach can be used (17).
The procedure is to pump through the prepacked column
the stationary phase which is dissolved at relatively
high concentrations (e.g., 20-40% by weight) in an
appropriate solvent. This purging step, conveniently
carried out at high carrier flow rates, is continued
until an equilibrium situation exists (e.g., a flat
detector baseline). Next, a second solvent, which is
miscible with the first carrier and immiscible with
the stationary liquid phase, is slowly pumped through
the column bed. Under these conditions, the stationary
phase is homogeneously precipitated in the pores of the
support. If high concentrations of the stationary
phase in the initial solvent are used, the pores of the
support can be almost completely filled with the precipitated stationary phase.

With viscous stationary phases, it is sometimes
necessary to pass mobile phase through the in situ
coated columns for several hours to ensure the elimination of excess stationary phase between the support
particles. Increasing the velocity of the carrier
assists in this process. If the support consists of a
very wide pore structure, use of high carrier velocities can result in some loss of stationary liquid within the pores. It is desirable that the column be

conditioned at flow rates higher than those which will
be employed for actual separations.

A significant practical advantage of in situ coating of supports in LLC is that the same prepacked column of support can be recoated several times before replacement or repacking. Should the column become unus-
able, the stationary phase can be stripped from the
support by elution with a polar solvent (e.g., dichloromethane) and the support recoated in situ with the
stationary phase. Similarly, a single column of prepacked support may be recoated in turn with different stationary liquids, and such a column usually can
be used several times in this manner before replacement is required.

Two other techniques for preparing LLC packings
are available, but not widely used. The solvent filtration technique, also successfully used in gas chromatography, produces efficient packings (18); however,
it is more difficult to predict exactly how much stationary liquid will be deposited on the support with
this approach. The equilibration technique is a modification of the in situ method and is based on the fact
that an equilibrium concentration of the stationary
phase will be adsorbed onto a support from the carrier
saturated with the stationary phase (19). However,
there is a danger when attempting the equilibrium
approach: If the concentration of liquid phase loaded
onto the support is less than the equilibrium amount,
tailing peaks can occur as a result of the combination
retention effects of adsorption plus liquid-liquid
partition.

In theory, for LLC it is desirable to fill the
porous structure of the support with the stationary
liquid. There should be little difference between solute mass transfer into a volume filled with only the
stationary phase, as compared with one filled with a
mixture of stationary phase and stagnant mobile phase
(viscosity being the same for both liquids). Therefore,

a support that is essentially filled with the stationary phase should maintain its efficiency, but the sample capacity of the column would be measureably increased, along with increased k' values. As indicated
in Figure 7.4, plate-height values obtained on two similar solutes with essentially the same capacity factors
occur on the same plot even though only 1% stationary
liquid was used in a precoated column, while 3% stationary phase is present in an in situ loaded column.

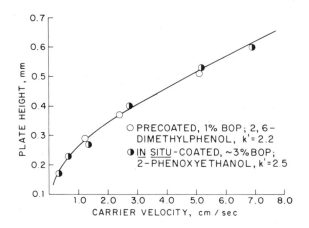

Figure 7.4   Effect of concentration of stationary
phase.   Column, 1000 x 2.1 mm i.d., Zipax® (< 37 µ);
1% β,β'-oxydipropionitrile; carrier, hexane; temperature, 27°C.   J. J. Kirkland and C. H. Dilks, Jr.,
Anal. Chem., 45, 1788, 1973. Reprinted by permission
of editor.

These data substantiate the contention (Section 2.3)
that stationary phase mass transfer is generally not
limiting column efficiency in modern LC, since the
plate heights are equivalent for these two columns
with different concentrations of stationary phase
liquids.

7.3   THE PARTITIONING PHASES

There are some basic characteristics of the mobile
phase which must be considered when designing an LLC
system.   First, the mobile phase must be immiscible
with the stationary phase.   Both partitioning phases,
but especially the mobile phase, should have as low a
viscosity as possible for higher column permeability
and/or efficiency (Section 2.3 and Appendix II).   The
detector can also limit the phases which can be used.
For instance, strongly UV-absorbing solvents should be
avoided with an ultraviolet photometric detector.   The
cost, toxicity, purity, and stability of a solvent also
should be taken into account.   Most important is the
selectivity of a liquid-liquid system for a given sample.   The choice of the mobile phase in LLC is largely
empirical, although some general guidelines are given
below.   A more detailed discussion of the requirements
for solvents in LLC is presented in Section 13.2.

   In LLC, the k' values of solutes are generally
controlled by changing the mobile phase.   While it is
possible to vary k' values somewhat by selection of
the stationary phase, this is often inconvenient in
actual operation.   Of course, the combination of the
two liquid phases (mobile and stationary phases) controls the overall range of k' values for solutes which
can be separated in the particular system.   Thus, polar
compounds may be eluted at optimum k' values using polar stationary phases and nonpolar carriers.   Converse-
ly, nonpolar solutes are optimumly eluted with a system consisting of a nonpolar stationary phase and a
polar carrier.   However, once the selection of the
basic LLC system is made, adjustment of k' values is
normally carried out by varying the mobile phase composition.

   How do we measure solvent polarity so that the
mobile phase can be systematically varied to optimize
the k' values and the selectivity of the system?

Empirical scales of relative solvent polarities have
been useful in this connection (20). However, a more
quantitative scale of solvent polarities is defined by
the Hildebrand solubility parameter δ (see also Section 13.2). Table 7.5 is a partial listing of some
solvents with their δ values. The parameter δ is a
good measure of what is commonly called polarity. Nonpolar solvents have low values of δ, while polar solvents have large values. The solvents in the first
column of Table 7.5 are listed in order of increasing
δ values and polarity. In conventional LLC (more polar
stationary phase), the strength of the carrier increases
with increasing solvent polarity. Thus, by increasing
the solvent δ values, we increase solvent strength and
decrease sample k' values.

The δ values in the first column of Table 7.5 can
be subdivided according to specific intermolecular interactions: dispersion, dipole orientation, and hydrogen bonding (see discussion of ref. 42). Thus the
parameter δ_d is a measure of the ability of the solvent to interact with sample molecules via dispersion
or London forces. The parameter δ_o measures the ability of the solvent to interact by dipole interaction.
The parameters δ_a and δ_h represent the ability of a
solvent to interact as hydrogen acceptor or donor, respectively. Solvents with large values of δ_d tend to
interact preferentially with sample compounds of higher polarizability or refractive index, for example,
aromatic compounds, higher molecular weight homologs,
compounds containing Cl, Br, I, S, and the like.
Therefore, compounds of higher refractive index will
distribute preferentially into that liquid phase which
has a higher value of δ_d. Solvents with large values
of δ_o tend to interact preferentially with sample molecules having large dipole moments (e.g., nitro compounds, nitriles, sulfoxides, amides). Solvents with
large δ_a values are good proton acceptors, and tend to
interact preferentially with hydroxylic sample

TABLE 7.5.   EXTENDED SOLUBILITY THEORY FOR THE CLASSI-
FICATION OF SOLVENTS

Solvent	I δ	II δ_d	III δ_o	IV δ_a	V δ_h	VI ϵ^o	VII η
Perfluoroalkanes[a]	6.0	6.0	0	1	0	-0.25	
CFCl$_2$-CF$_3$	6.2	5.9	1.5	0	0		
Isooctane[b]	7.0	7.0	0	0	0	0.01	0.50
Diisopropyl ether	7.0	6.9	0.5	0.5	0	0.28	0.37
n-Pentane	7.1	7.1	0	0	0	0.00	0.23
CCl$_3$-CF$_3$	7.1	6.8	1.5	0.5	0		
n-Hexane	7.3	7.3	0	0	0	0.01	0.32
n-Heptane	7.4	7.4	0	0	0	0.01	
Diethyl ether	7.4	6.7	2	2	0	0.38	0.23
Triethylamine	7.5	7.5	0	3.5	0		
Cyclopentane	8.1	8.1	0	0	0	0.05	0.47
Cyclohexane	8.2	8.2	0	0	0	0.04	1.00
Propyl chloride	8.3	7.3	3	0	0	0.30	0.35
CCl$_4$	8.6	8.6	0	0.5	0	0.18	0.97
Diethyl sulfide	8.6	8.2	2	0.5	0	0.38	0.45
Ethyl acetate	8.6	7.0	3	2	0	0.58	0.45
Propylamine	8.7	7.3	4	6.5	0.5		
Ethyl bromide	8.8	7.8	3	0	0	0.37	
m-Xylene	8.8	8.8	0	0.5	0	0.26	
Toluene	8.9	8.9	0	0.5	0	0.29	0.59
CHCl$_3$	9.1	8.1	3	0.5	3	0.40	0.57
Tetrahydrofuran	9.1	7.6	4	3	0	0.45	
Methyl acetate	9.2	6.8	4.5	2	0	0.60	0.37
Benzene	9.2	9.2	0	0.5	0	0.32	0.65
Perchloroethylene	9.3	9.3	0	0.5	0		
Acetone	9.4	6.8	5	2.5	0	0.56	
CH$_2$Cl$_2$	9.6	6.4	5.5	0.5	0	0.42	0.44
Chlorobenzene	9.6	9.2	2	0.5	0	0.30	0.80
Anisole	9.7	9.1	2.5	2			

TABLE 7.5 - Cont'd.

Solvent	I δ	II δ_d	III δ_o	IV δ_a	V δ_h	VI ϵ^o	VII η
1,2-Dichloroethane	9.7	8.2	4	0	0	0.49	0.79
Methyl benzoate	9.8	9.2	2.5	1	0		
Dioxane	9.8	7.8	4	3	0	0.56	1.54
Methyl iodide	9.9	9.3	2	0.5	0		
Bromobenzene	9.9	9.6	1.5	0.5	0		
CS$_2$	10.0	10.0	0	0.5	0	0.15	0.37
Propanol	10.2	7.2	2.5	4	4	0.82	2.3
Pyridine	10.4	9.0	4	5	0	0.71	0.94
Benzonitrile	10.7	9.2	3.5	1.5	0		
Nitromethane	11.0	7.3	8	1	0	0.64	0.67
Nitrobenzene	11.1	9.5	4	0.5	0		
Ethanol	11.2	6.8	4.0	5	5	0.88	1.20
Phenol	11.4	9.5					
Dimethylformamide	11.5	7.9					
Acetonitrile	11.8	6.5	8	2.5	0	0.65	0.37
Methylene iodide	11.9	11.3	1	0.5	0		
Acetic acid	12.4	7.0				1.0	1.26
Dimethylsulfoxide	12.8	8.4	7.5	5	0	0.6	2.2
Methanol	12.9	6.2	5	7.5	7.5	0.95	0.60
1,3-Dicyanopropane	13.0	8.0	8	3	0		
Propylene car- bonate	13.3						
Ethanolamine	13.5	8.3	Large	Large	Large		
Ethylene glycol	14.7	8.0	Large	Large	Large		
Formamide	17.9	8.3	Large	Large	Large		
Water	21	6.3	Large	Large	Large		

I. Solubility parameter (calculated from the boiling

Table 7.5. - Cont'd.

---

point).
II.   Dispersion solubility parameter.
III.   Orientation (polar) solubility parameter (approx.
values).
IV.   Proton-acceptor solubility parameter (approx.
values).
V.   Proton-donor solubility parameter (approx. values).
VI.   Solvent strength parameter for adsorption chromatography on alumina; see Section 8.3.
VII.   Viscosity (cP, 20°).
a Average values for different compounds; the fluoro-
chemicals (FC-75, FC-78) sold by 3M Company have similar properties and are considerably cheaper.
b 2,2,4-Trimethylpentane.

---

molecules (acids, phenols).   Solvents with large δ_h
values are good proton donors, and interact preferentially with basic samples (amines, sulfoxides, etc.).
     While solvent strength is governed by the parame-
ter δ, solvent selectivity is controlled by the sub-
parameters δ_d, δ_o, δ_a, and δ_h.   Thus after sample k'
values have been adjusted into the optimum range (1≤
k' ≤10) by choosing a solvent of the right δ value,
selectivity can be varied by choosing another solvent
of similar δ value, but different values of δ_d, δ_o, δ_a
and/or δ_h.   In this connection it should be noted that
solvent mixtures will have values of δ, δ_d, δ_o, and so
on, that are the averages of the corresponding values
for each of the pure solvents making up the solvent
mixture.   Table 7.6 lists the composition of some binary liquid-liquid systems which are suitable for obtaining a range of selectivities utilizing various
interactions.
     In addition to the normal two-liquid systems

TABLE 7.6.   ILLUSTRATIVE LIQUID–LIQUID CHROMATOGRAPH-
IC SYSTEMS

Stationary Phase	Carriers
Conventional LLC	
β,β'-Oxydipropionitrile	Pentane, cyclopentane, hexane, heptane, isooctane
1,2,3-Tris(2-cyanoethoxy)-propane	
Carbowax® 600	Same, modified with up to 10-20% chloroform, dichloromethane, tetrahydrofuran, acetonitrile, dioxane, etc.
Triethylene glycol	
Trimethylene glycol	
Ethylene glycol	Di-n-butyl ether
Dimethyl sulfoxide	Isooctane
H_2O/ethylene glycol	Hexane/CCl_4
Ethylenediamine	Hexane
Water	n-Butanol
1,2-Ethanediol	Nitromethane
Nitromethane	CCl_4/hexane
Reversed-Phase LLC	
Cyanoethylsilicone	Methanol/water
Dimethylpolysiloxane	Acetonitrile/water
Heptane	Aqueous methanol
Hydrocarbon polymer	Methanol/water

utilized widely in LLC, ternary liquid-liquid systems
are also useful. Ternary LLC systems are obtained from
two-liquid systems by adding a third component which is
miscible with both of the components of the two-liquid
system. With ternary systems, either of the two phases
can be used as the stationary liquid in an LLC system,
with the other phase as the mobile liquid. Thus, conventional or reversed-phase liquid-liquid chromatographic separations can be carried out by using the appropriate phase of a ternary system as the stationary liquid.
Selectivity and k' values can be varied over a wide
range in such a ternary system, permitting the separation of a large range of compound types. However, as
the two phases become quite similar, α values become
small and there is a greater tendency for the stationary phase to be displaced by the mobile phase. Similarly, as the two phases become quite dissimilar, α
values become large but the k' values of most compounds
fall outside the optimum range of 1-10. The α values
of a number of pesticides, steroids, and metal chelates
have been determined for the ternary system: water,
ethanol, and 2,2,4-trimethylpentane, which appears to
be a versatile medium for separating a wide range of
compounds by LLC (22).

A special type of LLC, called <u>ion-pair partition</u>
<u>chromatography</u>, is useful for the separation of ionic
or ionizable compounds such as quaternary ammonium
ions, sulfonates, organic sulfonates, amino acids, and
aminophenols (21). Ion pairs are formed by the association of the ionic or ionizable solute with a suitable counter ion (e.g., tetraalkylammonium compounds/
picric acid) in aqueous medium. This ion pair may then
be partitioned into various organic phases of moderate
solvating ability (e.g., chloroform) which gives rise
to high selectivity. This form of LLC is versatile
and can be adapted to different types of solutes by
the proper choice of kind and concentration of counter
ion.

To prevent the loss of mechanically held stationary phases from LLC columns, it is necessary to <u>pre-</u>
<u>saturate</u> the carrier with the stationary phase. This
pre-saturation step is particularly important if the
column is to be maintained constant over a long period
of time, or if the two phases used in the LLC system
have relatively high mutual solubility. Solvents of
relatively poor mutual solubility are difficult to
equilibrate by a simple shaking in a separatory funnel.
An effective procedure for preparing LLC partitioning
systems is to place a 50-100-fold excess of carrier
liquid in a large closed vessel with the stationary
phase. After thoroughly sparging these phases with
nitrogen to eliminate dissolved oxygen, the mixture in
the closed container is then rapidly mixed for several
hours or overnight by magnetic stirring under nitrogen
atmosphere, so that a turbulent vortex is produced in
the container. After this thorough mixing operation,
the phases are allowed to separate, and the equilibrated carrier is separated and maintained under a nitrogen atmosphere until use.

It is sometimes desirable to <u>degas</u> the mobile
phase before it is used with an LC column (see Section
4.2). This operation removes dissolved gases so that
small bubbles will not form in the detector; it also
eliminates oxygen from the carrier, which reduces the
possibility of changes in the partitioning system due
to oxidation.

To ensure continuing equilibration between the
mobile and stationary phases in the LLC columns, a <u>pre-</u>
<u>column</u> should be placed in series prior to the sam-
pling inlet. A convenient pre-column consists of a
50-cm length of 3/16 in. i.d. tubing packed with 20-
30% of the stationary phase impregnated on a relatively coarse diatomaceous earth support (e.g., 120-140
mesh). This pre-column does not need to be packed
carefully since it is not involved in the separation
process. A pre-column is easily prepared, and it

should be replaced frequently to ensure complete protection of the carefully prepared analytical column.
It is necessary that the incoming mobile phase, the
pre-column and the analytical column should all be
maintained at the same temperature. A variation of
less than 0.1° C within these columns is desirable.

As with the other methods of LC, it is important
that purified solvents be used in LLC. Many of the
commonly used solvents are commercially available in
satisfactory purity (see Appendix V). Solvents may
also be purified by treatment with activated adsorbents as described in Section 8.3.

The question is often asked, how long should a
LLC column last? The lifetime of a LLC column depends
on the particular partitioning system and the way it
is used. Generally, the higher the mutual solubility
of the partitioning phases, the more difficult it is
to maintain true equilibrium between these phases.
This situation usually results in shorter column life.
For systems of high mutual solubility (i.e., phases
that are difficult to maintain at equilibrium), a column life of 1 month is acceptable. For other columns,
3-6 months is a typical lifetime, and some columns
have been used for over 15 months without significant
change in column characteristics. The lifetime of a
LLC column with mechanically held stationary phases
depends on how carefully the system is operated. Good
lifetime is promoted by using purified solvents (both
the stationary and mobile phases), by excluding oxygen
from the system to prevent unwanted reactions, and by
using pre-saturated solvents and an appropriate pre-
column (12).

7.4   OTHER SEPARATION VARIABLES

Sample size effects were discussed in general terms in
Chapter 2. However, some sample injection effects

peculiar to LLC should be mentioned. The techniques
which are used to place the sample on the LLC column
to a large degree determine how well and how long the
column operates in a repeatable fashion. The maximum
volume of sample that can be injected into an LLC column varies greatly, and is highly dependent on the internal diameter of the column and the partitioning
system used. For 2-3-mm i.d. LLC columns, 50-100 μl
samples can often be introduced without significantly
decreasing column efficiency.

The weight of solute that can be introduced into
a column without overloading also greatly depends on
the system; however, 2-mm i.d. columns of superficially
porous supports with mechanically held liquid phases
will often accept from 25-200 μg of solute before severe band broadening is evident. Experience has shown,
however, that somewhat smaller sample weights generally
must be used with bonded-phase packings.

Sample solutions to be injected into LLC columns
with mechanically held liquid phases should be prepared
with the carrier (preferably saturated with the stationary phase) to prevent stripping of the stationary
phase. If a solvent stronger than the carrier is used
to dissolve the sample, each sample injection results
in a small amount of the stationary phase being bled
from the column. Therefore, with repeated injections,
the characteristics of the column will change significantly. Another potential problem arising from dissolving samples in a solvent stronger than the carrier
is that solutes of relatively low solubility may pre-
cipitate in the less polar carrier when injected into
the column. This precipitation results in severe peak
tailing and catastrophic band broadening.

7.5 SPECIAL PROBLEMS

While column deactivation is normally not a problem in
LLC, continued injection of samples with strongly

retained (noneluted) components can eventually affect
column performance. This is then seen as changes in
k' and H values and sometimes in altered selectivity
for the separation of a particular pair of compounds.
To prolong column life in such a situation, it is desirable to use a guard column between the injection
port and the analytical column. The guard column, consisting of a short length (5-10 cm) equivalent to the
analytical column, captures these strongly retained
compounds and prevents them from contaminating the analytical column. The guard column should be replaced
at regular intervals to ensure constant performance of
the analytical column. As indicated in Section 7.2,
complete renewal of the stationary phase sometimes can
be accomplished by washing the column with a very
strong solvent and replacing the stationary liquid by
the in situ coating approach.

In reversed-phase LLC, loss of the nonpolar stationary phase can occur as a result of preferential
adsorption of water to the surface of the support, with
consequent desorption of the nonpolar stationary phase
into the mobile phase as the small droplets. To eliminate this difficulty, very nonpolar polytetrafluoro-
ethylene polymers have been used as supports (23).
However, a more generally satisfactory procedure is
to silanize the commonly used siliceous supports before coating them with nonpolar stationary phase. Silanization can be carried out by first heating the
support with concentrated nitric acid for 1-2 hr on
a steam bath, washing to neutrality and drying at
150°C for 1 hr. This acid-washed support is then re-
fluxed with a 5-10% solution of freshly distilled di-
methyldichlorosilane in toluene, or the neat reagent,
for 2 hr, followed by a thorough washing with methanol,
and drying the support (24).

In the recovery of sample fractions in LLC, it
should be recalled that the mobile phase is saturated
with stationary phase. Therefore, high-boiling

stationary phases may be difficult to remove from separated sample fractions. In this case, volatile stationary phases are clearly preferred.

7.6   APPLICATIONS

Modern LLC has been used to separate a wide range of
sample types, some of which are given in Tables 7.1
and 7.4. A general review of the types of samples
that can be handled by LLC was given in Section 7.1.
To illustrate the wide range of applicability of LLC,
some actual separations carried out by this technique
will now be cited.
  Modern LLC has been used for several interesting
separations of compounds containing metal atoms, including isomeric arene tricarbonylchromium complexes
(25), isomers of cobalt complexes involved in the synthesis of vitamin B-12 (26), and metal-β-diketonates
(27). An LLC separation of some metal-β-diketonates
is shown in Figure 7.5.
  Steroids and related synthetically prepared compounds have been separated by LLC, including cortisol
(28), corticosteroids (29), derivatized urinary 17-
ketosteroids (30), fluocinolene acetonide (31), and
various steroids and derivatized steroids (32). Figure 7.6 shows the LLC separation of a mixture of some
steroids in which the column was monitored by two different detectors, refractive index and ultraviolet
absorption.
  Various pesticides have been separated by LLC,
and the technique has been particularly useful for the
trace analysis of these materials in naturally occurring backgrounds. Studies have included the analysis
of methylprednisolone in milk (33), Abate® larvicide
in salt ponds (34), and phosphorus-containing insecti-
cides in crop tissues (35). The analysis of methylparathion and parathion residues on lettuce using LLC
and a selective polarographic detector is demonstrated

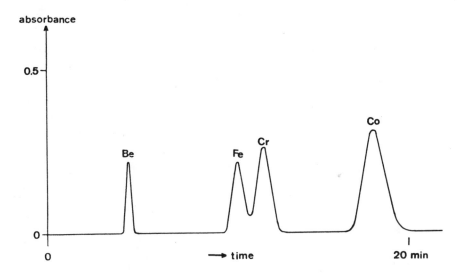

Figure 7.5   Separation of metal-β-diketonate complexes by LLC.   Column, 500 x 2.7 mm; particle size,
5-10 μm; water, ethanol, 2,2,4-trimethylpentane ternary system with less polar phase as mobile phase;
fluid velocity, 3.0 mm/sec; pressure drop 28 bar; theoretical plate number for Co-(AA)$_3$, 960. J. F. K. Huber,
J. C. Kraak, and H. Veening, Anal. Chem., 44, 1554
(1972).   Reprinted by permission of editor.

in Figure 7.7.
    Reversed-phase LLC has been used to separate mix-
tures of aromatics, paraffins, olefins, and diolefins
(13), chlorinated benzenes (13), and fused-ring aromatics (36).   A gradient elution, reverse-phase LLC
separation of some fused-ring aromatic standards is
shown in Figure 7.8.   Mixtures of fat-soluble vitamins
have also been separated by reversed-phase LLC (37),
as illustrated in Figure 7.9.
    A large variety of other materials has also been
separated by modern LLC, including benzodiazepine

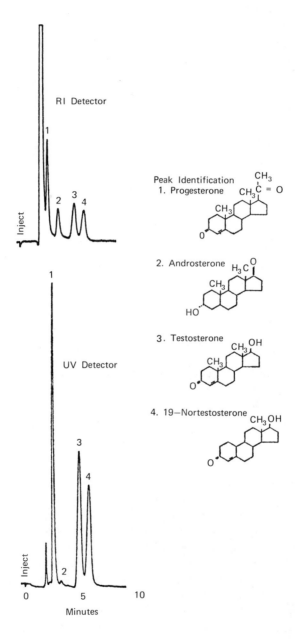

Figure 7.6   Separation of steroids by LLC.   Column, 1%
β,β'-oxydipropionitrile on Zipax®; mobile phase, heptane; pressure, 600 psig; flow, 1 ml/min; temperature,
ambient; detector sensitivity, UV – 0.32 AUFS, RI – 8
x 10^{-5} RIFS; limit of detection, RI – μg for all
peaks, UV – 1 μg for peak 2, 10 ng for peaks 1, 3, and
4.   R. A. Henry, J. A. Schmit, and J. F. Dieckman, J.
Chromatog. Sci., 9, 513 (1971).   Reprinted by permission of editor.

drugs (38), antioxidants and plasticizers (39), alkaloids (40), and various synthetic organics (41).   Figure 7.10 shows the LLC separation of a mixture of hydroxylated aromatics using a short column of very fine
(5-6 μ) particles operated at a high carrier velocity.
This separation illustrates the excellent speed and
selectivity that can be obtained by LLC in an optimized system.

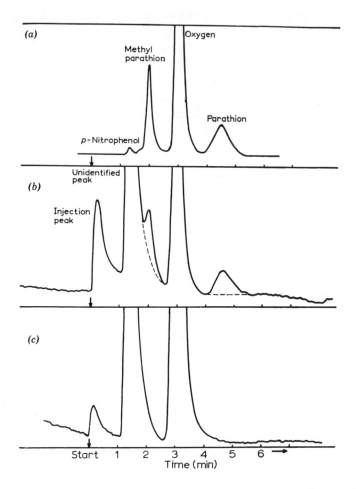

Figure 7.7 Analysis of Pesticide Residues by LLC.
(a) Chromatogram of test mixture; methylparathion,
200 mg/l; parathion, 265 mg/ml; 0.9 µl; full scale,
2×10^{-7} a. (b) Lettuce extract; methyl parathion,
0.8 ppm; parathion, 1.0 ppm; 100 µl. (c) Control extract. Column, 18 x 0.27 cm glass with 28–40 µ silan-
ized diatomaceous earth coated 2,2,4-trimethylpentane;
mobile phase, 60.1% water, 38.8% ethanol, 0.80% acetic
acid, 0.21% sodium hydroxide, and 0.09% potassium

chloride (w/w) saturated with isooctane; flow, 3.4
ml/hr (35).   Reprinted by permission of editor.

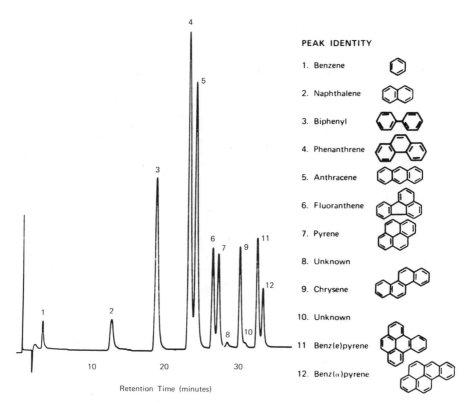

Figure 7.8   Separation of fused-ring aromatics by
reversed-phase LLC.   Operating conditions: column, ODS-
Permaphase®; column temperature, 50°C; linear gradient,
from 20% CH_3OH/H_2O to 100% CH_3OH at 2%/min; column
pressure, 1000 psi; flow rate, 1 cc/min; detector, UV
photometer.   J. A. Schmit, R. A. Henry, R. C. Williams,
and J. F. Dieckman, J. Chromatog. Sci., 9, 645 (1971).
Reprinted by permission of editor.

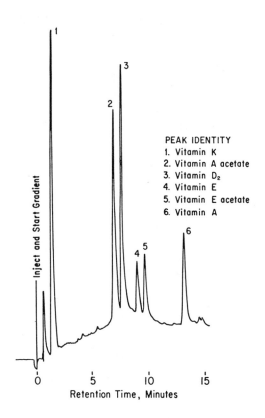

Figure 7.9 Gradient elution separation of fat-soluble vitamins by reversed-phase LLC. Operating conditions: instrument, Du Pont 820 liquid chromatograph; column, Permaphase ODS; mobile phase, gradient from H_2O to CH_3OH at 5%/min; column temperature, 70°C; column pressure, 1200 psi; flow rate, 2 cc/min; detector, UV photometer. R. C. Williams, J. A. Schmit, and R. A. Henry, J. Chromatog. Sci., 10, 494 (1972). Reprinted by permission of publisher.

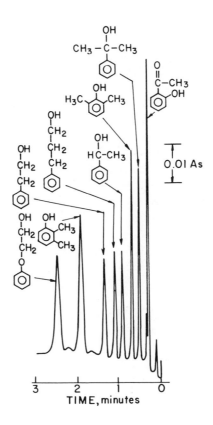

Figure 7.10 Separation of hydroxylated aromatics.
Column, 250 mm x 3.2 mm i.d., 5-6 μ porous silica
microspheres (∼ 350 Å); stationary liquid, β,β'-oxy-
dipropionitrile, ∼30% by weight; carrier, hexane;
pressure, 5000 psi; flow rate, 7.7 ml/min; temperature, 27°C; sample, 15 μl of solution.   J. J. Kirkland,
J. Chromatog. Sci., 10, 593 (1972).   Reprinted by permission of editor.

REFERENCES

1.   R. A. Henry, J. A. Schmit, and R. C. Williams,
     paper at the 163rd National American Chemical So-
     ciety Meeting, Boston, Mass., April, 1972.
2.   J. L. Papa and L. P. Turner, J. Chromatog. Sci.,
     10, 747 (1972).
3.   J. F. K. Huber, F. F. M. Kolder, and J. M. Miller,
     Anal. Chem., 44, 105 (1972).
4.   J. Halasz, H. Englehardt, J. Asshauer, and B. L.
     Karger, Anal. Chem., 42, 1461 (1970).
5.   B. L. Karger and L. V. Berry, Clin. Chem., 17,
     757 (1971).
6.   H. Englehardt and N. Weigand, Anal. Chem., 45,
     1149 (1973).
7.   R. E. Majors, Amer. Lab., 4, 27 (1972).
8.   R. E. Leitch and J. J. DeStefano, J. Chromatog.
     Sci., 11, 105 (1973).
9.   J. H. Knox and G. Vasvari, J. Chromatog., 83, 181
     (1973).
10. J. J. Kirkland, J. Chromatog. Sci., 9, 206 (1971).
11. R. A. Henry, J. A. Schmit, and J. F. Dieckman, J.
     Chromatog. Sci., 9, 513 (1971).
12. R. E. Leitch, J. Chromatog. Sci., 9, 531 (1971).
13. J. A. Schmit, in Modern Practice of Liquid Chro-
     matography, J. J. Kirkland, ed., Wiley-Inter-
     science, New York, 1971, Chapter 11.
14. Liquid Chromatography Application Sheet, Number
     15, Chromatronix, Berkeley, California, 1972.
15. Du Pont Liquid Chromatography Methods Bulletin,
     820M10, March 23, 1972, E. I. du Pont de Nemours
     & Co., Inc., Instrument Products Division,
     Wilmington, Del.
16. J. F. K. Huber, C. A. M. Meijers, and J. A. R. J.
     Hulsman, in Advances in Chromatography, 1971, A.
     Zlatkis, ed., Chromatography Symposium, Depart-
     ment of Chemistry, University of Houston, Houston,
     Texas, 1971, p. 230.

17. J. J. Kirkland and C. H. Dilks, Jr., Anal. Chem., 45, 1788 (1973).
18. J. J. Kirkland, in Modern Practice of Liquid Chromatography, Wiley-Interscience, New York, 1971, Chapter 5.
19. B. L. Karger, H. Engelhardt, and K. Conroe in Gas Chromatography, 1970, R. Stock and S. G. Perry, eds., Inst. of Petroleum (Elsevier), London, 1971, p. 124.
20. J. M. Hais and K. Macek, Paper Chromatography, Academic Press, New York, 1963, p. 115.
21. S. Eksborg, P. Lagerström, R. Modin, and G. Schill, J. Chromatog., 83, 99 (1973).
22. J.F.K. Huber, J. Chromatog. Sci., 9, 72 (1971).
23. S. Siggia and R. A. Dishuran, Anal. Chem., 42, 1223 (1970).
24. J. J. Kirkland, in Gas Chromatography, 1963, Academic Press, New York, 1963, p. 77.
25. J. M. Greenwood, H. Veening, and B. R. Willeford, J. Organometal. Chem., 38, 345 (1972).
26. J. Schreiber, Chimia, 25, 405 (1971).
27. J. F. K. Huber, J. C. Kraak, and H. Veening, Anal. Chem., 44, 1554 (1972).
28. C. A. M. Meijers, J. A. R. J. Hulsman, and J. F. K. Huber, Anal. Chem., 261, 347 (1972).
29. J. Mollica and R. Strusz, J. Pharm. Sci., 61, 444 (1972).
30. F. A. Fitzpatrick, S. Siggia, and J. Dingman, Anal. Chem., 44, 2211 (1972).
31. F. Bailey and P. N. Brittain, J. Pharm. Pharmacal., 24, 425 (1972).
32. R. A. Henry, J. A. Schmit, and J. F. Dieckman, J. Chromatog. Sci., 9, 513 (1971).
33. L. F. Krzeminski, B. L. Cox, P. N. Perrel, and R. A. Schlitz, J. Ag. Food Chem., 20, 970 (1972).
34. R. A. Henry, J. A. Schmit, and J. F. Dieckman, Anal. Chem., 43, 1053 (1971).

35. J. G. Koen and J. F. K. Huber, Anal. Chim. Acta,
     51, 303 (1970).
36. J. A. Schmit, R. A. Henry, R. C. Williams, and J.
     F. Dieckman, J. Chromatog. Sci., 9, 645 (1971).
37. R. C. Williams, J. A. Schmit, and R. A. Henry, J.
     Chromatog. Sci., 10, 494 (1972).
38. C. G. Scott and P. Bommer, J. Chromatog. Sci., 8,
     445 (1970).
39. R. E. Majors, J. Chromatog. Sci., 8, 338 (1970).
40. C. Wu and S. Siggia, Anal. Chem., 44, 1499 (1972).
41. J. J. Kirkland, J. Chromatog. Sci., 10, 593 (1972).
42. R. A. Keller, B. L. Karger, and L. R. Snyder, in
     Gas Chromatography 1970, R. Stock and S. G. Perry,
     eds., Institute of Petroleum, London, 1971, p.125.

BIBLIOGRAPHY

J. C. Giddings and R. A. Keller, in Chromatography, 2nd
   ed., E. Heftman, ed., Reinhold Publishers, New York,
   1967. (Theoretical basis of partition chromatography.)
E. Heftmann, Chromatography, 2nd ed., Reinhold Publishing Corp., New York, 1967. (Theory and application
   of classical liquid chromatography and paper chromatography.)
J. F. K. Huber, in Comprehensive Analytical Chemistry,
   Vol. 2B, C. L. Wilson and D. W. Wilson, eds.,
   Elsevier, Amsterdam, 1968. (Theory and practice of
   liquid chromatography.)
J. J. Kirkland, J. Chromatog. Sci., 7, 7 (1969); 7,
   361 (1969). (Practice of LLC with controlled surface porosity supports.)
J. J. Kirkland, in Modern Practice of Liquid Chromatography, J. J. Kirkland, ed., Wiley-Interscience,
   New York, 1971, Chapter 5. (Practice of modern LLC.)
J. J. Kirkland and J. J. DeStefano, J. Chromatog. Sci.,
   8, 309 (1971). (LLC with chemically bonded-phase
   packings.)

E. Lederer and M. Lederer, Chromatography, 2nd ed., Elsevier Publishing Co., New York, 1957. (Theory and application of classical liquid chromatography and paper chromatography.)

D. C. Locke, in Advances in Chromatography, Vol. 8, J. C. Giddings and R. A. Keller, eds., Marcel Dekker, New York, 1969. (Thermodynamics of liquid-liquid chromatography.)

K. Macek, ed., Pharmaceutical Applications of Thin-Layer and Paper Chromatography, Elsevier Publishing Co., New York, 1972. (Applications of partitioning systems using TLC and PC.)

F. A. v. Metzsch, Angewandte Chemie, 65, 586 (1953). (Extensive list of solvent pairs for liquid-liquid partition.)

L. R. Snyder, in Modern Practice of Liquid Chromatography, J. J. Kirkland, ed., Wiley-Interscience, New York, 1971. Chapter 4. (Detailed discussion of solvent selection.)

CHAPTER EIGHT

LIQUID-SOLID CHROMATOGRAPHY

8.1 INTRODUCTION

Liquid-solid or <u>adsorption</u> chromatography (LSC) is the oldest of the four basic LC methods. In its classical form--so-called <u>open-column</u> chromatography as in Figure 1.1--LSC was conceived and developed by Tswett at the turn of the century. <u>Thin-layer chromatography</u> (TLC), the open-bed version of adsorption chromatography (cf. Figure 1.1), was introduced in the early 1950s by Kirchner and later popularized by Stahl. Modern LSC, featuring automated operation and fast, high-resolution separation, has developed mainly since 1967. Figure 8.1 provides a comparison of the application of each of these three versions of LSC to the separation of a mixture of polyphenyls. While the advantage of modern LSC is clearly apparent in this example, the development of even better columns has further decreased separation times by at least another order of magnitude, as can be seen in the example of Figure 1.2.

Despite the superiority of modern LSC with respect to improved separation and convenience, both open-column and thin-layer chromatography are widely practiced. It is sometimes convenient to carry out preparative separations with open-column chromatography,

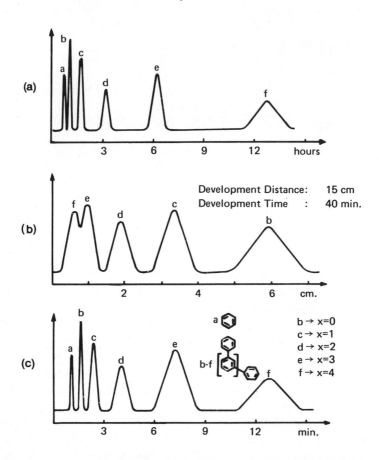

Figure 8.1   Comparison of different LSC techniques:
classical open-column (a), TLC (b), and modern LSC
(c) (1a).   Reprinted by permission.

in the case of easily separated samples.   TLC can be
used to great advantage, particularly in conjunction
with modern LSC.   Because of its simplicity and flexi-
bility, TLC is well suited for exploring the best combination of mobile phase and stationary phase for a
given sample.   Once the right solvent and column packing have been thus identified, they can be used in

modern LSC for improved separation and greater convenience. Certain precautions are required when transfer-
ring experimental conditions from TLC to modern LSC
(see, e.g., ref. 1, and following discussion), but this
is usually a straightforward procedure.

There are further advantages to the preliminary
use of TLC in scouting unknown samples. The use of
corrosive spray reagents plus charring (e.g., ref. 2)
--a sensitive, universal technique for the visualization of separated TLC bands--reveals many compounds
that might not be picked up by a single LC detector,
because of detector specificity or insensitivity. Also,
every band is observable in the final TLC chromatogram,
whereas strongly retained bands often escape detection
in isocratic elution from columns. Thus, TLC serves to
keep track of compounds that might otherwise be missed.
The literature of TLC (2,3) includes many applications
that can serve as guides for similar separations by
modern LSC, provided we know something about the composition of our sample.

LSC gives best results when applied to certain
kinds of samples, or when a certain kind of separation
selectivity is required. Generally the sample should
be organic-soluble, of intermediate molecular weight,
and nonionic. Water-soluble samples can often be
handled satisfactorily by LSC, but usually it is more
difficult to find the right separation conditions;
better results are usually obtained with one of the
other three LC methods. Compounds with molecular
weights greater than 1000-2000 are normally not well
separated by LSC, while compounds with molecular
weights of less than 100-200 are best separated by gas
chromatography. Ionic samples generally give badly
tailing bands in LSC, unless special precautions are
taken. Usually such samples should be separated by one
of the other LC methods.

Concerning separation selectivity, compounds of
different chemical type are easily separated on the

polar adsorbents (see Section 8.2), as illustrated in Table 8.1. Table 8.1 also shows that LSC can provide easy resolution of compounds with differing numbers of functional groups. Finally, LSC is unique among the four LC methods in providing maximum differentiation among isomeric mixtures; a few examples are provided in Table 8.1, and many more could be cited. However, LSC is not noted for its ability to separate homologs, or other mixtures differing in the extent of aliphatic substitution.

The basis of selectivity in LSC can be understood in terms of the adsorption process (for details, see ref. 4). Retention of a solute S in LSC requires displacement of an equivalent number of adsorbed solvent molecules E, as illustrated in Figure 8.2a. Because nonpolar hydrocarbon groups are weakly attracted to the polar adsorbent surface, most solvents tend to displace such hydrocarbon groups from the surface, as in Figure 8.2b for n-butyl phenol. Therefore, hydrocarbon substituents contribute little to solute retention, and there is usually little difference in k' values among molecules differing only in their aliphatic substituents. Polar functional groups, on the other hand, are strongly attracted to the adsorbent surface, so that compounds with substituents of differing polarity--or a differing number of such groups--are readily separated.

Another feature of LSC is the presence of discrete adsorption sites on the surface of the adsorbent--these are illustrated by the points A in Figure 8.2c. Optimum interaction between a solute molecule and the adsorbent surface occurs when solute functional groups exactly overlap these adsorption sites--as a hand fits a glove, or a key is matched to its lock. This overlapping is readily possible for monofunctional solutes such as phenyl-X or phenyl-Y in Figure 8.2c, but not for polyfunctional solutes such as phenyl-X,Y. However, certain polyfunctional solutes will be better matched to the adsorbent surface than their isomeric counterparts, resulting in preferential retention of

TABLE 8.1.   SEPARATION SELECTIVITY IN LIQUID-SOLID
CHROMATOGRAPHY (ALUMINA)

	Value of k'
Compound Type[a]	
2-methoxy naphthalene	0.6
1-nitro naphthalene	1.8
1-cyano naphthalene	2.7
1-aceto naphthalene	5.5
Number of functional groups[a]	
2-methoxy naphthalene	0.6
1,7-dimethoxynaphthalene	1.4
1-nitro naphthalene	1.8
1,5-dinitro naphthalene	6.1
acetophenone[b]	1.1
3-nitro acetophenone[b]	1.6
Isomers	
m-dibromobenzene[c]	3.8
p-dibromobenzene[c]	6.9
quinoline[d]	5.4
isoquinoline[d]	18.6
1,2,3,4-dibenzanthracene $(C_{22}H_{14})$[e]	0.6
picene $(C_{22}H_{14})$[e]	12.0

[a]23%v CH_2Cl_2/pentane (5).   [d]Benzene (7).
[b]60%v CH_2Cl_2/pentane (5).   [e]CH_2Cl_2 (6).
[c]Pentane (6).

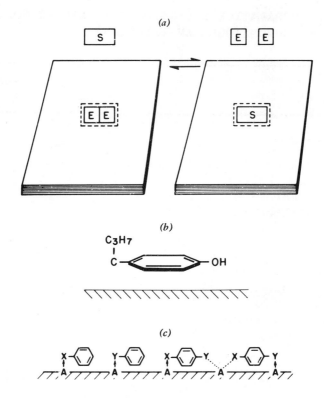

Figure 8.2   The adsorption process and separation selectivity in LSC.

the compound that best fits the surface. In practice
it is rare that an isomeric pair of compounds cannot
be separated by LSC, and often the separation factors
are surprisingly large; for example, α equal to 10 or
more.

Apart from gel chromatography, where the basis of
separation is quite simple and relative retention is
easily predicted for compounds of different molecular
structure, retention in LSC is easier to understand and
predict than for the other LC methods. Semiquantitative estimates of k' are often possible, knowing only

the experimental conditions and the structure of the migrating compound (4). In some cases predictions of k' can be made within a few percent (5,8) over a wide range in experimental conditions. This means that the control of separation in LSC (via changes in k' or α) is relatively straightforward, if we have a good understanding of the basis of LSC separation. For a detailed discussion of the fundamental basis of LSC, including its experimental optimization, see refs. 4,9, and 10.

8.2   COLUMN PACKINGS

Packings for LSC are described in terms of the adsorbent or stationary phase, the type of particle (porous versus pellicular), and particle size. Each of these particle characteristics has an important effect on the performance and use of a given packing material.

Adsorbents are usually defined by their chemical composition; for example, silica, alumina, charcoal. While significant differences in separation selectivity exist from one adsorbent to another, there has been little tendency to deliberately select a particular adsorbent for a given sample. Instead, the composition of the moving phase is usually altered to provide needed changes in k' and α, as discussed in Section 8.3. For reasons that will become apparent, silica is used much more often than other adsorbents. Aside from the generally superior performance of silica in LSC, most of the examples from the TLC literature feature silica as adsorbent. Finally, a much wider range of high performance silicas are commercially available than is the case for other adsorbents.

Although we should consider silica first in modern LSC, there will be occasional samples that are simply not well separated on silica. Or some very complex samples may require the use of more than one adsorbent for complete separation (for a good example, see ref. 11). Therefore, it is useful to know some of

the distinguishing features of the various chromatographic adsorbents. Adsorbents can be divided into so-
called <u>polar</u> and <u>nonpolar</u> types. Polar adsorbents include the various inorganic oxides: silica, alumina,
magnesia, Florisil® (magnesium silicate), molecular
sieves, and some less commonly used materials. Charcoal is the only widely used nonpolar adsorbent. Polar
sample molecules are preferentially retained on polar
adsorbents such as silica, and separation order is
roughly in the following sequence:

> saturated hydrocarbons (small k') < olefins
> < aromatic hydrocarbons ≈ organic halides <
> sulfides < ethers < nitro compounds < esters
> ≈ aldehydes ≈ ketones < alcohols ≈ amines <
> sulfones < sulfoxides < amides < carboxylic
> acids (large k').

The relative retention of different compounds on
nonpolar adsorbents follows quite different rules.
Molecular polarizability is the dominant consideration,
so that compounds of higher refractive index are held
more strongly on charcoal. These include higher-
molecular-weight homologs, aromatic compounds, and
molecules containing atoms such as chlorine, sulfur,
bromine, and iodine. While charcoal appears uniquely
useful for separating aromatic compounds as a class
from corresponding aliphatic derivatives (see, e.g.,
ref. 11), high-performance charcoal for modern LSC has
only recently become commercially available. Its general utility is therefore still somewhat unclear.
The polar adsorbents can be further divided into
<u>acidic</u> and <u>basic</u> adsorbents. Acidic adsorbents include
silica and Florisil®, while adsorbents such as alumina
and magnesia are basic (unless they have been acid
treated). The acidic adsorbents preferentially retain
bases such as aliphatic or aromatic amines. Similarly,
basic adsorbents such as alumina retain acids more

strongly--for example, pyrrole derivatives, phenols,
thiophenols, and carboxylic acids. Occasionally the
matching of an acidic adsorbent to a basic sample--or
vice versa--leads to pronounced band tailing or irreversible retention of the sample.

Apart from these general differences among the
various classes of adsorbents, individual adsorbents
also exhibit unique selectivity for certain compounds.
Thus, alumina is particularly useful for its ability to
separate isomers of aromatic hydrocarbons, as well as
corresponding halogen-substituted derivatives. Magnesia appears to retain planar molecules more strongly
than corresponding less planar compounds, so that poly-
cyclic aromatic hydrocarbons are strongly retained, and
cholesterol esters (more planar) are easily separated
from fatty alcohol esters (less planar) (12). Other
differences in adsorbent selectivity are summarized in
(4).

Table 8.2 summarizes a number of useful adsorbents
that are commercially available. This listing is divided into three different groups. Porous, high-performance adsorbents offer greater capacity (Section 2.2),
thereby permitting larger sample sizes on preparative
separations, and corresponding increases in detection
sensitivity. Porous, low-performance adsorbents are
inexpensive (e.g., $1.00/lb), and can be used for puri-
fying solvents or carrying out large-scale preparative
separations of easily resolved mixtures. Pellicular,
high-performance adsorbents offer greater column efficiency (larger N values) and convenience, compared to
porous adsorbents, but are more expensive and have
lower capacity.

The surface areas of the adsorbents of Table 8.2
are also listed. Adsorbent surface area is a critical
property, and its value for a given adsorbent should
always be known. Surface area in LSC is equivalent to
the percent liquid loading in liquid-liquid chromatography; the larger the surface area, the greater is the

TABLE 8.2.   SOME COLUMN PACKINGS FOR LIQUID-SOLID
            CHROMATOGRAPHY (13)

	d_p (μ)	Surface Area (m^2/g)	Shape[a]	Supplier[b]
A. Porous, high-performance				
Silica				
Porasil® A[c]	37-75	350-500	S	Waters (11)
Porasil® B	37-75	125-250	S	
Porasil® C	37-75	50-100	S	
Porasil® D	37-75	25-45	S	
Porasil® E	37-75	10-20	S	
Porasil® F	37-75	2-6	S	
Porasil® T	15-25	300	I	
Sil-X®[d]	36-45	300	I	Perkin-Elmer (7)
Biosil®	20-44		I	Bio-Rad Labs (62)
LiChrosorb®	10 20 30 40	200+	I	EM Labs (64) E.Merck (63)
Zorbax®-Sil	6-8	300	S	DuPont (3)
Alumina				
Woelm alumina	18-30	200+	I	Waters (11)
Bio-Rad ®AG	74	200+	I	Bio-Rad Labs (2)

TABLE 8.2. (Continued)

	d_p (μ)	Surface Area (m^2/g)	Shape[a]	Supplier[b]
B. Porous, low-performance				
Silicas				
Davison Code 12	150-	800	I	W.R. Grace (66)
Davison Code 62	150-	350	I	W.R. Grace (66)
Alumina				
Alcoa F-20	150-	200	I	Alcoa (67)
C. Pellicular, high-performance				
Silica				
Corasil®-II	37-50	14	S	Waters (3)
Pellosil®-HC	37-44	8	S	Reeve-Angel (46)
Perisorb®-A	30-40	10	S	EM Labs(64) E.Merck(63)
Vydac®[d]	30-44	12	S	Separations Group (65)
Alumina				
Pellumina® HC	37-44	8	S	Reeve-Angel (46)
Charcoal				
Pellicarb®			S	Reeve-Angel (46)

[a]S, spherical; I, irregular
[b]See Appendix V for complete addresses.

TABLE 8.2.   (Continued)

---

^{c}Also called Spherosil®.
^{d}Deactivated (permanently); requires no added water.

capacity of the adsorbent, and the more strongly retained are all compounds. Porous adsorbents should
have surface areas in the 200-400 m^2/g range, while
pellicular adsorbents usually have surface areas of
5-20 m^2/g. Duplication of a given adsorbent requires
matching its surface area, and the surface area of the
adsorbent determines how much water should be added to
that adsorbent for optimum performance (see below).
 The efficiency of columns packed with various adsorbents depends mainly on particle size and type. As
an example, for otherwise similar experimental conditions (i.e., column pressure P, separation time t, solvent viscosity η , etc.) column efficiency increases
as follows:

Absorbent	Particle Size (μ)	Column efficiency, N_{eff} (plates per sec)
Porous silica	100-125	1.6
Porasil®-A	35-75	2.3
Porous silica	20	4
Corasil®-II, Sil®-X	37-50	4
Vydac®	30-44	6
LiChrosorb®	10	12
Zorbax®-SIL	5-6	25

The considerable advantage of the small-particle (d_p
≤ 10 μ) columns is unfortunately accompanied by more
difficult column preparation (Chapter 6), and requires
an LC unit of quite small extra-column hold-up (small
V_i values, as in Section 3.3). Some studies have shown

(16)   that silica gives more efficient columns than alumina.

The packing of LSC columns has been discussed in general terms in Chapter 6.   The 30-μ and larger particles are packed dry, with tapping.   Smaller particles are packed by the balanced-density-slurry technique, under high pressure (14,15).   Ready-packed columns of 5-10-μ particles for LSC can also be purchased (e.g., Du Pont, Varian Aerograph, Applied Science).

One last property of the adsorbent should be cited: its water content.   Dry or fully activated adsorbents have a number of undesirable properties, compared to adsorbents that have been deactivated by water or other strongly adsorbed material:

- Much lower linear capacities   (see Figure 2.11) (4).
- Greater tendency toward sample alteration and loss (4).
- Lower column efficiency, in some cases (16).
- Variable retention properties from batch to batch.

Therefore, it is important to add a certain quantity of water to the adsorbent before it is used in LSC. Since the k' values of all compounds decrease as more water is added to the adsorbent (e.g., Figure 8.3), the preparation of repeatable or standardized absorbents requires close control over the amount of added water.

The amount of water used to deactivate the adsorbent can be controlled by initial addition of liquid water to the adsorbent, and/or by equilibration with water-wet solvents (Section 8.3).   The initial-addition technique is illustrated in Figure 8.4.   We begin in A by selecting an adsorbent of a given type (e.g., Porasil®-A) from a particular supplier.   This ensures that we start with a product that is nominally similar

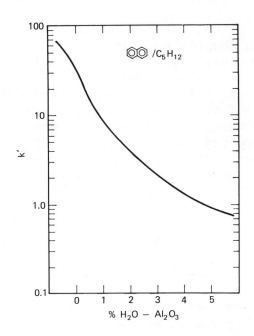

Figure 8.3   Variation of k' with adsorbent water content; naphthalene eluted by pentane from alumina.

from batch to batch.   At this point the adsorbent will contain an unknown amount of water, which is removed (Figure 8.4B) by heating in air for 8-16 hr.   The temperature of activation varies with adsorbent type, being 125-150°C for silica.   After activation or drying of the adsorbent, it is allowed to cool in a stoppered glass bottle (Figure 8.4C).   Enough liquid water is now added to the adsorbent (Figure 8.4D) to give 50-75% of a monolayer; that is, 0.02-0.03 g water per 100 m^2 of adsorbent surface.   This corresponds, for example, to 8-12 g water per 100 g of Porasil®-A (Table 8.2). The glass container is again stoppered, vigorously shaken until all lumps of wet adsorbent have disappeared, and stored for a short time.   At this point

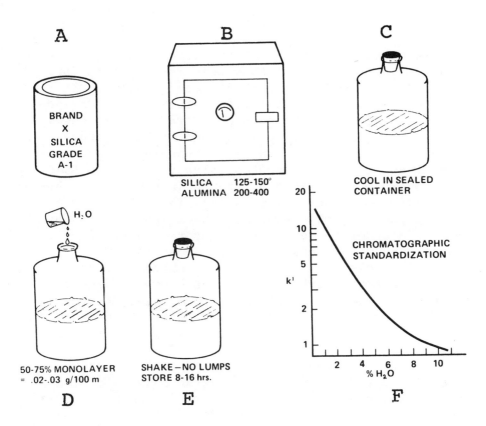

Figure 8.4   Preparation of standardized (controlled water content) adsorbents.

(Figure 8.4E) adsorbent of the correct water content has been prepared.

Small differences in the retention characteristics of the adsorbent may remain at the end of step E of Figure 8.4, due to batch-to-batch variations in the starting adsorbent. These can be eliminated, for the most part, by chromatographic standardization. For a given batch of adsorbent, a standard curve of k' versus adsorbent water content is measured, as in

Figure 8.3 or Figure 8.4F. A standard solute and solvent are used for this purpose, as in Figure 8.3. A
value of k' (same solute, solvent) is now measured for
our fresh batch of adsorbent from step E. This k' value should be the same as that read from the standard
curve for the desired adsorbent water content. If this
k' value is significantly different, water or dry adsorbent is added to the adsorbent from step E, the mixture is reequilibrated as in E, and k' is redetermined.* In this manner the adsorbent is adjusted to
make k' equal to the expected value from Figure 8.4F.

Adsorbents may be deactivated in other ways than
by addition of water. Very polar liquids such as ethylene glycol and glycerine have been substituted for
water, in order to reduce the dehydration of water-
deactivated adsorbents by dry solvents (see Section
8.3). Alcohols such as isopropanol have also been
added directly to the solvent to achieve adsorbent deactivation (Section 8.3). Some commercial adsorbents
such as Vydac® and Sil®-X are claimed to be permanently
deactivated by a proprietary process, and therefore require no added water.

Repeatable separations using different adsorbent
batches are seen to require comparable water deactivation (i.e., Figure 8.3). Therefore, if we use TLC as
a guide in selecting conditions for a corresponding
separation by modern LSC, the water content of the TLC
plate should match that of the adsorbent used in the
column. Typically the TLC adsorbent will be somewhat
drier, unless special precautions are taken, so that
k' values tend to be somewhat smaller in the column

---

*This technique is now used less often. It is more
convenient to maintain adsorbent activity (and k' values) constant through variation of the solvent water
content, as described in Section 8.3.

separation. This is commonly taken care of by using
a slightly weaker solvent in the column separation
than was found optimum for the TLC separation. Equipment is commercially available for the precise control
of adsorbent water content in TLC (Camag), however.

8.3 MOBILE PHASES

The choice of solvent or mobile phase is all-important
in LSC. As we have seen, water-deactivated silica is
usually selected as adsorbent. The solvent is then
varied to give k' values in the optimum range ($1 < k' < 10$), and in some cases to improve the α values of one
or more band pairs.

Solvent strength, which controls the k' values of
all sample bands, is easily predicted in LSC. It can
be defined quantitatively by the solvent strength parameter $\epsilon°$, which is listed for several pure solvents in
Table 8.3. These values are for alumina as adsorbent,
but a similar series of values is observed also for
silica and the other polar adsorbents. A series of
solvents, arranged in order of increasing strength as
in Table 8.3, is referred to as an eluotropic series.
If an initial solvent is too strong (k' values too
small), then a weaker solvent is substituted (smaller
value of $\epsilon°$). Similarly, if the initial solvent is
too weak (k' values too large), a stronger solvent is
substituted. Thus, an eluotropic series can be used
to find the right solvent strength by a rapid trial-
and-error approach. Often the right solvent strength
is determined by preliminary TLC separations of the
sample of interest, keeping in mind that a slightly
weaker solvent should generally be used in the corresponding separation by modern LSC.

Solvent mixtures, particularly binary solvents,
are used more often in LSC than pure solvents. There
are several advantages to the use of binary solvents;
for example, solvent strength changes continuously with

TABLE 8.3.   SOLVENT STRENGTH DATA; ALUMINA AS ADSORBENT (4,7)

Solvent	$\epsilon°$	nb	Solvent	$\epsilon°$	nb
Fluoroalkanes[a]	-0.25		Ethylene dichloride	0.44	4.8
n-Pentane	0.00	5.9	Methyl ethyl ketone	0.51	4.6
Isooctane	0.01	7.6	1-Nitropropane	0.53	4.5
Petroleum ether	0.01	6.7	Triethyl amine	0.54	7.5
n-Decane	0.04	10.3	Acetone	0.56	4.2
Cyclohexane	0.04	6.0	Dioxane	0.56	6.0
Cyclopentane	0.05	5.2	Tetrahydrofuran	0.57	5.0
1-Pentene	0.08	5.8	Ethyl acetate	0.58	5.7
Carbon disulfide	0.15	3.7	Methyl acetate	0.60	4.8
Carbon tetrachloride	0.18	5.0	Diethyl amine	0.63	7.5
Xylene	0.26	7.6	Nitromethane	0.64	3.8
i-Propyl ether	0.28	5.1	Acetonitrile	0.65	10.0
i-Propyl chloride	0.29	3.5	Pyridine	0.71	5.8
Toluene	0.29	6.8	Dimethyl sulfoxide	0.75	4.3
n-Propyl chloride	0.30	3.5	i-, or n-Propanol	0.82	8.0
Benzene	0.32	6.0	Ethanol	0.88	8.0
Ethyl bromide	0.35	3.4	Methanol	0.95	8.0
Ethyl sulfide	0.38	5.0	Ethylene glycol	1.1	8.0
Chloroform	0.40	5.0			
Methylene chloride	0.42	4.1			

[a]Fluorochemical FC-78 (3-M Company) is equivalent to the fluoroalkanes with respect to strength, but is considerably cheaper (27).

composition, as illustrated in Figure 8.5 for several
solvent binaries. Thus, it is possible to obtain just
the right solvent strength, by the proper choice of
binary composition. If pentane as solvent (ϵ° = 0)
is too weak, and chloroform is too strong (ϵ° = 0.40),
then some mixture of these two solvents will have just
the right strength. Normally we can zero in on this
composition quite rapidly. Another advantage of binary solvents is that solvent viscosity can be kept low,
by selecting a nonviscous solvent as the major component of the mixture. We have seen in Section 2.3 that
low solvent viscosities favor high column efficiencies.
    Table 8.4 lists ϵ° values for several useful bi-
nary solvents on alumina. Figure 8.6 provides similar

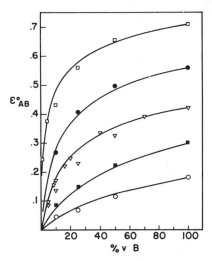

Figure 8.5 Variation of solvent strength ϵ° in LSC
as a function of binary composition (4). Binary solvent ϵ° values versus solvent composition for adsorption on H_2O-Al_2O_3. O , Pentane (A)-CCl_4 (B); ■
pentane (A)-n-propyl chloride (B); △ , pentane
(A)-CH_2Cl_2 (B); ● , pentane (A)-acetone (B); □ ,
pentane (A)-pyridine (B); solid line, calculated values. Reprinted with permission.

TABLE 8.4. THE STRENGTHS OF SOME BINARY SOLVENTS; ALUMINA AS ADSORBENT

| | Values of Volume Percent B Shown for Given $\epsilon°$ Value | | | | | | | |
| | Solvent A: | | | | Pentane | | | |
$\epsilon°$	Solvent B:	CS_2	i-PrCl[a]	Benzene	Ethyl ether	$CHCl_3$	CH_3Cl_2	Acetone	Methyl acetate
0.00		0	0	0	0	0	0	0	0
0.05		18	8	3.5	4	2	1.5		
0.10		48	19	8	9	5	4	1.5	
0.15		100	34	16	15	9	8	3.5	2
0.20			52	28	25	15	13	6	3.5
0.25			77	49	38	25	22	9	5
0.30				83	55	40	34	13	8
0.35					81	65	54	19	13
0.40						100	84	28	19
0.45								42	29
0.50								61	44
0.55								92	65
0.80									100

TABLE 8.4 - Cont'd.

| | Solvent A: | Pentane | Benzene | | | | | |
| | Solvent B: | Diethyl-amine | Acetone | Methyl acetate | Diethyl-amine | Aceto-nitrile | i-PrOH[b] | MeOH[c] |
$\epsilon°$								
0.30		2.5						
0.35		5	6	4	2			
0.40		8	18	12	7			
0.45		13	36	24	14	1.5		
0.50		22	60	42	26	6		
0.55		38	93	66	45	14	4	
0.60		73		100	77	36	7	
0.65						100	12	
0.70							21	
0.75							40	4
0.80							75	8
0.85								18
0.90								44
0.95								100

[a]Isopropyl chloride
[b]Isopropanol
[c]Methanol

MIXED SOLVENT STRENGTHS ON SILICA GEL

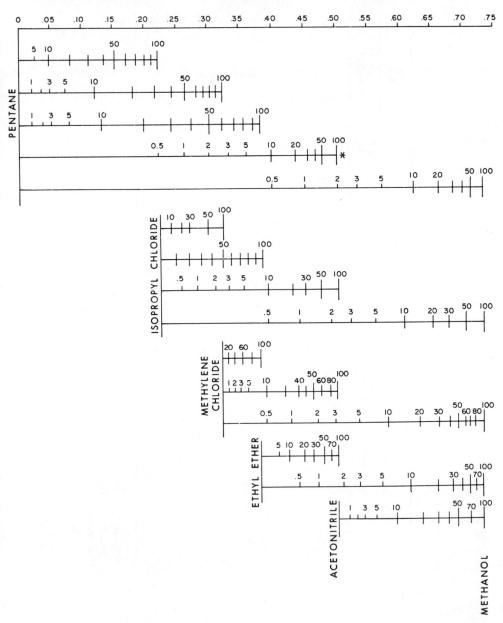

Figure 8.6 Solvent strength for various binaries as a function of binary composition (adsorbent, silica) (17). Note for acetonitrile/pentane solutions that a small amount of cosolvent (e.g., CCl$_4$, benzene) must be added to intermediate compositions for miscibility.

data for silica as adsorbent, in a slightly different format. Here solvent strength is plotted horizontally at the top of Figure 8.6, while corresponding solvent compositions are plotted below; for example, pentane/isopropyl chloride, pentane/methylene chloride, pentane/ethyl ether, and so on. For example, 50 vol % methylene chloride/pentane has $\epsilon°$ = 0.27. For additional data on solvent strength see Ref. 4.

Solvent selectivity in LSC is varied by holding solvent strength constant, while varying solvent composition. In practice this is achieved through the use of data such as those of Table 8.4 or Figure 8.6. Typically a trial-and-error substitution of different solvents of equal strength is tried. For example, if 16 vol % benzene/pentane is found to have the right strength, but α is close to one for some pair of bands, then from Table 8.4 (alumina as adsorbent) we might choose 100% carbon disulfide, 34 vol % isopropyl chloride/pentane, or 2 vol % methyl acetate/pentane as alternative solvents. The strength of each of these solvents is the same ($\epsilon°$ = 0.15), but the different compositions should lead to differences in α values. Or if 32 vol % ethyl ether/pentane as solvent has the right strength for a separation on silica ($\epsilon°$ = 0.25, Figure 8.6), and α equals ~1 for some pair of bands, we might substitute any of the following solvents from Figure 8.6 for the purpose of increasing α: 0.8 vol % acetonitrile/pentane, 42 vol % methylene chloride/pentane, or 10 vol % ethyl ether/isopropyl chloride.

Such a trial-and-error approach to solvent selectivity or the optimization of α values is often

successful, but generally tedious。 What we want--when
α must be increased--is a set of simple rules that predict the variation of α with solvent composition. We
will now examine a few rules of this type which are
well documented for LSC. A rather general dependence
of solvent selectivity on solvent composition is illustrated by the data of Table 8.5. Here α values for two
representative solutes, dinitronaphthalene and aceto-
naphthalene, are shown for a series of solvents of similar strength. All of these solvents are blends of
different strong solvents with the weak solvent pentane.
Because the <u>pure</u> strong solvents vary in strength, the
concentration of each of these strong solvents must
also vary, since $\epsilon°$ for each solvent <u>mixture</u> is held
roughly constant. Thus the concentration of strong solvent varies from 50 vol % benzene ($\epsilon°$ for benzene

TABLE 8.5.   SOLVENT SELECTIVITY IN LSC; CONCENTRATED
VERSUS DILUTE SOLUTIONS OF THE STRONG SOLVENT COMPO-
NENT (ALUMINA ADSORBENT) (7)

Solvent (Pentane Solution) (vol %)		k' COCH₃ (acetonaphthalene)	k' dinitronaphthalene	α
50	Benzene	5.1	2.5	0.5
25	Diethyl ether	2.5	2.9	1.16
23	CH$_2$Cl$_2$	5.5	5.8	1.05
4	Ethyl acetate	2.9	5.4	1.8
5	Pyridine	2.3	5.4	2.4
0.05	CH$_3$SOCH$_3$	1.0	3.5	3.5

0.32), to 4 vol % ethyl acetate ($\epsilon°$ for ethyl acetate equal 0.58), to 0.05 vol % dimethyl sulfoxide ($\epsilon°$ for dimethyl sulfoxide equal 0.75). As the concentration of the strong solvent decreases in going down the data of Table 8.5, we see a regular increase in α. As a result, maximum values of α (recalling that α is equivalent to $1/\alpha$) occur for either large or small concentrations of the strong solvent. Intermediate solvent concentrations give less favorable values of α. Thus in Table 8.5, for $\epsilon° = 0.25$, we would expect maximum values of α for either 77 vol % isopropyl chloride/pentane as solvent, or for 5 vol % methyl acetate/pentane. Such solvents as 38 vol % ethyl ether/pentane are predicted to give intermediate (less favorable) α values. A more detailed discussion of this dependence of α on solvent composition is given in ref. 5.

Another useful rule in guiding the choice of solvents for maximum selectivity is that hydrogen bonding between solvent and sample usually results in large changes in α. Thus solvents containing alcohols as components (e.g., 4 vol % isopropanol/benzene in Table 8.4) will often give different α values, relative to other solvents (e.g., 14 vol % acetonitrile/benzene in Table 8.4). Similarly, if the sample contains compounds that are hydrogen donors (e.g., alcohols, amines, pyrroles, etc.), a change in solvent basicity can lead to striking changes in α. Solvent basicity has already been defined in terms of the solvent parameter δ_a (Chapter 7 and Table 7.5). In Table 8.6 we see an example of this, the separation of N-methyl aniline from 2-chloroquinoline by solvent mixtures of different basicity. Here δ_a for the stronger solvent is shown (pentane is nonbasic), and a good correlation between α and solvent basicity δ_a is shown. Thus another useful rule in choosing solvents for maximum α is: Choose a solvent containing either a very basic (large δ_a) or nonbasic (small δ_a) strong solvent. Further examples of solvent selectivity in LSC are discussed in ref. 4.

TABLE 8.6.   SOLVENT SELECTIVITY IN LSC; BASIC VERSUS
NONBASIC SOLVENT COMPONENTS FOR SAMPLES CONTAINING
PROTON DONORS (ALUMINA ADSORBENT) (7)

Solvent (Pentane Solution) (vol %)		$\bigcirc\!\!-\!NHCH_3$ / 2-chloroquinoline		
		k'^a	α	δ_a
30	$CHCl_3$	2.6	0.8	0.5
1	CH_3NO_2	1.8	1.1	1
28	Benzene	1.8	1.2	0.5
0.6	CH_3COCH_3	3.2	1.6	2.5
5	Et_3N	2.0	2.1	3.5
0.05	CH_3SOCH_3	0.6	3.0	5

[a]For 2-chloroquinoline.

The use of binary solvents which contain a small
concentration (e.g., $\leq 2\%$) of a strong solvent B in a
weaker solvent A can lead to the phenomenon of solvent
demixing. When such a mixture is passed through an
LSC column for the first time, there is a tendency for
preferential adsorption of the component B at the front
of the column. This leaves pure A as the mobile phase
leaving the column. Gradually the adsorbent becomes
saturated with B, until eventually the mobile phase
compositions entering and leaving the column are the
same. It is undesirable to carry out separations during the period the column is equilibrating with B, so

enough solvent should be passed through the column for equilibration to occur. The extent of equilibration can be followed by frequent injections of some standard compound with a convenient k' value. When the value of k' becomes constant for repeated injections, column equilibration is complete and the column can be used. Solvent demixing occurs frequently in TLC when binary solvents are used, because only a limited amount of solvent contacts the adsorbent. Therefore, separations by TLC with dilute solutions of B (e.g., ≤2%) as mobile phase often lead to different results than would be obtained by comparable column separations. This complication should be kept in mind when using TLC to explore useful mobile phase compositions for column chromatography. Devices are available (Camag; see also ref. 1) which allow the pre-equilibration of TLC plates with mobile phase, prior to separation. Under these conditions, separations by TLC more closely parallel those by column chromatography.

If water is used to deactivate the adsorbent, as is usually the case, then water-free solvents cannot be used for the mobile phase. The reason is that the solvent will then pick up water from the adsorbent, and the adsorbent will gradually dry out with continued passage of solvent through the column. This is illustrated by the following experiment, summarized in Figure 8.7.

A column of 4% water-alumina was packed, the column filled with solvent, and k' determined for a given solute. After elution of the solute from the column, a second sample injection was made and k' redetermined. The column was then discarded. This experiment was repeated (starting with fresh columns) for solvents of differing water contents, and the initial and final k' values were plotted as in Figure 8.7. For dry solvent (0% water) the initial k' value is seen to be about 2.7, and the final value is greater than 25. In this case continued passage of solvent through the column

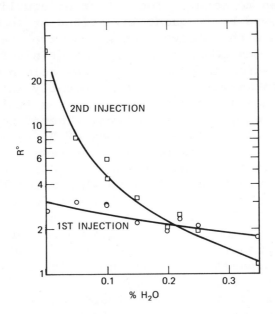

Figure 8.7 The importance of solvent water-content; elution of benzaldoxime from 4% water-alumina by 5 vol % isopropanol/ethyl ether (18). Reprinted by permission.

(between first and second sample injections) removed a significant amount of water from the adsorbent, with a resulting increase in k' (cf. Figure 8.3). When the experiment was repeated with solvent containing 0.35% water, the second value of k' (1.3) is now seen to be less than the initial value (1.9). In this case, the thermodynamic activity of water in the solvent is greater than on the adsorbent surface, so that water is given up to the adsorbent by the solvent. The two curves of Figure 8.7 cross for a water value of 0.22%, which means that solvent containing this amount of water is in equilibrium with starting adsorbent, neither

removing nor adding water to the adsorbent, so that k'
values do not change with time. Experiments such as
those of Figure 8.7 therefore allow the correct solvent
water content to be determined for a given water-deactivated adsorbent.

   A more convenient way to control adsorbent water
content--and the necessary solvent water content--is
to vary solvent water content and observe k' for some
standard solute. This is illustrated in Figure 8.8,
for elution of quinoline from alumina by methylene
chloride/pentane as solvent. At the beginning of this
experiment, the column had been pre-equilibrated with
solvent that was 50% water-saturated; that is, a 50/50
blend of water-saturated and dry (water-free) solvents

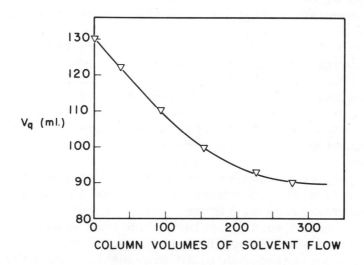

Figure 8.8   Adjusting adsorbent water-content in the
column; elution of quinoline from alumina by 35 vol %
CH_2Cl_2 pentane (19). Reprinted by permission.

was passed through the column until solvent and adsorbent were in equilibrium. At this point the water content or activity of the adsorbent was such that the
retention volume V_q of quinoline equalled 130 ml. Now
the solvent water content was changed from 50% to 75%
saturation, and flow through the column was continued.
At intervals quinoline was injected and its retention
volume determined. As can be seen in Figure 8.8, this
resulted in a gradual decrease in V_q, with a final, constant value being found after about 300 column volumes
of solvent had passed through the column: V_q = 90 ml.
The change to a wetter solvent (75% water saturation)
results in a gradual increase of the adsorbent water
content, until after 300 column volumes the adsorbent
has become re-equilibrated with the new solvent. At
this point no further change in adsorbent water content
or V_q values occurs.

Several aspects of this experiment (Figure 8.8)
deserve comment. First, if adsorbent water content is
controlled via the water content of the solvent, solute
retention volume or k' measurements can be used to follow the course of column equilibration. When retention
becomes constant, equilibration has occurred and the
column is ready to use. Note that an unequilibrated
column (e.g., 100 column volumes in Figure 8.8) should
never be used for actual separation, since the resulting separation will be poor relative to an equilibrated column.

Second, retention volume or k' values can be used
to standardize the adsorbent in situ, as in the similar standardization of Figure 8.4F. If the k' value
of some standard solute is less than that measured previously for a similar column, the water content of the
solvent is decreased enough to provide the necessary
increase in k'.

Third, determining the necessary water content of
a series of solvents, so that they will be in equilibrium for a given column, would be a tedious procedure

if each solvent had to be examined individually (recall
that we will usually use several different solvents
with a given column, in order to control k' and α for
different samples). For water-immiscible solvents, however, the work involved can be much reduced. We simply
equilibrate the column with any solvent of some specified water content, measured as percent water saturation (e.g., 50% water-saturated chloroform). Now any
other (water-immiscible) solvent that is 50% water-
saturated can be used directly, since its water activity will be approximately the same as the adsorbent
activity; for example, 50% water-saturated benzene, 50%
water-saturated ethyl ether, etc.

   Fourth, and finally, we see in Figure 8.8 that
equilibrating a column with a relatively nonpolar solvent (i.e., pentane/methylene chloride) can be a relatively slow process. The large solvent volume required
here to re-equilibrate the column is the result of the
low solubility of water in this particular solvent.
Therefore, for addition of some amount of water to the
adsorbent via the solvent, a large amount of solvent
must pass through the column. We can effect re-equilibration of the column with water much more rapidly, if
we select an equilibrating solvent that has a higher
solubility for water; for example, 50% water-saturated
ethyl ether.

   The example of Figure 8.8 can also be extended to
the use of solvents that contain small amounts of very
strong components; for example, 0.1% isopropanol/pen-
tane. As with water-containing solvents, the initial
use of such solvents requires that we pass enough of
the solvent through the column to achieve equilibration, as evidenced by constant k' values for some standard solute (see preceding discussion of solvent demixing).

   The preparation of water-saturated solvents, for
blending with dry solvent to achieve a given percent
saturation, deserves comment. Particularly in the case

of nonpolar solvents of low water solubility (e.g., hydrocarbons and their mixtures), it can be surprisingly difficult to achieve complete saturation by simply shaking or stirring the solvent with water. An alternative is to coat a low-cost porous adsorbent with an excess of water (e.g., 30% water on Davison Code 62 silica; Table 8.2) and pack a large glass column with the resulting packing. Now dry solvent is passed through the column, yielding 100% water-saturated solvent as effluent. This technique should not be used with solvent mixtures which contain water-miscible components (e.g., 1% isopropanol/pentane), since these will be extracted by the water left on the packing. However, such solvents can be readily water-saturated by shaking with a _small_ excess of water (shaking such solvents with a large excess of water would be equivalent to passing them through a water-coated silica column, with loss of part of the solvent to the water, and a resulting change in the solvent composition).

What percent water-saturation is optimum in LSC? The desired addition of half a monolayer of water to the adsorbent (Section 8.2) is usually achieved by about 25% water-saturated solvents for alumina as adsorbent, or 50% water-saturated solvents for silica as adsorbent.

From the above discussion it should be apparent that a certain effort is required in the use of water-deactivated adsorbents. Some workers prefer to add a few tenths of a percent of isopropanol directly to the solvent, in place of water. This avoids the difficulty of having to prepare water-saturated solvents, as described above. While some workers have obtained good results with isopropanol-deactivated solvents, other studies (20) have shown that this leads to peak distortion in some cases. On balance, water-deactivation appears to be preferable.

A final word on solvent purification: Many commercial solvents can be used directly in LSC, without

further purification. This includes such solvents as
benzene, methylene chloride, ethyl ether (freshly
opened containers), and the lower alcohols. Other
solvents, although sufficiently pure for preparative
separations (no contamination of recovered sample fractions), may contain traces of volatile UV absorbers
that must be removed if we are to use a UV detector
(especially at 254 nm). These include aliphatic hydrocarbons (e.g., pentane, hexane; 99% pure Philips Petroleum grade), isopropyl chloride, dioxane, and some
other solvents. Finally, some solvents may contain
impurities that are deliberately added as stabilizers,
but should be removed (e.g., ethanol in chloroform,
BHT in tetrahydrofuran, etc.). In most cases, highly
purified ("spectro-grade") solvents can be purchased
at a premium price (Appendix V).

Whether a solvent should be purified must be determined by the user. Generally, solvent purification is
best achieved by passing the solvent through a column
of dry, porous silica; for example, Davison Code 12
silica (Table 8.2), preactivated at 125°C. Further
purification of the solvent can be achieved within the
LC unit, by placing a pre-column between the pump and
the sample injection unit. The pre-column should be
packed with the same deactivated adsorbent used in the
separation column, but can be larger and need not be
well packed (i.e., its N value is unimportant). Note,
however, that the pre-column serves as a buffer between
the pump and the column. Therefore re-equilibration
of the separation column with a new solvent (as in Figure 8.8) will be further delayed by a pre-column (which
must itself be re-equilibrated).

8.4 OTHER SEPARATION VARIABLES

Sample size effects were discussed in general terms in
Sections 2.3 and 2.4. In the present discussion we
need to distinguish between the sample per se, and

added solvent or other nonretained sample components.
Normally the total volume of sample plus solvent injected onto the column will be 30-100 µl or less in analytical separations. A much larger volume can be tol-
erated in preparative separations, or if the solvent
used to contain the sample is much weaker (smaller ϵ°
value) than the mobile phase used for the separation.

The weight of sample (not including solvent or unretained components) that can be charged to an LSC column varies with several factors. The linear capacity
(Section 2.6) of the adsorbent will be less for dry
adsorbents, and greater for water-deactivated adsor-
bents--particularly for samples with larger k' values.
Thus, for naphthalene eluted by pentane, the linear
capacity of dry alumina is only 1 µg of sample per gram
of adsorbent. However, 4% water-alumina has a linear
capacity 100 times greater (0.1 mg/g). Silicas have a
generally higher linear capacity than alumina; for example, values of 0.5-3 mg/g for water-deactivated silicas. Pellicular adsorbents normally have about one-
fourth the linear capacity of corresponding porous
sorbents; for example, 0.3 mg/g for Corasil®-II (14).
Larger weights of sample can be charged, when a single
component does not comprise most of the sample, or when
the sample is dissolved in a larger volume of solvent
(see Figure 12.11).

Temperature effects in LSC are generally unimportant, except as the detector requires thermostatting
for maximum sensitivity. Separation is usually carried
out at room temperature, without thermostatting of the
column. Values of k' normally decrease with increasing
temperature, by about 2% per °C change. This assumes
the water content of the adsorbent is not changed as a
result of separation, or that the same percent water
saturation is maintained (note that 50% water-saturated solvent, prepared at 25°C, will have a lower percent water saturation at higher temperatures).

8.5   SPECIAL PROBLEMS

Sample loss or alteration during separation is more of
a problem with LSC than for the other LC methods, but
does not occur often.   Such effects are observed frequently in classical LSC and thin-layer chromatography.
However, modern LSC features both speed and controlled
separation, and these factors greatly reduce the like-
lihood of sample reaction during separation.   Sample
reaction in LSC can occur as a result of oxidation, or
acid- or base-induced catalysis by the adsorbent.   Sample oxidation can be minimized, in the case of oxygen-
sensitive samples, by purging all solvents with nitrogen before use (Section 7.3), and/or by adding 0.05%
of HBT (hydroxylated butyl toluene or 2,6-di-t-butyl-
p-cresol) to the solvent.   HBT does not interfere with
separation (it is weakly adsorbed in LSC), and it is
easily removed from recovered sample fractions by evaporation.   However, HBT absorbs strongly below 285 nm,
and cannot be used with most UV detectors.
      To avoid acid or base catalysis by the adsorbent,
acid-sensitive samples should not be separated on acidic adsorbents (e.g., Florisil), and base-sensitive
samples should not be separated on basic adsorbents
(e.g., alumina, magnesia).   Pure silica is one of the
gentlest adsorbents, and seldom induces reaction of
the sample.   However, it is sometimes contaminated with
traces of strong acids left over from its preparation;
this can be removed by a water wash.
      Some adsorbents (notably charcoal, Florisil®) irre-
versibly adsorb some or all of certain sample components, leading to low recoveries of separated bands.
This also occurs occasionally with other polar adsorbents, but can be greatly reduced by using water-deactivated adsorbents.   Water deactivation also seems
to cut down on the extent of other adsorbent-related
reactions of the sample.
      The batch-to-batch duplication of adsorbents can
be an occasional problem.   Normally we standardize the

adsorbent by varying its water content, so as to dupli-
cate the k' value for a given compound (and solvent)
from one adsorbent batch to the next. Sometimes, however, it is found that the k' values of other sample
bands have shifted (relative to our standard compound),
and the original separation is no longer possible with
the new batch of adsorbent and the original separation
conditions (i.e., solvent, column length, etc.). Until
the suppliers of LSC packings improve the batch-to-
batch uniformity of their materials, there is little
that can be done to avoid this difficulty. If a routine LSC assay is being set up for application over a
period of time to a large number of samples, it is
best to stockpile a sufficient quantity of the adsorbent (from a single or combined batch) for the entire
series of analyses. Since a single column can often
be used for a hundred or more separations, this does
not represent a very large inventory of adsorbent.

Column deactivation is a problem in LSC, when injecting "dirty" samples or using solvents that have not
been carefully prepurified. Strongly adsorbing contaminants from either sample or solvent gradually accumulate on the column, and are not washed off. One then
sees a gradual decrease in k' values for a given solute (and solvent), accompanied by a decrease in column
plate number. There are two solutions to this problem,
other than to replace the column after its apparent de-
terioration. A guard column (Section 7.5) can be
coupled between the sample injection system and the
column. This will pick up chemical "garbage" and protect the main column. The guard column can then be
replaced at regular intervals. A second approach is
to occasionally regenerate the column. This involves
washing the column with a very strong solvent, such as
50 vol % benzene/methanol, to remove these strongly adsorbed contaminants. The methanol is then washed from
the column by 5-10 column volumes of two or three successively weaker solvents; for example, 10 vol %

isopropanol/pentane, 100% ethyl ether, and so on, end-
ing up with the solvent for the next separation. Alternatively, the starting solvent can be used to remove the methanol, but larger volumes of solvent are
then required. The water content of the adsorbent
must also be restored to its original value, by using
appropriately treated solvents during regeneration (i.
e., solvents with the correct water saturation).

8.6 APPLICATIONS

A general review of the types of samples that can be
handled by LSC was given in Section 8.1, with emphasis
on the value of previous applications by TLC (2,3) as
guides to modern LSC applications. Modern LSC has now
been applied to a wide range of sample types, many of
which are reported in the literature of Table 1.1. A
brief review with some actual chromatograms will now
be given.
     A number of challenging separations of mixtures of
synthetic organics have been reported, including some
vitamin B-12 precursors (21), substituted ureas and
carbamates (22), and synthetic dyes (23). A urea-
carbamate separation is shown in Figure 8.9.
     Mixtures of antioxidants and pesticides have been
separated by several workers using LSC (5, 21). Figure 8.10 shows a pesticide separation. The analysis
of reaction products is another fruitful application
of LSC, as illustrated by the hydrogenated-quinoline
analysis described in Chapter 11.
     The drugs of abuse are being determined by LSC,
including mixtures of barbiturates, alkaloids, and
certain analgesics (24, 25). Figure 8.11 illustrates
a barbiturate separation.
     A variety of natural products and compounds of
biomedical interest are now being assayed by LSC.
These include mixtures of aflatoxins, vitamins, and

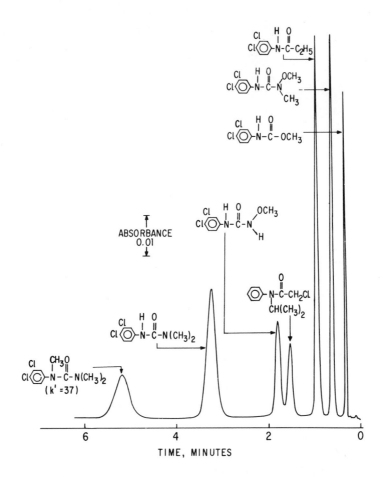

Figure 8.9 High-speed separation of substituted ureas and carbamates (2). Reprinted by permission of editor.

steroids (24-26). Figure 8.12 shows the separation of a mixture of vitamins. These and other compounds have been determined in such samples as raw produce, prepared food, body fluids such as urine, and so on. For examples see recent issues of the Journal of Chromatography, Journal of Chromatographic Science, and

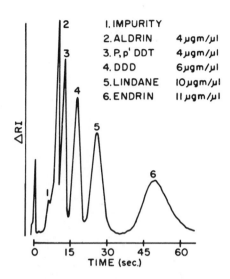

Figure 8.10   Separation of pesticides on silica.
Sample size, 2 µl; column, 20 cm x 2.3 mm i.d.; packing, 37–50 µ Corasil®-II; carrier and flow rate, n-
hexane at 3.0 ml/mm.   Reprinted by permission of Waters
Associates.

Analytical Chemistry (1973 on).

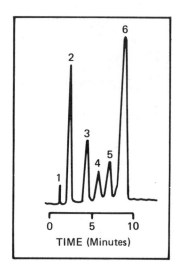

Figure 8.11   Separation of barbiturates on a 1 m x 2 mm i.d. stainless steel column packed with Vydac® Adsorbent.   Peak identification: (1) solvent front, (2) hexobarbital, (3) secobarbital, (4) amobarbital, (5) barbital, (6) phenobarbital; mobile phase, 2% methanol in heptane at 1.1 ml/min; pressure, 500 psig; detector, UV at 16 x 10^{-2} AUFS; temperature, 25°C. Reprinted by permission of Applied Science Laboratories.

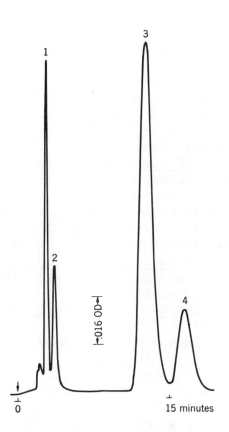

Figure 8.12   Separation of vitamin A acetate (1), vitamin E (2), vitamin D (3), and vitamin A (4).   Method,
adsorption; packing, Corasil®-II; solvent, 10% chloroform in isooctane; column, 2 x 1000 mm glass; flow
rate, 1 ml/min; pressure, 380 psi; temperature, ambient; sample size, 40 µl; UV at 254 nm, 16 X; 2-5 µg
samples.   Reprinted by permission of Chromatronix, Inc.

REFERENCES

1. H. Schlitt and F. Geiss, J. Chromatog., 67, 261 (1972).
1a. F. Eisenbeiss, Chemiker Zeitung, 95, 237 (1971).
2. E. Stahl, Thin-Layer Chromatography. A Laboratory Handbook, 2nd ed., Academic Press, New York, 1969.
3. J. G. Kirchner, Thin-Layer Chromatography, Wiley-Interscience, New York, 1967.
4. L. R. Snyder, Principles of Adsorption Chromatography, Marcel Dekker, New York, 1968.
5. L. R. Snyder, J. Chromatog., 63, 15 (1971).
6. L. R. Snyder, J. Chromatog., 20, 463 (1965).
7. L. R. Snyder, J. Chromatog., 8, 319 (1962).
8. L. R. Snyder, Advan. Chromatog., 4, 3 (1967).
9. L. R. Snyder, in Modern Practice of Liquid Chromatography, J. J. Kirkland, ed., Wiley-Interscience, New York, 1971, Chapter 6.
10. L. R. Snyder, in Gas Chromatography 1970, R. Stock and S. G. Perry, eds., Inst. of Petroleum (Elsevier), London, 1971, p. 83.
11. L. R. Snyder, Acct. Chem. Res., 3, 290 (1970).
12. N. Nicollaides, J. Chromatog. Sci., 8, 717 (1970).
13. R. E. Leitch and J. J. DeStefano, J. Chromatog. Sci., 11, 105 (1973).
14. R. E. Majors, Anal. Chem., 44, 1722 (1972).
15. J. J. Kirkland, J. Chromatog. Sci., 9, 206 (1971).
16. L. R. Snyder, Anal. Chem., 39, 698 (1967).
17. D. L. Saunders, Anal. Chem., in press.
18. L. R. Snyder, J. Chromatog., 16, 55 (1964).
19. L. R. Snyder, J. Chromatog., Sci., 7, 595 (1969).
20. J. J. Kirkland, J. Chromatog., 83, 149 (1973).
21. Waters Associates, Application Sheets.
22. J. J. Kirkland, in Gas Chromatography 1972, S. G. Perry, ed., Applied Science Publishers, Essex, England, 1973, p. 39.
23. Varian Aerograph, Application Sheets.
24. Applied Science Labs, Application Sheets.

25. Separations Group, Application Sheets.
26. Chromatronix, Application Sheets.
27. L. R. Snyder, J. Chromatog., <u>36</u>, 476 (1968).

CHAPTER NINE

ION-EXCHANGE CHROMATOGRAPHY

9.1 INTRODUCTION

At the present time, more separations are probably carried out by ion-exchange chromatography than by any of
the other LC methods. Ion-exchange chromatography came
of age as a separation tool with the development of the
cross-linked polystyrene resin in the early 1940s. During this period, the use of ion-exchange chromatography
for separating the rare earths and various fission products was extremely important in the development of atom-
ic energy. More recently, the application of ion-
exchange chromatography has resulted in the solution
of many very difficult and important problems in bio-
chemistry (e.g., structure of proteins).
   Ion-exchange chromatography is carried out with
packings having charge-bearing functional groups. The
most common mechanism is <u>simple ion exchange</u>:

(a) $X^- + R^+Y^- \rightleftharpoons Y^- + R^+X^-$ (anion exchange),

(b) $X^+ + R^-Y^+ \rightleftharpoons Y^+ + R^-X^+$ (cation exchange),

$$(9.1)$$

where

   X = sample ion
   Y = mobile phase ion

283

R = ionic sites on exchanger.

As shown in Eq. 9.1a, in anion-exchange chromatography the sample ion, X^-, is in competition with the mobile phase ion, Y^-, for the ionic sites on the ionic exchanger, R^+. Similarly, in cation-exchange chromatography (Eq. 9.1b), sample cations, X^+, are in competition with the mobile phase ions, Y^+, for the ionic sites, R^-, on the ion exchanger. Simple ion-exchange separations are based on the different strengths of the solute ion/resin (ion-pair) interactions. Solutes interacting weakly with the resin in the presence of the mobile phase ions are poorly retained on the column and elute early in the chromatogram, while solutes that interact strongly with the resin are more strongly retained and elute later.

Both acids and bases can be separated by ion exchange. In Eq. 9.2a, we see the dissociation of a weak acid, HA, into its ionic form, H^+ and A^-:

$$\text{(a)} \quad HA \;\rightleftharpoons\; H^+ + A^-,$$

$$\text{(b)} \quad B + H^+ \;\rightleftharpoons\; BH^+. \tag{9.2}$$

The degree of this dissociation can be controlled by the pH of the system. By increasing the concentration of hydrogen ions (decreasing pH), the dissociation equilibrium can be forced to the left, thus decreasing the concentration of A^- ions which are available for interaction with the anion-exchange resin and decreasing the k' value for solute HA. As indicated in Eq. 9.2b, change in pH can also control the availability of ionic forms of bases for ion-exchange separations. In this case, increasing the concentration of hydrogen ions (decreasing pH) shifts equilibrium to the right, thereby increasing the concentration of the charged (BH^+) ions available for ion-exchange interaction and increasing the k' value for basic solute B.

Various ligands can have a marked effect on ion charge and ion/resin interaction, for example,

$$Fe^{+3} + 4Cl^- \rightleftharpoons FeCl_4^- \qquad (9.3)$$

In this illustration, the presence of chloride ions greatly influences the concentration of positively charged ferric ions because of the formation of the negatively charged $FeCl_4^-$ complex. The formation of such complexes can be useful analytically, and many heavy metals can be separated as anion complexes by anion-exchange chromatography.

Ionic complexes of neutral molecules can sometimes be used for ion-exchange separations. Sugar in the presence of borate ions forms a sugar/borate ion:

$$Sugar + BO_3^{=} \rightleftharpoons Sugar \cdot BO_3^{=} \qquad (9.4)$$

complex mixtures of various neutral sugars have been separated in this manner on anion exchange resins (see Section 9.6).

Ion exchangers in the acid or base form can also be used to carry out separations via <u>acid-base reactions</u>:

(a) $HA + R^+OH^- \rightleftharpoons R^+A^- + H_2O,$

(b) $B + R^-H^+ \rightleftharpoons R^-BH^+,$ $\qquad (9.5)$

where R = ion-exchange resin.

This technique is useful for separating weak acids and bases in nonaqueous systems. In Eq. 9.5a, we see that a weak acid, HA, undergoes interaction with the anion exchanger, R^+OH^-, to form the resin anion salt, R^+A^-, plus water. Note that this is not a simple ion exchange such as that shown in Eq. 9.1, but a reversible acid-base reaction. Equation 9.5b shows the similar reaction of a weak base, B, reacting with the cationic resin, R^-H^+, to form a salt, R^-BH^+.

Ion-exchange packings can also be involved in <u>ligand exchange</u> reactions:

$$RM-L + X \rightleftharpoons RM-X + L, \qquad (9.6)$$

where

     RM = metal-resin ion pair,
      L = mobile phase ligand,
      X = sample component (another ligand).

In this approach, a resin modified with a metal ion, RM, to which is attached a mobile phase ligand, L, comes in contact with a sample component, X, to form a new complex, RMX. Since this new complex with the sample is more stable, the mobile phase ligand is displaced from the resin metal ion site. This type of interaction has been successfully used for the separation of amino acids and other amino compounds (1). In this application, a cation-exchange resin is reacted with a heavy metal, such as copper or nickel, to form the metal-modified resin. The various amino acids then compete with the mobile phase ligand (e.g., ammonium ions) to form new complexes with the metal atom attached to the resin. Selective amino acid-metal interactions result in different retention for the various compounds.

Ion exchangers can also involve separations via a <u>partitioning mechanism</u>:

$$X \ (org) \ \rightleftharpoons \ X \ (H_2O\text{-}R) \qquad\qquad (9.7)$$

This form of retention is not due to an ion-exchange reaction, and could be included in the discussion of partition chromatography in Chapter 7. In this case, a partitioning system is established between water immobilized in a swollen ion-exchange resin and the aqueous mobile phase containing some organic solvent (e.g., ethanol), to form a different medium which can be utilized in a partitioning process. Mixtures of sugars have been separated in this manner.

A combination of various forms of retention mechanisms can frequently occur in separations with ion exchangers. The possibility of more than one type of interaction described above, in addition to possible matrix adsorption effects, makes it difficult to predict the selectivity of separations with ion-exchange materials.

9.2   COLUMN PACKINGS

Many different types of ion exchangers are available;
however, the cross-linked polystyrene ion exchangers
are most widely used.   Figure 9.1a shows a cation resin structure: a polystyrene backbone cross-linked with
divinylbenzene to form an insoluble matrix.   In this
instance, the aromatic rings have been sulfonated to
form the strong sulfonic acid cation exchanger.   Figure
9.1b shows an anion-exchange resin structure with the
same cross-linked polystyrene matrix, but with a tetra-
alkylammonium functional group.   Dry polystyrene-based
resins tend to <u>swell</u> when placed in solvents.   Water
penetrates into the resin exchanger and hydrates the
ionic functional groups.   These ions may then be considered to be dissolved in their water of hydration,
thus forming a very concentrated solution within the
resin.   Osmotic pressure tends to drive more water into
the resin and the bead swells (i.e., increases in size).
The amount of water taken up by the resin depends on
the exchangeable ion of the resin and decreases with
increasing resin crosslinkage.   Ion-exchange resins
also swell in organic solvents, but the swelling generally is less than in water.
     Other types of materials have been used for ion-
exchange chromatography, including cross-linked poly-
dextran exchangers (used for the separation of macromolecules, such as proteins) and cellulose derivatives.
There are a few inorganic exchangers (e.g., zirconium
phosphate), but these materials have not found widespread application.   A new form of organic ion exchanger is the bonded-phase materials.   The structure of
one of these packings is shown in Figure 9.2.   In this
material the ion-exchange functional groups are chemically attached to a siliceous support by means of a

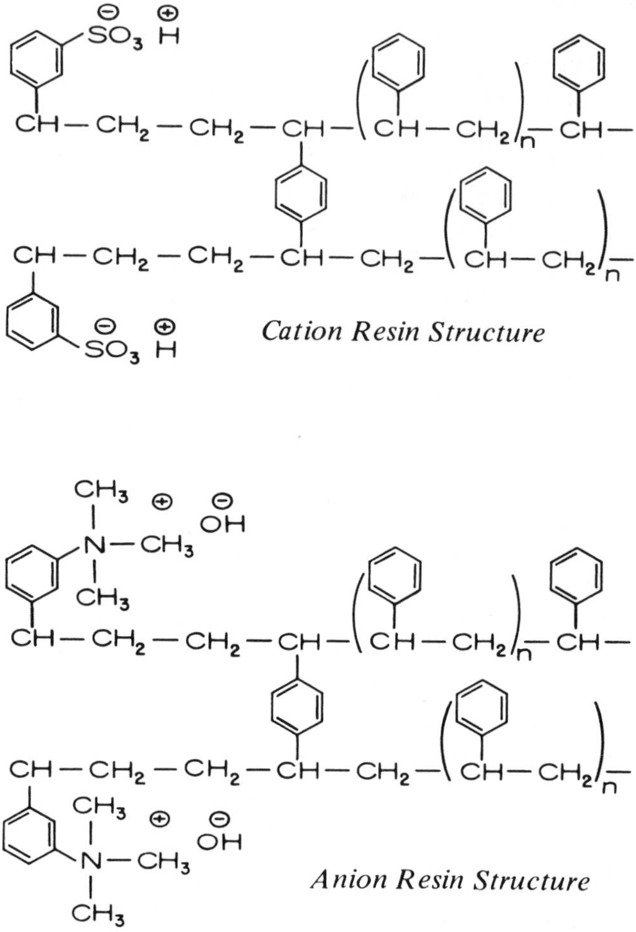

Cation Resin Structure

Anion Resin Structure

Figure 9.1   Structure of typical cross-linked polystyrene ion exchangers.   Reprinted by permission of Varian
Aerograph.

covalently bonded cross-linked silicone network.
   Ion-exchange packings are available with several
different particle geometries, as shown in Figure 9.3.
The most commonly used material for ion-exchange

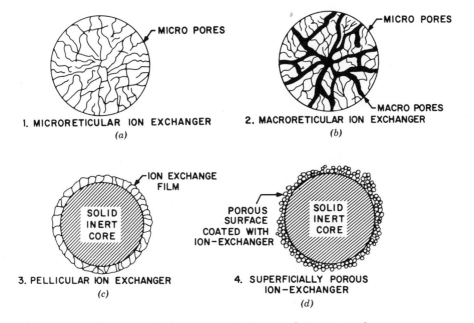

Figure 9.2   Structure of bonded-phase ion exchanger.

Figure 9.3   Particle geometries of ion exchangers.
R. E. Leitch and J. J. DeStefano, J. Chromatog. Sci.,
11, 105 (1973).   Reprinted by permission of editor.

separations is the <u>microreticular resin</u> shown in Figure 9.3a.   These are spherical particles of a divinyl-
benzene-cross-linked polystyrene containing micropores
throughout the structure.   The porosity of these

microreticular resins is largely obtained by swelling the material in the mobile phase prior to use. This type of ion exchanger is widely used for separating small molecules and is characterized by relatively small openings in the cross-linked structural network and a relatively high exchange capacity. Figure 9.3b illustrates the porous polystyrene-based macroreticular resin. In addition to micropores this resin also con- tains rigid pores that may be as large as several hun- dred Å. These larger pores allow easier access of large molecules to the interior of the particle for chromatographic interaction. Macroreticular ion ex- changers are also useful for separations in nonaqueous solvents, such as the acid-base interactions (Eq. 9.5) involving organic-soluble weak acids and bases.

The pellicular and superficially porous materials are the newest forms of ion-exchange packings. The pellicular ion-exchange particle (Figure 9.3c) has a solid inert core with a thin coherent film of ion- exchange material on the surface. The superficially porous ion exchanger (Figure 9.3d) is similar to the pellicular materials in that it also has a solid inert core. However, the porous surface in this particle is composed of very small solid particles, which in turn are coated with a very thin film of either mechan- ically held or chemically bonded ion exchanger. Both the pellicular and superficially porous particles have only a relatively small amount of ion-exchange material available at the surface, and therefore they have rela- tively lower exchange capacity. As suggested by the discussion in Section 6.2, the solid-core ion exchang- ers exhibit higher performance than the totally porous resin particles of the same size, particularly at high carrier velocities; faster equilibration of the solute between the stationary and mobile phases is permitted by these materials. Rigid, solid-core ion exchangers swell very little when contacted by liquids. There- fore, these particles are particularly desirable for gradient elution separations involving large changes in the composition of the mobile phase. Equilibrium appears still faster with the superficially porous particles, since thinner ion-exchange films are used; mass transfer in ion exchange normally is limited by stationary phase effects associated with the highly viscous, swollen resin.

A special form of ion-exchange materials are the liquid ion exchangers. In most processes of liquid ion exchange in which anions or cations are involved, one species of ions moves across a liquid-liquid interface, while an equivalent amount of similarly charged ion moves in the opposite direction across the same boundary. Substances such as water-insoluble, long-chain amines or quaternary ammonium compounds (e.g., $R_3CH_3N^+Cl^-$, where $R = C_8-C_{10}$ aliphatic groups) have been used for anion-exchange chromatography, and di-alkyl esters of phosphoric acid and the monoalkyl esters of alkanephosphonic acids have been found useful as cation exchangers. Desirable properties of liquid ion exchangers include: low solubility in water, large specific exchange capacity, high stability, and small surface activity.

When using liquid ion exchangers in conjunction with siliceous supports, it is usually desirable to silanize the support to increase the stability of the liquid ion-exchange film (see Section 7.5 for silanization procedure). Some liquid exchangers are moderately soluble in water; hence, the use of a pre-column is desirable (Section 7.3). Columns of liquid ion exchangers appear to be more stable when mobile phases of higher ionic strength are used. Instability of the liquid phase generally results in a loss from the support by emulsification.

A variety of ionic substituents are used in ion exchangers. Some of these are shown in Table 9.1. Strongly acidic cation exchangers contain strong sulfonic acid groups, while substituted phosphoric acids are employed as moderately strong acidic ion exchangers.

TABLE 9.1.   SOME FUNCTIONAL GROUPS IN RESIN ION EX-
CHANGERS

Type	Matrix Structure	Functional Group[a]	pH Limits[b]
Strong acid	Cross-linked polystryene	$-SO_3H$	0-14
Moderately strong acid	Cross-linked polystyrene	$-PO(OH)_2$	4-14
Weak acid	Polymerized acrylic acid	$-COOH$	6-14
Chelating resin	Cross-linked polystyrene	$\overset{\displaystyle CH_2COONa}{\underset{\displaystyle CH_2COONa}{-CH_2N}}$	6-14
Strong base	Cross-linked polystyrene	$-CH_2NMe_3Cl$ (type 1) $\underset{-CH_2NMe_2Cl}{\overset{CH_2CH_2OH}{\vert}}$ (type 2)	0-14
Weak base	Cross-linked polystyrene	$-CH_2NHMe_2OH$ and $-CH_2NH_2MeOH$	0-7
Bifunctional	Phenol-for-maldehyde polymer	phenolic groups and $-SO_2Na$ or $-CH_2SO_3Na$	c

[a]The exchangeable ion in the functional group indicates the form in which the resin is generally sold.
[b]The pH limits within which the resin ionizes; useful
exchange range somewhat less (see Figure 9.4).
[c]The sulphonate group functions as an exchanger at any
pH, whereas the phenolate group can exchange cations

only at very high pH.
(Taken in part from ref. 2.)

Weakly acidic cation exchangers use carboxylic acid
functional groups. Strongly basic anion exchangers
utilize tetraalkylammonium groups for ionic interaction, while weakly basic ion exchangers normally contain tertiary amine groups. A combination of functional groups is used in ion exchangers designed for special purposes.

   Some of the specially prepared, high-performance
LC ion exchangers and their characteristics are listed
in Table 9.2. Table 9.3 lists the commercially available porous and pellicular ion exchangers. New packings of this type are now being made available in rapid
order; thus the current commercial literature should
be consulted for the latest information on the newest
ion-exchange packings.

   As mentioned in Section 9.1, divinylbenzene is used
as a cross-linking agent in conventional polystyrene
ion-exchange resins. High percentages of divinylbenzene increase the strength of the resin particle as a
result of the additional cross-linking, but decrease
its tendency to swell in the mobile phase, increase
selectivity, and also decrease the permeability of
the particle. Conventional resins commonly contain
4-12% of divinylbenzene, with the 8% cross-linked resins being the most popular. If a very high pressure
drop (5000 psi) or gradient elution is to be employed,
it is desirable to use an ion-exchanger resin with a
high degree of cross-linking (e.g., 8-12% divinylbenzene).

   The capacity of ion-exchange materials is determined by the concentration of measureable ionic groups
within the structure. The capacity of ion exchangers
is a function of pH:

TABLE 9.2. POROUS ION-EXCHANGE RESINS FOR HIGH-PERFORMANCE LIQUID CHROMATOGRAPHY

Type	Name	Supplier[a]	Particle Size (μ)	Strength[b]	Functional Group[c]	Form	Ion Exchange Capacity[d] (dry)	Description
Anion	Durrum Type DA-X8	Durrum (68)	X8 20±2 X8A 8±2	SB	$-NR_3^+$	Cl	4 milliequiv./g	Polystryene (PS)-divinyl-benzene (DVB); 8% cross-linked.
	Durrum Type DA-X4	Durrum (68)	X 20±5	SB	$-NR_3^+$	Cl	2	PS-DVB; 4% cross-linked.
	Aminex®A-Series	Bio-Rad Labs (62)	A-14 17-23 A-25 17.5±2 A-27 12-15 A-28 7-11	SB	$-NR_3^+$	Cl	3.4 A-14 3.2 others	PS-DVB; A-14: 4% cross-linked; others: 8% cross-linked.
Cation	Durrum Type DC-A Series	Durrum (68)	1A 18±3 2A 12±3 4A 8±2	SA	$-SO_3H$	Na	5	PS-DVB; 8% cross-linked.
	Aminex A-Series	Bio-Rad Labs (62)	A-4 16-24 A-5 13±2 A-6 17.5±2 A-7 7-11	SA	$-SO_3H$	Na	5.0	PS-DVB; 8% cross-linked.
	Beckman AA Series PA Series	Beckman (20)	AA-15 22±6 PA-28 16 PA-35 13	SA	$-SO_3H$	Na	5	PS-DVB; AA-15 8% cross-linked; others 7.5% cross-linked.
	Hamilton HP-Series H-Series	Hamilton (27)	AN90 22±6 B80 15±5 H-70 24±6	SA	$-SO_3H$	Na	5.2	PS-DVB; AN90 7% cross-linked; B80 7.75% cross-linked; H-70 8% cross-linked.

[a]See Appendix V for complete names and addresses.
[b]SB = strongly basic, WB = weakly basic, and SA = strongly acidic.
[c]NR_3^+ is the tetra-alkylammonium group, R = alkyl group.
[d]Capacity often depends on the compound used and the mechanism of separation.
The values in the table are only approximate.
Taken in part from R. E. Majors, American Lab., May, 27 (1972).

TABLE 9.3. SUPERFICIALLY POROUS OR PELLICULAR ION-EXCHANGE RESINS FOR HIGH-PERFORMANCE LIQUID CHROMATOGRAPHY

Type	Name	Supplier[a]	Particle Size (μ)	Strength[b]	Functional Group[c]	Particle Type[e]	Ion-exchange Capacity[d](dry) ~10 μ-equiv./g	Description
<u>Anion</u>	Pellicular Anion	Varian (10)	~40	SB	$-NR_3^+$	P		Available in tested prepacked columns; can be used in organic solvents.
	AS-Pelliomex®-SAX	Reeve-Angel (46)	44-53	SB	$-NR_3^+$	P	10	Resin is styrene-divinylbenzene copolymer; can be used with organic solvents.
	Zipax®-SAX	Du Pont (3)	25-37	SB	$-NR_3^+$	SP	12	Can be used with small amounts of nonaqueous solvents.
	Zipax®-WAX	Du Pont (3)	25-37	WB	$-NH_2$	SP	12	Can be used with small amounts of nonaqueous solvents.
	Zipax®-AAX	Du Pont (3)	25-37	SB	$-NR_3^+$	SP; B	10	Can be used with any solvent; bonded substituted silicone.
	Vydac®-Anion	Applied Science, Separations Group (28, 65)	30-44	SB	$-NR_3^+$	SP; B	NA[f]	Can be used with any solvent; bonded substituted silicone.
	Bondapak®/AX/ Corasil®	Waters (11)	37-50	SB	$-NR_3^+$	SP; B	10	Can be used with any solvent; bonded substituted silicone.
	Sepcote®-AN	Separation Technology (70)	~40	SB	$-NR_3^+$	P; B	NA	Bonded polystyrene-based resin; benzyl trimethyl quaternary.

TABLE 9.3. (Continued)

Type	Name	Supplier[a]	Particle Size (μ)	Strength[b]	Functional Group[c]	Particle Type[e]	Ion-Exchange Capacity[d](dry)	Description
	Sepcote®-WAN	Separation Technology (70)	~40	WB	$-NH_2$	P; B	NA	Bonded polystyrene-based resin; polyamine groups.
Cation	Pellicular Cation	Varian (10)	~40	SA	$-SO_3H$	P	~10	Available in tested pre-packed columns; can be used with organic solvents.
	HC-Pellionex®-SAX	Reeve-Angel (46)	44-53	SA	$-SO_3H$	P	60	Resin is styrene-divinyl-benzene copolymer; can be used with organic solvents.
	HS-Pellionex®-SCX	Reeve-Angel (46)	44-53	SA	$-SO_3H$	P	8-10	Resin is styrene-divinyl-benzene copolymer; can be used with organic solvents.
	Zipax®-SCX	Du Pont (3)	25-37	SA	$-SO_3H$	SP	5	Can be used with small amounts of nonaqueous solvents.
	Vydac® Cation	Applied Science, Separations Group(28,65)	30-44	SA	$-SO_3H$	SP; B	0.1	Can be used with any solvent.
	Sepcote® CA	Separation Technology (70)	~40	SA	$-SO_3H$	P; B	NA	Bonded polystyrene-based resin.

[a]See Appendix V for complete names and addresses.
[b]SB = strongly basic, WB = weakly basic, and SA = strongly acidic.
[c]NR_3^+ is the tetra-alkylammonium group.
[d]Capacity often depends on the compound used and the mechanism of separation. Capacity values in this table are only approximate.
[e]P = pellicular; SP = superficially porous or porous layer bead type; B = bonded phase material.
[f]NA = not available.
(Taken in part from R. E. Majors, American Lab, May, 27 (1972), and R. E. Leitch and J. J. DeStefano, J. Chromatog. Sci., $\underline{4}$, 105 (1973).)

$$\text{(a)} \quad R_cH \; \rightleftharpoons \; R_c^- + H^+,$$
$$\text{(b)} \quad R_aOH \; \rightleftharpoons \; R_a^+ + OH^-, \tag{9.8}$$

where

R_c = cation exchanger,
R_a = anion exchanger.

In Eq. 9.8a, we see that the ionization of a cation-exchange resin, R_cH, to produce the resin ion, R_c^-, and H^+, is influenced by pH. Thus, at low pH (high concentration of the hydrogen ion), the ionization of the acidic resin is inhibited and the exchange capacity is diminished. In Eq. 9.8b, the ionization of a basic ion exchanger is inhibited at high pH, thereby reducing the exchange capacity of this resin. The effect of pH on the capacity of ion exchangers is more quantitatively shown in Figure 9.4. The top curves show that strongly acidic cation exchangers have useful capacities (totally ionized) above about pH = 2. Similarly, the strongly basic anion exchangers are useful up to about pH = 10. In the lower curve of Figure 9.4, we see that the weakly acidic cation exchangers have a useful exchange capacity only above about pH = 8, while the weakly basic anion exchangers are useful only to about pH = 6. These plots show that the usable pH range of the strong ion exchangers is much greater than the useful pH range of the weakly acidic and basic exchangers. This effect is the main reason for the more widespread use of the strong acid and base materials in ion-exchange chromatography. However, the weak ion-exchangers do have unique application. For instance, weak resins are often used for peptide and protein separations to avoid denaturation of these materials. Cellulose-based ion exchangers are also employed for these separations.

The exchange capacity of the various ion exchangers also depends on the type of the particle and its geometry. Conventional porous resins usually have a

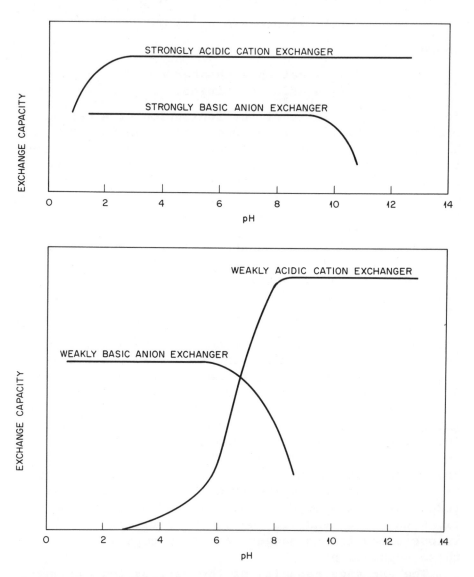

Figure 9.4   Effect of pH on capacity of ion exchangers.
C. D. Scott, in Modern Practice of Liquid Chromatography,

J. J. Kirkland, ed., Wiley-Interscience, New York,
1971. Reprinted by permission of publisher.

capacity of about 1-10 milliequiv/g, while the pellicular or superficially porous beads, having only thin
films of ion exchange on the surface of the particle,
have exchange capacities of about 5-50 microequiv./g.
     The packing of ion-exchange columns has been dis-
cussed in general terms in Chapter 6. The solid-core
materials are normally packed by dry-packing techniques, while the conventional polystyrene-based resins are slurry-packed. Packing procedures appear to
be somewhat less critical for conventional resin exchangers with narrow size ranges. Columns of the
smaller sizes give the best performance, and ion exchangers in the 3-7-μ range have been successfully
used. However, columns of these very small particles
are more difficult to pack, and most workers prefer
to use columns of ion exchangers in the 10-40-μ range
(but with narrow particle-size fractions). The resin
ion exchangers in the 10-20-μ range afford a good compromise between column efficiency and convenience. At
present, pellicular and superficially porous materials
are only available with particles larger than 20 μ.

9.3 MOBILE PHASES

Most ion-exchange chromatographic separations are carried out in aqueous media because of the desirable
solvent and ionizing properties of water. In some instances, mixed solvents, such as water-alcohol, and
even totally organic solvents have been used. In ion-
exchange chromatography with aqueous mobile phases,
peak retention is controlled by the total salt concentration (or ionic strength), and by the pH of the mobile phase. An increase in the concentration of the
salt in the mobile phase decreases the retention of

sample components. This is due to the decreased ability of the sample ions to compete with the mobile phase
counter ions for the ion-exchange groups in the resin.
This effect is illustrated in Figure 9.5, which shows
the influence of the concentration of sodium nitrate
in a mobile phase on the relative retention of a number of food additives (3).

The type of ions in the mobile phase can significantly affect the retention of sample molecules, due
to the varying ability of these different mobile phase
ions to interact with the ion-exchange resin. The retention sequence of various anions for conventional
cross-linked polystyrene anion-exchange resins is as
follows:

$$citrate > SO_4^{-2} > oxalate > I^- > NO_3^- > CrO_4^{-2} >$$

$$Br^- > SCN > Cl^- > formate > acetate > OH^- > F^-.$$

This retention sequence varies somewhat for different
commercially available resins, but it is a qualitative
indication of the ability of various anions to interact with strong anion exchangers. In this sequence,
citrate is very strongly bonded to the resin, while
fluoride ions are very weakly bound. Thus, sample
molecules generally are more quickly eluted with citrate ions than with fluoride ions. There is a similar
retention sequence for cation-exchange resins:

$$Ba^{+2} > Pb^{+2} > Sr^{+2} > Ca^{+2} > Ni^{+2} > Cd^{+2} > Cu^{+2} >$$

$$Co^{+2} > Zn^{+2} > Mg^{+2} > UO_2^{+2} > Te^+ > Ag^+ > Cs^+ >$$

$$Rb^+ > K^+ > NH_4^+ > Na^+ > H^+ > Li^+.$$

Generally, there is less variation in the retention
of sample ions within this cation series, since there
is less difference in the ability of cations to compete with the cation exchanger than for various anions
to complex with anion exchangers. This is mainly due to

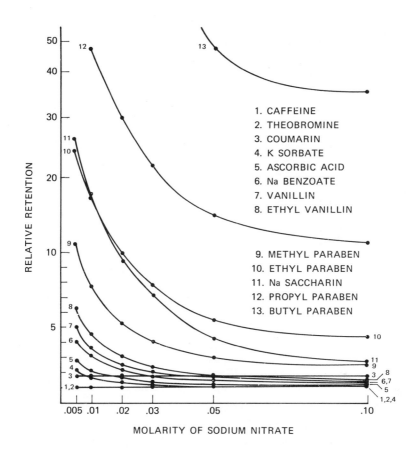

Figure 9.5 Effect of ionic strength on retention.
Column, 1 m x 2.1 mm i.d. Zipax®-SAX (anion exchange);
pressure, 1000 psi; temperature, 24°C; mobile phase,
0.01 M sodium borate (pH = 9.2) with sodium nitrate;
flow, 0.85 ml/min. J. J. Nelson, J. Chromatog. Sci.,
<u>11</u>, 28 (1973). Reprinted by permission of editor.

the smaller variations in size and charge characteristics of various cations. Figure 9.6 shows the effect

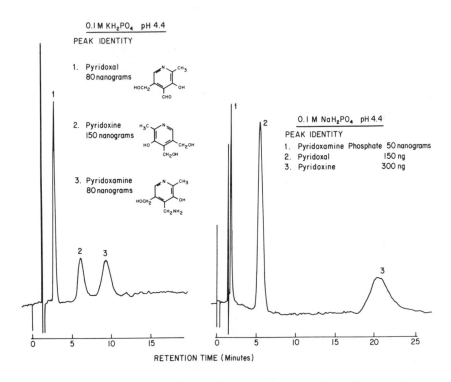

Figure 9.6 Effect of cation on retention in cation-exchange chromatography. R. C. Williams, R. A. Henry, J. A. Schmit, accepted by J. Chromatog. Sci., 1973. Reprinted by permission of editor.

of varying the nature of the cation (Na$^+$ versus K$^+$) in the eluent on the retention of sample components in a cation-exchange separation. It should be noted that because of the smaller exchange capacity of pellicular packings (usually a factor of 50-100 compared to conventional resin exchangers), the concentration of salt in the mobile phase must generally be decreased by this factor to obtain solute retentions comparable to those with the conventional porous resin ion exchangers.

There are some general guidelines for predicting the ability of an exchanger to complex with solute or mobile phase ions. The ion exchanger tends to prefer:

- The ion of higher valence (charge).

- The ion with a smaller (solvated) equivalent volume.

- The ion with the greater polarizability.

- The ion which interacts more strongly with the fixed ionic groups or with the matrix.

- The ion which interacts weakest with other materials in the mobile phase.

Since sample retention in ion-exchange chromatography is often not limited to simple ion exchange (see Section 9.1), these ion-exchange retention sequences can show significant variations. Variations are particularly apparent with pellicular or superficially porous ion exchangers.

Peak retention in ion exchange chromatography can also be controlled by changes in mobile phase pH. Thus, an increase in pH decreases sample retention in cation-exchange chromatography and increases retention in anion-exchange chromatography. The influence of pH on retention can be explained by the effect of hydrogen ion concentration on available sample ions.

$$
\begin{aligned}
&\text{(a)} \quad {>}COOH \; \rightleftharpoons \; COO^- + H^+ \quad \text{(anion exchange)}\\
&\text{(b)} \quad {>}NH + H^+ \; \rightleftharpoons \; \overset{+}{N}H_2 \quad \text{(cation exchange)}
\end{aligned}
\qquad (9.9)
$$

In Eq. 9.9a, we see that a free carboxylic acid is in equilibrium with the carboxylic ion plus a proton. At higher pH (lower concentration of H^+) the concentration of carboxylic ions is higher; thus, the retention of this sample ion is greater because of its increased

ability to compete for the anionic sites of the resin.
Similarly, in Eq. 9.9b, the formation of the charged
cationic form of a weak base is enhanced at higher hydrogen ion concentrations. Thus, a decrease in pH results in increased retention due to the increased competition of these charged sample ions for the cation-
exchange sites. The influence of pH on the retention
of various mononucleotides has been shown in Figure
3.19. Variation in pH can also cause changes in the
relative retention of sample components such as that
illustrated in this figure. Therefore, changes in pH
can be used not only to control peak retention (k'),
but can also be useful in changing selectivity. Unfortunately, changes in selectivity as a function of
pH are difficult to predict and usually have to be
found by a trial-and-error process. When using anion-
exchange resins, control of pH is normally carried out
with the use of cationic buffers (e.g., containing
ammonia, pyridine, etc.); with cation-exchange resins,
buffers containing ions such as acetate, formate, and
citrate often are used to fix the pH in the system.

As discussed in Section 9.1, the control of selectivity in ion-exchange chromatography is difficult to
predict accurately because of the tendency of the ion
exchanger to engage in more than one mechanism of retention. Section 9.4 contains a discussion of some
of the separating conditions controlling the separation of amino acids. This treatment furnishes some
insights which are useful for predicting which
carrier system might provide the desired selectivity
in an ion-exchange separation. More information on
the selectivity of ion exchangers is presented in
materials given in the bibliography at the end of
this chapter.

Just as thin-layer chromatography (TLC) can be
used to pilot the development of many LSC systems,
this technique sometimes is used in establishing satisfactory conditions for ion exchange chromatography.

Commercially available TLC plates of strong cation or anion exchangers can be used to determine the approximate strength and selectivity of the mobile phase required to cause differential migration of sample components for the desired separation. However, this approach is only a rough guide, and significant adjustment in the final conditions for separations in the column may have to be made. Difficult visualization of the solutes on the ion-exchange plate limits the use of TLC as a pilot technique.

The solvents used in the mobile phase can also provide changes in selectivity. As previously indicated, an aqueous carrier is normally used in ion-exchange chromatography; however, small amounts of organic solvents, such as ethanol, tetrahydrofuran, and acetonitrile, are sometimes added to these aqueous systems to increase the solubility of sample components, and these solvents also can provide changes in selectivity. The addition of organic solvents to aqueous buffers sometimes decreases the tendency of certain sample components to form tailing peaks.

As mentioned in Section 9.1, weak acids and bases can sometimes be separated with conventional ion-exchange resins, using nonaqueous carrier systems. In this form of chromatography, decreasing the polarity of the organic solvent increases the retention of sample components so that optimum k' values can be obtained. Changing from methanol to benzene, then to pentane, greatly increases the retention of sample components with conventional macroreticular ion-exchange resins (4).

Gradient elution is widely used in ion-exchange chromatography with aqueous mobile phases. Both salt and pH gradients are used to control peak retention and selectivity. In Figure 9.7, the separation of the ultraviolet absorbers in urine by strong ion-exchange chromatography was carried out using a change in the concentration of acetate ion in the mobile phase to generate the gradient. Salt gradients should not be

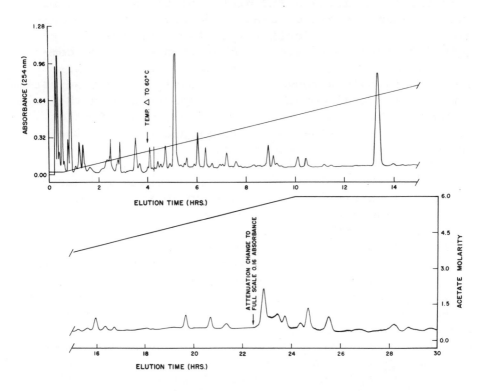

Figure 9.7 Gradient elution in ion-exchange chroma-
tography. UV-absorbing constituents of human urine.
Column, 100 cm, 2.4 mm i.d., strong anion-exchange
resin, 12-15 μ; gradient elution. D. R. Gere, in
Modern Practice of Liquid Chromatography, J. J.
Kirkland, ed., Wiley-Interscience, New York, 1971,p.417.
Reprinted by permission of publisher.

utilized with the less highly cross-linked resins (i.e.,
(≤ 8% divinylbenzene), so that changes in the dimensions of the ion-exchange bed due to the changes in
the size of the resin particle during the gradient elution process are minimized.
    Each ion-exchange column must be pre-equilibrated

with the mobile phase before it can be used for reproducible separations. Fresh mobile phase should be
pumped through the column until the effluent has the
same pH as the initial mobile phase. With conventional
cross-linked polystyrene resins of the larger particle
sizes, this pre-equilibration period may require considerable time, perhaps hours. However, equilibration
is much faster with the very small particles of conventional resins. Superficially porous or pellicular ion
exchangers are equilibrated very rapidly, usually in a
few minutes. This rapid equilibration represents a
considerable advantage in high-speed gradient elution
ion-exchange chromatography, which requires rapid changes in the ionic strength or pH of the mobile phase.
Regeneration of the ion-exchange column is required
after a gradient elution run, and this regeneration
also is much faster with the superficially porous ion
exchangers.

9.4   SOLVENT SELECTIVITY IN AMINO ACID SEPARATIONS

(H. J. Adler)

The analysis of amino acids was the first example of
modern liquid chromatography to be extensively applied
on a large scale. Over the past 15 years literally
millions of these analyses have been carried out, and
a great deal of attention has been given to the problem of optimizing the resolution of amino acids. During that period the time required for the analysis of
a single protein hydrolysate sample has been reduced
from 22 hr (5) to 1 hr (6). Figure 9.8 (6) is a chromatogram of the amino acids from a protein hydrolysate
and Figure 9.9 (7) shows a chromatogram of some of the
amino acids found in physiologic fluids.
    Here, we will examine some of the variations in
separation conditions which have been used to control
selectivity, or the spacing of individual bands. As

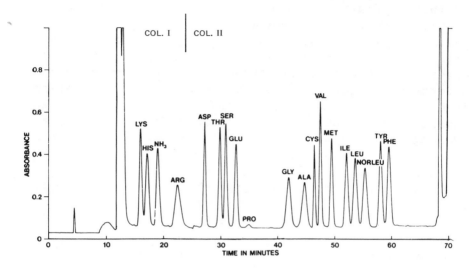

Figure 9.8   Synthetic mixture of protein–hydrolysate amino acids (5).   Two-column system: column 1, basic amino acids (lysine-arginine); column 2, acidic and neutral amino acids (aspartic acid-phenylalanine).

we noted in Section 3.6, this is a special problem in the routine analysis of complex samples containing (in the case of the amino acids) 20 to 40 different compounds.

The selective behavior of amino acids as they pass through a column of ion-exchange resin is not predictable in terms of fundamental theory.   However, through experimental observation, sufficient data have been gathered to allow general predictions to be made.

The variables most commonly used to achieve selectivity in amino acids separations are the pH and ionic strength of the mobile phase (usually an inorganic buffer), addition of an organic solvent to the mobile phase, the column operating temperature, and the mobile phase cation.

Table 9.4 shows the range of operating conditions typically used to separate mixtures of amino acids. Several different buffers are usually used for each analysis. They may be switched in a step-wise fashion going directly from one buffer to the next (5,8), or pre-loaded into a gradient device (9) to give a smooth gradual change in the mobile phase. The step-wise method of buffer switching allows the user to select the most efficient conditions for each part of the analysis to achieve both selectivity and speed.

TABLE 9.4.   TYPICAL OPERATING CONDITIONS

Resin	Spherical cation exchange	
	Nominally 8% cross-linked, poly-styrene-divinylbenzene copolymer	
	Particle diameter: 10-20 μ range (usual tolerance of $\pm$2-3 μ)	
Buffers	**Protein Hydrolysate**	**Physiologic Fluid**
Composition	Sodium citrate	Sodium or lithium citrate
Concentration	0.2 M-1.5M sodium	0.2M-1.5M sodium or 0.3M-0.8M lithium
	0.03M-0.1M citrate	0.05M-0.20M citrate
pH	2.8-6.0	2.5-6.0
Temperature	55°C-65°C	35°C-65°C

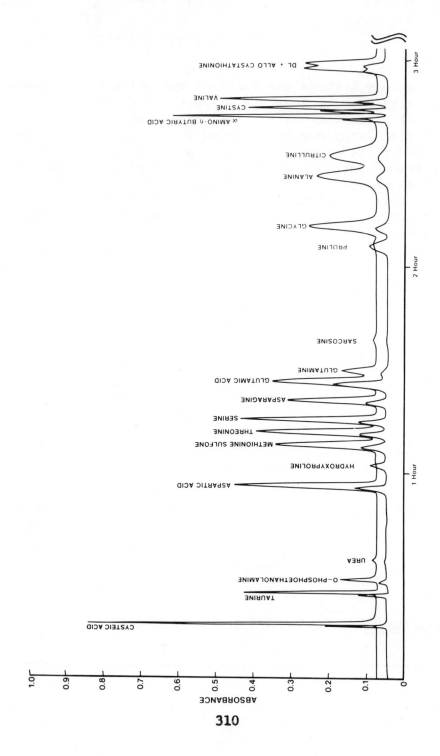

310

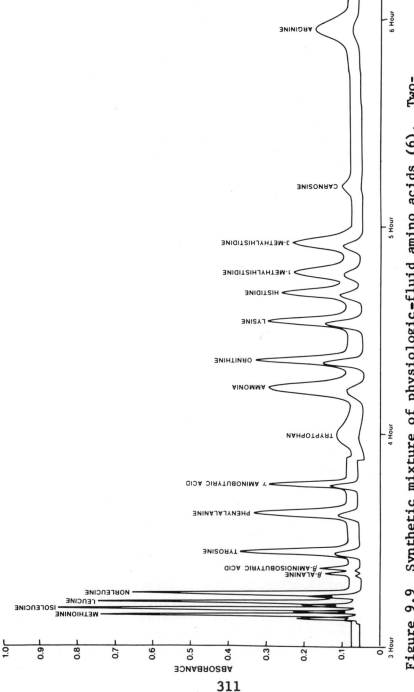

Figure 9.9 Synthetic mixture of physiologic-fluid amino acids (6). Two-column system: column 1, acidic and neutral amino acids (cysteic acid—γ-aminobutyric acid); column 2, basic amino acids (tryptophan—arginine).

311

Figure 9.10 is from a classical study by Hamilton (8) and shows the effect of changes in some of these variables on the elution order of the basic amino acids. As can be seen from this figure, changes in pH have a much greater effect on some amino acids than on others, and it illustrates why changes in buffer pH are most often used to control selectivity. For instance, for the basic amino acids found in a protein hydrolysate we have:

Elution Order at	pH 5.00	pH 5.25	pH 6.00
1	Ammonia	Lysine	Histidine
2	Lysine	Histidine	Lysine
3	Histidine	Ammonia	Ammonia
4	Arginine	Arginine	Arginine

With the basic amino acids (see Figure 9.10), the increase in temperature and cation concentration improves resolution (band width) and decreases k' values. However, changes in k' are nearly parallel, so these variables are not of much help in selectively moving one peak with respect to the others.

Of the acidic and neutral amino acids, the elution times of aspartic acid and particularly cystine are shifted easily, with respect to their adjacent peaks, by changes in pH. Two peaks of interest in the analysis of physiologic fluids, β-alanine and β-aminoisobutyric acid, are also differentially moved by buffer pH changes. They are eluted in the same region of the chromatogram as another pair of amino acids, tyrosine, and phenylalanine. Due to the interaction of their aromatic groups with the resin polystyrene matrix, tyrosine and phenylalanine are retained on the column longer than would be expected and are eluted at a pH considerably above their pK values (7). The variation in elution order with pH of these four amino acids is:

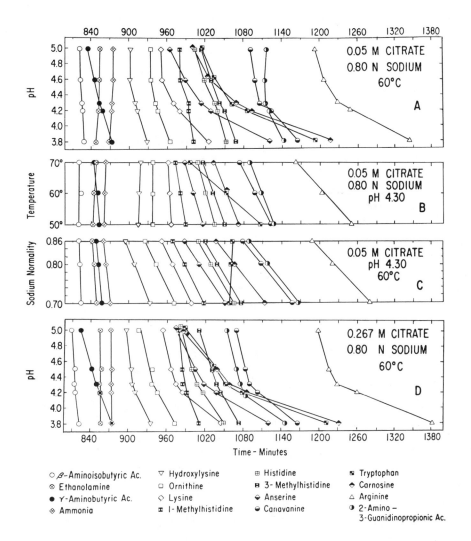

Figure 9.10   Resolution of basic amino acids as function of pH, temperature, sodium concentration, and buffer composition.

Elution Order at	pH 4.15	pH 4.80	pH 5.25
1	Tyrosine	Tyrosine	β-Alanine
2	Phenylalanine	β-Alanine	β-Aminoiso-butyric acid
3	β-Alanine	β-Aminoiso-butyric acid	Tyrosine
4	β-Aminoiso-butyric acid	Phenylalanine	Phenylalanine

The differential solubility of threonine and serine in various organic solvents (methanol, thiodiglycol, methyl cellosolve) is an aid in separating this pair of amino acids. The addition of up to 10% of one of these organics in the buffer makes a dramatic improvement in their separation. It has been observed (7,10), however, that the addition of these solvents has some effect on the elution pattern of other sample components. All other conditions remaining the same, several amino acids (methionine sulfone, glutamic acid, proline, and citrulline) are eluted sooner in the presence of an organic, and others (glycine, alanine, and α-amino-n-butyric acid) are delayed.

More than with the basic amino acids, the elution times of some of the acidic and neutral amino acids are selectively affected by the column operating temperature. Aspartic acid, glutamic acid, and especially proline and citrulline are easily shifted by temperature adjustments; higher temperatures cause elution times to decrease relative to other amino acids.

For many years, when sodium citrate buffers were used almost exclusively as the mobile phase, a complete analysis of physiologic fluids by ion-exchange chromatography was hampered by an inability to separate asparagine and glutamine from the other amino acids (8,11).

Either two separate analyses under different conditions
were required to separate all of the peaks, or two
analyses were carried out under the same conditions--
with one or two of the amino acids selectively de-
stroyed between runs (11,12) and a difference of peak
areas used as a means of quantitation. The use of
lithium citrate buffers instead of sodium citrate has
now made the complete separation of these two important amino acids possible (10). In this case, the
lithium ion is not bound as strongly to the resin,
yielding generally larger k' values (i.e., see retention sequence for cation-exchange resins in Section
9.3). This, in addition to specific characteristics
of the amino acids, gives some separation advantages
with asparagine, glutamine, and the other amino acids
in that region of the chromatogram.

9.5   OTHER SEPARATION VARIABLES

Because the effect of temperature on the selectivity
of ion-exchange separations is difficult to predict,
variations of temperature are generally not used as a
primary means of improving separations. However, as
in the other forms of LC, the number of theoretical
plates of a column generally increases at elevated
temperatures, due to the improved mass transfer of the
sample components resulting from the lowering of the
viscosity of the mobile phase and enhanced mobility in
and out of the ion exchanger. Increased temperature
also generally decreases the retention of the sample
components. If the stability of the sample compounds
allows, ion-exchange separations using aqueous mobile
phases are carried out at 60-80°C, to lower the viscosity of the eluent. Conventional anion-exchange
resins in the hydroxide form generally should not be
used above 50°C, because of instability. However, in
the salt form these resins may be used satisfactorily

up to 80°C. Some forms of pellicular or superficially porous anion exchangers are particularly sensitive to elevated temperatures and can be used only at room temperature. Cation exchangers usually can be used up to 60° for the superficially porous type and 80°C for the conventional resins.

The weight of samples that can be placed on an ion-exchange column depends on the exchange capacity of the packing. Normally, sample loads less than 5% of the ion-exchange capacity are used to ensure that the efficiency of the column is not degraded. The volume of the sample solution may be relatively large, if the sample components are strongly retained on the column from the initial solvent. For instance, in gradient elution ion-exchange chromatography, very large samples may be injected into the column when the initial mobile phase produces large k' values for all sample components. On injection, the sample components collect on the top of the column, and a gradient may then be applied to start the migration of the sample components for the desired separation. This technique is particularly useful for analyzing trace components.

9.6   SPECIAL PROBLEMS

On rare occasions, we may find that a sample is degraded during an ion-exchange separation. This difficulty sometimes can be overcome by operating the column at lower temperature. Alternatively, we can change the pH or operate with a carrier containing a higher ionic strength. If none of these steps is successful, it may be necessary to change to a different type of ion exchanger: use a weak instead of a strong ion exchanger. For example, changing from a strong to a weak anion exchanger eliminated the dehydrohalogenation of a ring-substituted β-bromoethylbenzene compound during separation (13).

Occasionally, deactivation of an ion-exchange column can occur due to either blockage of the ion-exchange bed by particulate matter, or as a result of
retained sample components. Such problems arise particularly in the analysis of extracts from naturally
occurring systems. Particulate matter can be conveniently removed from sample solution by filtration
through a 0.5-μ Millipore® filter held in a "Swinney"
adapter (see Appendix V). Deactivation of the analytical column by strongly retained components can be
prevented by the use of a guard column. This is a
short length of chromatographic column (3-5 cm) which
is attached to the inlet of a carefully made analytical column with a low-volume connector. The sample is
injected into this guard column which captures very
strongly retained components of no interest. Periodically, this short guard column is discarded and replaced
with a fresh unit. Ion-exchange columns deactivated
by very strongly retained components can very often be
regenerated by washing the column with a very strong
mobile phase to remove the strongly adsorbed contaminants. Superficially porous cation exchangers have
been regenerated by treatment with 1 M nitric acid
(14). Following regeneration, the column must be re-
equilibrated with the initial mobile phase before subsequent use for analysis.

The use of organic solvents in the mobile phase
must be restricted with certain of the superficially
porous ion-exchange packings. Some of these packings
contain thin films of the ion exchanger dispersed on
the superficially porous support. Consequently, only
moderate concentrations of organic solvents can be
used in aqueous mobile phases without disturbing the
ion-exchange film. Directions supplied by the manufacturer should be closely followed for best results.
Organic solvents generally can be used without difficulty with the bonded-phase ion exchangers.

Ion-exchange packing materials vary somewhat from

lot to lot. Consequently, there may be some problems
in reproducing a separation with various lots of ion
exchangers. As mentioned in Chapter 6, stockpiling a
relatively large lot of a particular packing is recommended for critical applications. The superficially
porous anion exchangers are generally less stable than
cation exchangers, so we can anticipate having to replace columns of these packings more frequently.

9.7 APPLICATIONS

The use of modern ion-exchange chromatography now cov-
ers a wide range of applications, particularly for organic compounds in biochemical systems. It should be
pointed out that many of the ion-exchange separations
described in the older literature can be greatly speeded up by use of modern techniques. Thus, information
on separations carried out by classical LC can form
the basis of a modern high-performance separation. For
instance, it is anticipated that separations of rare
earths, actinides, transplutonium elements, polyphos-
phates, and many other inorganic materials can be vast-
ly improved. An example of this is the high-speed
separation of radioactive inorganics, Bk(III), Es(III),
and Fm(III) using a liquid anion exchanger (15). Shown
in Figure 9.11 is the high-speed separation of a mixture of radioactive alkaline metals, using a conventional resin cation-exchanger and a radiometric detector
(16).
    A large number of organic compounds has been sepa-
rated by modern ion-exchange chromatography. Extensive
studies have been carried out on nucleic acids, nucleo-
sides, nucleotides, and related compounds (17). Figure 9.12 shows a separation of a mixture of various
nucleotides in a rat liver extract by anion-exchange
chromatography, using a bonded-phase superficially
porous bead-type ion exchanger (18). Modern ion-

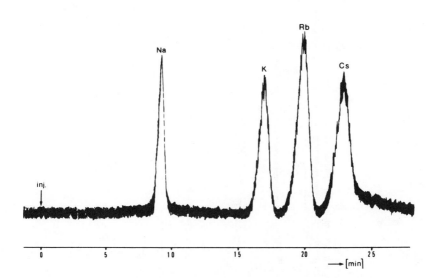

Figure 9.11   Separation of alkali metals by cation-exchange chromatography.   Separation of a mixture of $^{24}NaCl$, ^{42}KCl, $^{86}KbCl$, and $^{137}C_5Cl$.   Column, 48 cm x 5 mm; ion exchanger, Aminex Q-1505; particle size, 20-35 μm; eluant, 1.61 M HCl; flow rate, 128 cm^3/hr (fluid velocity, 1.86 mm/sec); sample size, 20 μl. J. F. K. Huber, A. M. Van urk-Schoen, Anal. Chim. Acta, 58, 395 (1972).   Reprinted by permission of editor.

exchange techniques are extensively used for the separation of amino acids, as discussed in Section 9.4. A large number of samples related to agricultural chemicals and drugs have been analyzed by ion-exchange chromatography, including substituted benzimidazoles (19), nitrile-substituted pyridine isomers (20), and a variety of drug formulations (21).   Figure 9.13 shows the separation of a mixture of sulfa drugs carried out using a superficially porous strong anion exchanger (22).   Ion-exchange chromatography has also been

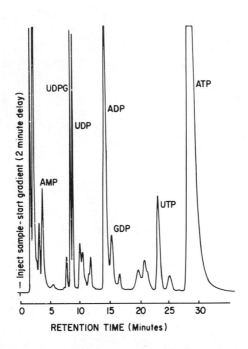

Figure 9.12   Nucleotides in rat liver extract.   Operating conditions: column, Permaphase®-AAX; sample and size, 5μl extract; temperature, 40°C; mobile phase, linear gradient, 0.003 M KH_2PO_4 (pH 3.3) to 0.5 M KH_2PO_4 at 3%/min; column pressure, 1000 psig; flow rate, 1 ml/min; detector, UV photometer at 254 nm; detector sensitivity, 0.04 AUFS.   Du Pont Liquid Chromatography Methods Bulletin 820M11, May 1, 1972.   Reprinted by courtesy of Du Pont Instrument Products Division.

extensively used for trace analysis of various compounds in soils and surface waters, as well as plant and animal tissues and fluids.   Figure 9.14 shows the trace analysis of three metabolites of benomyl fungicide in milk using a single cation-exchange separation (14).

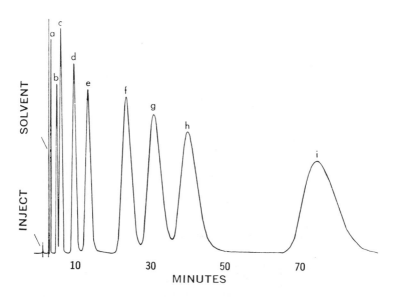

Figure 9.13   Ion exchange separation of sulfa drugs.
Identity: (a) sulfaguanidine, (b) sulfadiazine, (c)
sulfamerazine, (d) sulfamethazine, (e) sulfapyridine,
(f) sulfisoxazole, (g) sulfadimethoxine, (h) sulfa-
methizide, and (i) sulfathiazole; mobile phase, 0.01 M
borax-0.01 M $NaNO_2$; column, Zipax® SAX (strong anion
exchange), 1200 psig; flow rate, 0.80 ml/min.
T. C. Kram, J. Pharm. Sci., 61, 254 (1972).   Reprinted by permission of editor.

Modern ion-exchange chromatography is now being
utilized extensively in the analysis of food and food
additives.   Figure 9.15 shows a separation of some
common food additives carried out by anion-exchange
chromatography.   Many of the water-soluble vitamins
have been analyzed by strong cation-exchange chromatography, as illustrated in Figure 9.16 (23).   In

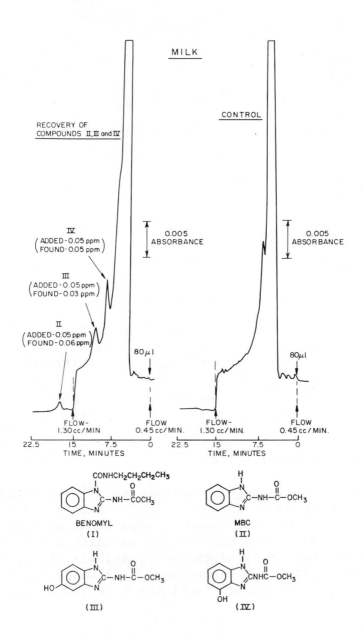

Figure 9.14   Recovery of benomyl metabolites in cow milk.   Column, 1 m x 2.1 mm strong cation exchange (Zipax®-SCX); carrier, 0.1 M acetic acid-0.1 M sodium acetate; temperature, 60°C.   J. J. Kirkland, J. Agr. Food Chem., 21, 171 (1973).   Reprinted by permission of editor.

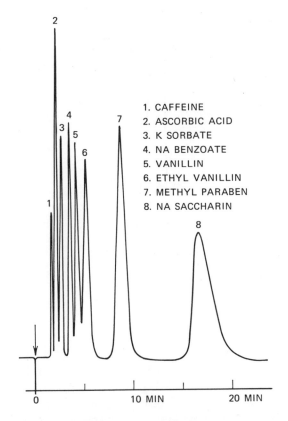

1. CAFFEINE
2. ASCORBIC ACID
3. K SORBATE
4. NA BENZOATE
5. VANILLIN
6. ETHYL VANILLIN
7. METHYL PARABEN
8. NA SACCHARIN

Figure 9.15   Separation of food additives.   Conditions: same as for Figure 9.5, except mobile phase, 0.01 M sodium borate + 0.02 M sodium nitrate.   J. J. Nelson, J. Chromatog. Sci., 11, 28 (1973).   Reprinted by permission of editor.

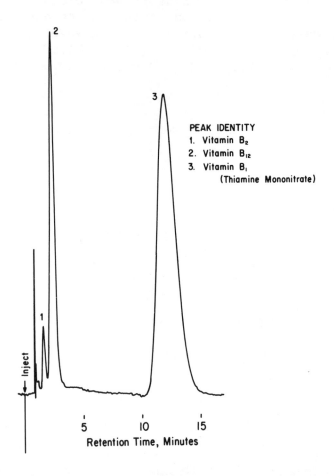

Figure 9.16   Separation of Vitamins B_{12} and B_1.   Conditions: column, Zipax®-SCX; mobile phase, H_2O at pH 9.2 with 0.6 M $NaClO_4$; column temperature, ambient; column pressure, 1500 psi; flow rate, 1 cc/min; detector, UV photometer (0.16 AUFS).   Du Pont Liquid Chromatography Methods Bulletin 820M10, March 23, 1972. Reprinted by permission of Du Pont Instrument Products Division.

Figure 9.17, we see a separation of various sugars
as borate anion complexes as carried out on an ion-
exchanger using gradient elution with a borate
buffer (24).

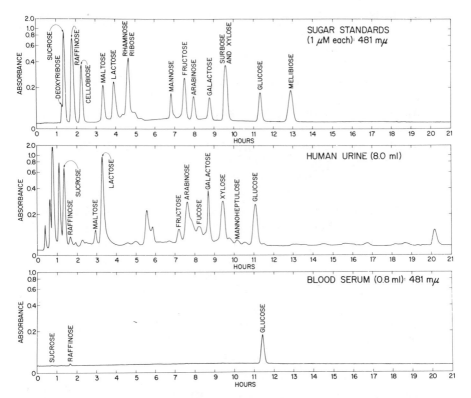

Figure 9.17   Separation of sugars by anion-exchange
chromatography of borate complexes. Column, 150 x
0.62 cm, 10 μ Aminex® A-27 anion-exchange resin; gradient elution of borate buffer with carbohydrate ana-
lyzer. C. D. Scott, in Modern Practice of Liquid
Chromatography, J. J. Kirkland, ed., Wiley-Interscience, New York, 1971, p. 287. Reprinted by permission
of publisher.

REFERENCES

1.   W. Reiman, III and H. F. Walton, Ion Exchange in
      Analytical Chemistry, Pergamon Press, New York,
      1970, p. 170.
2.   W. Reiman, III and H. F. Walton, Ion Exchange in
      Analytical Chemistry, Pergamon Press, New York,
      1970, p. 226.
3.   J. J. Nelson, J. Chromatog. Sci., 11, 28 (1973).
4.   L. R. Snyder and B. E. Buell, Anal. Chem., 40,
      1295 (1968).
5.   D. H. Spackman, W. H. Stein, and S. Moore, Anal.
      Chem., 30, 1190 (1958).
6.   G. Ertingshausen, H. A. Adler, and A. S. Reichler,
      J. Chrom., 42, 355 (1969).
7.   G. Ertingshausen and H. A. Adler, Am. J. Clin.
      Path., 53, 680 (1970).
8.   P. B. Hamilton, Anal. Chem., 35, 2055 (1963).
9.   E. A. Peterson and H. A. Sober, Anal. Chem., 31,
      857 (1959).
10.   J. V. Benson, M. J. Gordon, and J. A. Patterson,
      Anal. Biochem., 18, 228 (1967).
11.   J. C. Dickinson, H. Rosenbloom, and P. B. Hamilton,
      Pediatrics, 36, 2 (1965).
12.   S. Moore and W. H. Stein, J. Bio. Chem., 211,
      893 (1954).
13.   J. J. Kirkland, unpublished studies.
14.   J. J. Kirkland, J. Agr. Food Chem., 21, 171 (1973).
15.   E. P. Horwitz and C. A. A. Bloomquist, J. Inorg.
      Nucl. Chem., 35, 271 (1973).
16.   J. F. K. Huber, A. M. Van urk-Schoen, Anal. Chim.
      Acta, 58, 395 (1972).
17.   P. R. Brown, High-Pressure Liquid Chromatography;
      Biochemical and Biomedical Application, Academic
      Press, New York, 1973.
18.   Du Pont Liquid Chromatography Methods, Bulletin
      820M11, May 11, 1972.
19.   J. J. Kirkland, J. Chromatog. Sci., 7, 361 (1969).

20. C. P. Talley, Anal. Chem., **43**, 1512 (1971).
21. J. B. Smith, J. A. Mellica, and H. K. Govan, Amer. Lab., **4**, 13 (1972).
22. T. C. Kram, J. Pharm. Sci., **61**, 254 (1972).
23. Du Pont Liquid Chromatography Methods, Bulletin 820M10, March 23, 1972.
24. C. D. Scott, in <u>Modern Practice of Liquid Chromatography</u>, J. J. Kirkland, ed., Wiley-Interscience, New York, 1971, p. 287.

BIBLIOGRAPHY

F. Helfferich, <u>Ion Exchange</u>, McGraw-Hill Book Co., New York, 1972. (General information on ion exchangers.)

E. Heftmann, ed., <u>Chromatography</u>, 2nd ed., Reinhold Publishing Corp., New York, 1967. (Contains good discussion of ion-exchange chromatography, many applications.)

C. Horvath, B. Preiss, and S. R. Lipsky, Anal. Chem., **39**, 1422 (1967). (Description of pellicular ion exchangers used in separating nucleotides, etc.)

C. Horvath, in <u>Methods of Biochemical Analysis</u>, D. Glick, ed., Wiley-Interscience, New York, **21**, 1973, p. 79-154. (High performance ion exchange chromatography with narrow-bore columns; rapid analysis of nucleic acid constituents at the nanometer level.)

C. Horvath, in <u>Ion Exchange and Solvent Extraction</u>, J. Marinsky and Y. Marcus, eds., Marcel Dekker, New York, **5**, 1973, p. 207-260. (Pellicular ion exchangers in chromatography.)

J. Inczedy, <u>Analytical Applications of Ion Exchangers</u>, Pergamon Press, London, 1966. (Use of conventional ion-exchange resins.)

J. J. Kirkland, J. Chromatog. Sci., **7**, 361 (1969). (Description of porous layer bead ion exchangers, examples.)

R. Kunin, <u>Ion Exchange Resins</u>, 2nd ed., John Wiley & Sons, New York, 1958. (Classical discussion of ion exchangers.)

W. Reiman, III and H. F. Walton, Ion Exchange in Analytical Chemistry, Pergamon Press, New York, 1970.
(Source book on ion exchangers in analysis.)

O. Samuelson, Ion Exchange Separations in Analytical
Chemistry, John Wiley & Sons, New York, 1963. (Good
summary of classical ion-exchange chromatographic
separations.)

F. C. Saville, Comprehensive Analytical Chemistry, C.
L. Wilson and D. W. Wilson, ed., Elsevier Publishing
Co., New York, 1968, p. 212. (More recent discus-
sion of conventional exchangers.)

C. D. Scott, Modern Practice of Liquid Chromatography,
J. J. Kirkland, ed., Wiley-Interscience, New York,
1971, p. 287. (General discussion of ion-exchange
chromatography, techniques, equipment, some applications.)

CHAPTER TEN

GEL CHROMATOGRAPHY

10.1 INTRODUCTION

Gel chromatography, the newest of the four LC methods, is also referred to as exclusion chromatography, gel filtration, or gel permeation chromatography (GPC). It is the easiest of the various modern LC methods to understand and to use. Despite this simplicity, the technique is very powerful. It has been applied to a broad variety of sample types to solve widely different separation problems. No single, comprehensive book on modern gel chromatography has yet appeared, so the reader must refer to a number of sources for complete information; for example, refs. 1-8.

For what kinds of samples and separation problems is gel chromatography a logical first choice? Any kind of soluble sample can be handled, but the main applications of gel chromatography usually fall into one of three areas. First, gel chromatography is uniquely useful for separating high-molecular-weight species (mol. wt. > 2000), particularly those that are non-ionic. Aside from the resolution of individual macromolecules such as proteins and nucleic acids, gel chromatography is often used to obtain the molecular weight distribution of a synthetic polymer (e.g.,

329

Figure 1.4). Second, simple mixtures can be separated
easily and conveniently by gel chromatography, particularly when the components of the mixture are of widely
differing molecular weights. In such cases it is often
possible to handle larger sample sizes than by other LC
methods. Third, gel chromatography is nicely suited
for the initial, exploratory separation of unknown samples. Such a separation quickly provides an overall
picture of the total sample. It tells us whether we
are dealing with a simple or complex mixture, and
whether the sample components are of low, intermediate,
or high molecular weight. Initial separations by gel
chromatography often define which LC method--or combination of methods--will be required for a given sample.
An initial gel separation is also often an essential
first step in the resolution of a complex sample by the
successive application of more than one LC technique.
One example of this is provided in Section 11.3.

Consider next the basis of separation by gel chromatography. Figure 10.1 illustrates the relative retention of different sample molecules by a particle of
the gel packing material. A cross section of the spherical gel particle is shown, illustrating the presence
of pores of differing sizes. Sample molecules of varying size are shown as the dark circles outside the particle. Some of these molecules are too large to enter
any of the gel pores, and are totally excluded from the
particle. They therefore move directly through the
column, and appear first in the chromatogram. Other
molecules are so small that they can enter all the gel
pores and permeate the entire particle. These compounds are retained to the greatest degree, move
through the column most slowly, and appear last in the
chromatogram. Since solvent molecules are usually
quite small, they are eluted last (at t_0). As a result, the entire sample elutes before t_0, which is the
reverse of what is observed with the other three LC
methods (e.g., Figure 2.4). Finally, molecules of

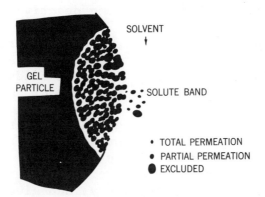

Figure 10.1 Representation of the gel chromatography process: differential permeation of sample molecules into the gel particle. Reprinted by permission of Waters Associates.

intermediate size can enter some pores but not others, and move through the column at intermediate speeds. Thus, separation in gel chromatography occurs strictly on the basis of molecular size. Sorption or adsorption onto the gel particle is undesirable, and it usually does not occur in properly designed systems.

The characteristic elution behavior of gel chromatography leads to several advantages in its application:

1. Narrow bands for easy detection.

2. Short separation times without gradient elution (no "general elution problem").

3. Predictable separation times.

4. Predictable values of t_R according to molecular size.

5. Freedom from sample loss or reaction during

separation.

6. Little problem in column deactivation.

Because all sample bands leave the column relatively quickly (before t_o), they are narrow and easily detected. The greater broadening of sample bands in the other LC methods can lead to problems of inadequate detector sensitivity--particularly for samples that do not absorb in the UV. In gel chromatography, however, all bands give good signals with refractometer detectors, even though these detectors are relatively insensitive. For this reason refractometers are generally used in gel chromatography, and give consistently good results.

A second advantage of gel chromatography is that the total sample is eluted quickly, without the need for gradient elution or other complicated procedure. This is not the case for the other LC methods, except for very simple mixtures and well chosen experimental conditions. A third advantage is that the beginning and end of the chromatogram are almost always predictable. All compounds elute between the value of t_R for total exclusion (i.e., for large molecules) and t_o. Therefore, a series of different samples can be injected over some predetermined series of times, without danger of one chromatogram overlapping another. This favors the use of automatic (unattended) sample injections. The predictable separation times in gel chromatography also mean less time wasted in waiting for the last bands in an unknown sample.

A fourth advantage in gel chromatography is the easier identification of unknown sample bands. Retention is determined strictly by molecular size, and, in fact, can be predicted for a compound of known molecular structure (Section 10.4). This feature can aid us in qualitative analysis. With the other LC methods a retention time value does not usually tell anything definite about molecular structure, because many different aspects of molecular structure can contribute to t_R. A fifth advantage of gel chromatography is the almost complete absence of sample loss or chemical alteration during separation. Such effects are not often found in any of the modern LC methods, but they do occur occasionally. However, the usual absence of sorptive forces in gel chromatography makes this one of the gentlest of separation techniques. A final advantage of gel chromatography is that the column packing does not tend to accumulate strongly retained compounds, from either impure solvents or dirty samples. Therefore, the column does not degrade with time because of the buildup of such compounds on the packing. Pre-columns and guard columns are often used with the other LC methods to protect the main column, but are unnecessary in gel chromatography. The only danger to a gel column is its clogging by samples that contain particulate matter. Such samples should, therefore, be filtered before injection (for example, through an $0.5~\mu$ filter; see Appendix V).

The unique elution behavior of gel chromatography also presents some disadvantages:

1. Limited peak capacity.

2. Inapplicable for samples whose components are of similar size.

3. Principles of separation unlike those for other LC methods.

The most serious disadvantage is limited peak capacity; that is, only a few separated bands can be accommodated within the total chromatogram, because the chromatogram is quite short (i.e., all compounds are eluted before t_o). With the other LC methods it is possible to resolve tens or even hundreds of compounds in a single separation. However, in gel chromatography we seldom see more than half a dozen distinct bands in a single chromatogram. This means that gel chromatography is usually incapable of completely resolving a complex, multicomponent sample, without further separation by other methods. However, polymeric mixtures need not be separated completely for determination of molecular weight distributions.

A second disadvantage of gel chromatography is that it cannot resolve compounds of similar size (e.g., molecular weights within 10% of each other). This rules out gel chromatography for many important separation problems, including the resolution of most isomers. A final difference in gel chromatography--but not a real disadvantage--is that the principles of separation differ somewhat from those applicable to the other LC methods. Thus, the concepts of separation factor α and capacity factor k' are not normally used. The composition of the moving phase is also relatively unimportant in gel chromatography, but plays a major role in the other LC methods.

It is useful to classify the various applications of gel chromatography, as these lead to quite different experimental approaches and/or use of the final results. First, gel chromatography is commonly divided into the techniques of gel filtration (aqueous solvents) and gel permeation chromatography or GPC (organic solvents), for respective application to water-soluble and water-insoluble samples. As we will see, the column packings and techniques used with these two methods can be quite different. Second, gel chromatography can be carried out either with rigid or nonrigid column packings. Rigid packings are required for high pressure (modern) LC, but some samples are better separated on the nonrigid gels. Finally, gel separations can yield either the resolution of individual compounds, or the partial separation of polymeric samples for calculating a molecular weight distribution (e.g., Figure 1.4).

In the present chapter we will emphasize the separations of individual compounds on rigid gels.

Separations on nonrigid gels are important, but fall
outside the realm of modern LC, because high-pressure
operation is precluded. These separations are also
adequately reviewed elsewhere (4-8). The determination of molecular-weight distributions for polymers by
gel chromatography is a large subject which demands a
more detailed treatment than can be presented here
(e.g., see ref. 2). We will give only a limited treatment of separations on the nonrigid gels, or the de-
determination of molecular-weight-distribution data.

10.2   COLUMN PACKINGS

Separation in gel chromatography is controlled largely
by the type of packing used in the column. Therefore,
the choice of an appropriate packing or gel for a given
separation is the major decision to be made when setting up a gel chromatography system.

   Commercially available packings for gel chromatography are summarized in Table 10.1. These materials are each available in a number of different pore
sizes, each suitable for the separation of compounds
of differing molecular weight (see below). Most separations by GPC are currently carried out with porous
polystyrene beads that have been cross-linked with
divinyl benzene. Poragel® and Bio-Bead®-S gels allow
the separation of relatively small molecules, up to
molecular weights of about 20,000. The Styragel®
packings allow the separation of still larger compounds, with molecular weights up to 20 million. The
porous vinyl acetate gels, recently introduced by
Merck, are similar to the polystyrene gels, but cover
a somewhat lower molecular-weight range. The smaller
pore Merckogels® are slightly better adapted to the
separation of compounds with molecular weights of less
than 2000, but are restricted to molecular weights of
less than one million.

TABLE 10.1   PACKINGS FOR GEL CHROMATOGRAPHY

	Source[a]
GPC (Organic Solvents)	
Polystyrene beads	
Poragel®, Styragel®	Waters (11)
Bio-Bead®-S	Bio-Rad Labs (62)
Vinyl acetate gels	
Merckogel®-OR	E. Merck (63), EM Labs (64)
Porous silica or glass	
Porasil®	Waters (11)
CPG Beads®	Electro-Nucleonics (71)
Bio-Glas®	Bio-Rad Labs (62)
Merckogel®-Si	E. Merck (63), EM Labs (64)
Gel filtration (Aqueous Solvents)	
Semirigid (low mol. wt.)	
Sephadex®	Pharmacia (72)
Bio-Gel®-P	Bio-Rad Labs (62)
Nonrigid (high mol. wt.)	
Sepharose®	Pharmacia (72)
Bio-Gel®-A	Bio-Rad Labs (62)
Special purpose gels	
Porous silica or glass	See above (GPC)
Deactivated Porasil®	Waters (11)
Aquapak®	Waters (11)

[a]See Appendix V for complete addresses.

GPC with organic solvents is also carried out with rigid packings of porous silica or glass. These materials are more stable than the organic gels, and somewhat more conveniently packed and handled. However, they often give undesired retention of sample by adsorption onto the silica surface. This problem and various ways of avoiding it are discussed further in Section 10.5.

Gel filtration in aqueous solvents is most often carried out on the semirigid or nonrigid gels listed in Table 10.1. Compounds with molecular weights up to several hundred thousand can be separated on one of the Sephadex® or Bio-Gel®-P gels. Sephadex® is a cross-linked dextran (polycarbohydrate), while Bio-Gel®-P is a cross-linked polyacrylamide. The smaller pore gels of this type, used for compounds with molecular weights up to a few thousand, can withstand column pressures of about 100-200 psi, but no more. For separating compounds with molecular weights of several hundred thousand and above, the larger pore gels can only be used at column pressures of a few psi. Greater pressures compress the gel and plug the column.

The Sepharose® and Bio-Gel®-A packings are composed of agarose, a poly-galactopyranose (carbohydrate). Pore size is controlled by varying the agarose concentration. Compounds with molecular weights between 10^4 and 10^8 can be separated on the agarose gels, but very low pressures (e.g., 0.5 psi) must be used for columns of the larger-pore gels.

From the preceding discussion, it is apparent that high-pressure, modern LC cannot be used in gel filtration with semirigid or nonrigid gels, except for the separation of relatively small molecules (or of small molecules from a single high-molecular-weight species). In an effort to improve the separation of larger, water-soluble compounds by gel filtration, the special purpose gels of Table 10.1 have been introduced. The same porous silicas and glasses used in GPC can also be

used in gel filtration, except that sample adsorption
is often a problem. Various techniques for minimizing
adsorption under these conditions have been developed
(Section 10.5). Waters Associates markets "deactivated
Porasil®", a porous silica whose surface has been chemically modified to reduce adsorption. Aquapak® is a
lightly sulfonated polystyrene. The sulfonation permits Aquapak® to swell in water, and hence to be used
in gel filtration. Since Aquapak® is also an ion exchanger, its use in gel filtration must be confined to
nonionic samples.

None of these special purpose gels is universally
satisfactory, even when used with the various techniques discussed in Section 10.5. Some samples can be
separated satisfactorily, but other samples give problems. Another approach to the goal of high-pressure
gel filtration is the use of glass beads to support the
gel within the column (e.g., ref. 9). However, this
technique is only moderately successful.

For a good general review of commercially available
packings for gel chromatography, see ref. 10. The manufacturers' literature should be consulted for detailed
information on the availability of specific gels for
different sample types and molecular-weight ranges, and
for the use of these materials.

Calibration plots for the various gel packings define the ability of a gel to separate samples of different molecular weights. A calibration plot (for a particular gel or column) is a graph of sample molecular
size versus relative retention. Figure 10.2 is a hypothetical example, where sample molecular weight is
plotted against retention volume V_R. Other measures of
molecular size and retention are used, and will shortly
be discussed. In Figure 10.2 we see that there is an
exclusion limit (point A) for the gel, corresponding
here to a molecular weight of 10^5. Molecules larger
than A are totally excluded from all gel pores (i.e.,
are too large to enter); they elute together as a single

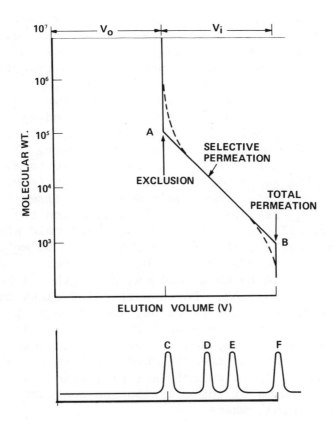

Figure 10.2 Hypothetical calibration plot for a given gel or gel column.

band C, with retention volume V_o. The interparticle volume V_o is equal to the volume of solvent within the column, but outside the gel particles.

We also see in Figure 10.2 a total permeation limit (point B, for a molecular weight of 1000 in this case), such that all compounds smaller than B are able to completely permeate the gel particles. These compounds elute as a single band F, with retention volume V_t. The latter quantity, V_t (which corresponds to retention time t_0) is equal to the total volume of solvent within the column (referred to as V_m in Eq. 2.4). The quantity V_i, equal to ($V_t - V_0$), is the volume of solvent held within the gel particles (<u>intraparticle volume</u>).

Compounds with molecular weights between A and B can enter some pores but not others, depending upon the size of the molecule. Therefore, these compounds elute in order of decreasing molecular size. This intermediate region (A < V_R < B) is referred to as the <u>fractionation range</u> of the gel. Compounds (e.g., D, E) can be separated when their molecular sizes are different <u>and</u> fall within the fractionation range.

As we will see in Section 10.4, molecule size in gel chromatography is actually determined by <u>molecular length</u>. Some manufacturers, therefore, report calibration plots in terms of molecular length (Å) rather than molecular weight. For a particular class of compounds (e.g., globular proteins, monodisperse polystyrenes, etc.), molecular weight increases regularly with molecular length, and calibrations in terms of molecular weight can then be used. Retention in gel chromatography (particularly gel filtration) is also measured in terms of the <u>distribution constant</u> K_0 (e.g., Figure 10.4), where

$$K_0 = \frac{V_R - V^0}{V_i} .$$

Values of K_0 therefore range between 0 (exclusion) and 1 (total permeation).

Manufacturers of different gel packings normally provide <u>approximate</u> calibration plots for a given gel. Some examples are given in Figure 10.3 for the various polystyrene gels sold by Waters for GPC. Similarly, Figure 10.4 shows calibration plots for the various Bio-Rad packings for gel filtration. <u>Exact</u> calibration curves should be determined by the user for every gel column, using polymer standards available from

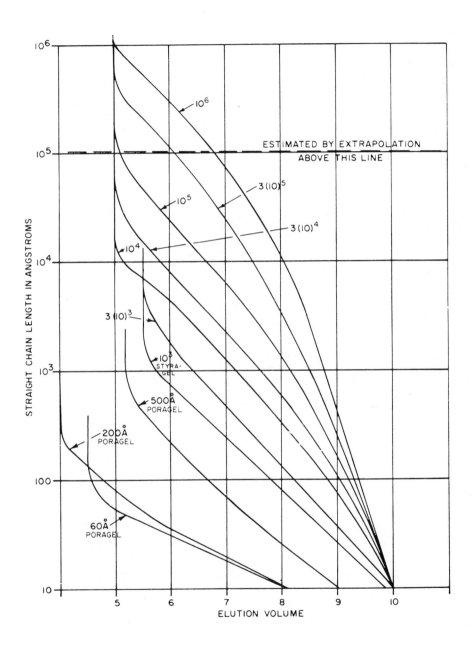

Figure 10.3 Calibration plots for various polystyrene
gels sold by Waters Associates for gel permeation.
The numbers on each curve refer to maximum pore size
in Å (e.g., 10^4, 10^6). Note, however, that these pore
sizes are "nominal" values, corresponding to molecular
length (L_m) values for maximally extended sample molecules. The <u>actual</u> pore sizes are considerably smaller
than those indicated values. Reprinted by permission
of Waters Associates.

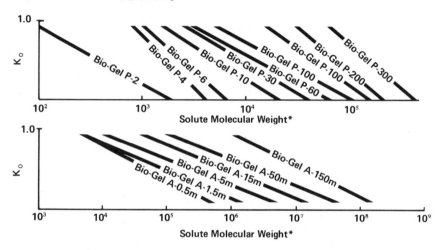

Figure 10.4. Calibration plots for various gels sold
by Bio-Rad Laboratories for gel filtration. Various
peptides and globular proteins were used as standards
to calibrate these gels. Reprinted by permission of
Bio-Rad Laboratories.

the suppliers of gel packings. This is discussed in
detail in Section 10.4.

Control of separation in gel chromatography is
simple and predictable. Band widths and resolution
vary with column plate number N, as described in Section 3.5. Separation is also determined by the <u>spac-
ing</u> of band centers; the farther apart are two bands,
the better is their resolution. Two compounds of

differing molecular size will be best separated when
they elute near the middle of the gel fractionation
range. This is illustrated in the hypothetical example
of Figure 10.5. Here we show calibration plots for
three different gels (A, B, C), where the separation
of two compounds (No. 1, mol. wt. 1000; No. 2, mol. wt.
10^4) is of interest. On gel A the two compounds elute
near the gel exclusion limit, so that the spacing between band centers (ΔV_A) is small and separation is
therefore poor. In this case a gel with a higher exclusion limit should be used. Gel C has a higher exclusion limit, but our two compounds now elute near
the total permeation limit. So, the peak spacing
(ΔV_C) is again small, and separation is still poor.
With gel B, however, the molecular sizes of compounds
No. 1 and No. 2 fall near the center of the fractionation range, the spacing of band centers (ΔV_B) is
large, and separation is much improved.

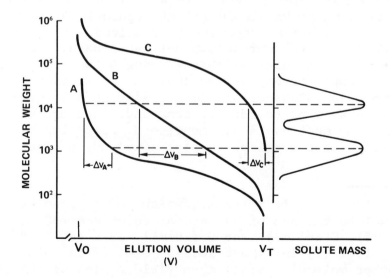

Figure 10.5 Selection of the right gel for maximum
resolution of two adjacent bands.

In the case of broad range samples (i.e., containing several species which cover a wide spread in molecular weights) we encounter a different sort of separation problem. In this case the fractionation range should span a wide range in molecular sizes, and this is most easily accomplished by combining several columns in series--each column containing a gel with a different exclusion limit. This approach is illustrated in Figure 10.6. Here we visualize four different gels, A, B, C, and D, whose individual calibration plots are given in (1) in Figure 10.6. Retention volumes for a series of connected columns are additive, so that the calibration plot for connected columns A, B, C, and D can be calculated from the data of (1) in Figure 10.6.* This calculated curve is shown in (2) in Figure 10.6 (solid curve). As expected, the fractionation range for this four-column set now spans a wide range of molecular sizes, and can, therefore, be used for separating samples with components of widely different molecular weight. This is illustrated for the hypothetical sample in (3) of Figure 10.6, where four separate bands appear to be resolved.

While such a four-column set can separate compounds with molecular weights varying between 10 and 10^5, it should be emphasized that the separation of any two adjacent bands of similar size is only marginal. When we want to resolve compounds that are close in size, we must use a column with a <u>narrow</u> fractionation range that overlaps the two compounds of interest.

---

*Thus, for a sample molecular weight of <u>10^3</u>, we have V_R equal to the sum of values for columns A (30 ml), B (48 ml), C (61 ml), and D (70 ml), or a total of <u>209 ml</u> (see (1) in Figure 10.6). Similar calculations for other molecular sizes then yield a plot of the overall composite calibration curve for columns A-D in series.

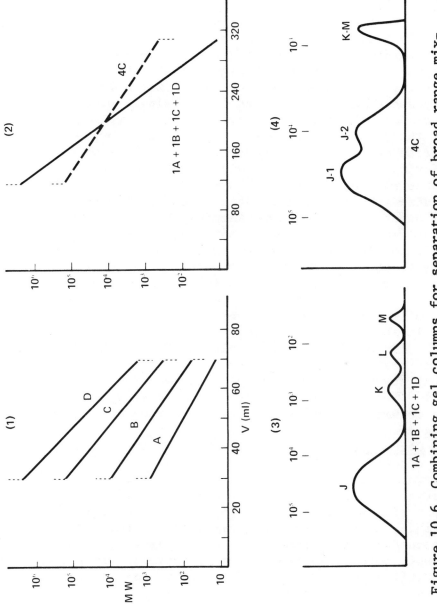

Figure 10.6 Combining gel columns for separation of broad range mixtures or two compounds of similar size.

For example, assume we suspect that band J in (3) of Figure 10.6 is really two bands that are unresolved. In this case we should reseparate the sample on a four-column set of gel C (band J falls near the center of the fractionation range for gel C). The resulting separation is shown in (4) in Figure 10.6. Now band J is split into two bands. At the same time, however, bands K, L, and M have been merged into a single band which elutes near the total permeation limit of gel C.

Thus, several columns in series can provide either moderate resolution for several compounds of widely differing size (different gels in each column), or greater resolution for two compounds of similar molecular size (the same gel in each column). For the initial separation of unknown samples, columns with different gels should normally be used, since some minimum resolution is needed over the entire molecular size range (because we do not know what size molecules are present in the sample). On the other hand, columns of the same gel are used when we must enhance the separation of two (or more) compounds of similar size.

It should be emphasized that different gels should not be mixed into the same column. First, this precludes the use of several (connected) columns in different ways (as in the examples (3) and (4) of Figure 10.6). Second, mixing different LC packings into the same column can result in particle sizing and the lowering of column efficiency.

Column packing techniques for gel chromatography were discussed in Chapter 6. The rigid polystyrene and vinyl acetate gels are packed by the balanced-slurry high-pressure method. For the packing of the porous silicas and glasses, and of the semirigid or nonrigid gels, see the manufacturers' recommendations. Prepacked columns of many of these gels (e.g., polystyrene, porous silica) can be purchased. Because of the difficulty in preparing high-efficiency polystyrene gel columns, and the importance of maximum column efficiency in gel chromatography (see below), beginners are advised to purchase prepacked columns of these materials.

Column efficiency in gel chromatography is based on the same factors as in the other LC methods (e.g., Chapters 2, 6). Until recently, highest column efficiencies were achieved with the polystyrene packings, although some workers now claim comparable or better efficiencies with porous glass or silica. Small-particle porous silicas of the Zorbax® type appear to give the highest column efficiencies of all (11). It should be noted that column efficiency is more important in gel chromatography, than for the other LC methods, because of the low peak capacity of gel chromatography. Thus, marginal columns in gel chromatography are not useful and should be discarded. Some idea of the plate numbers that are achievable in gel chromatography can be obtained from the Waters Associates catalog (see Appendix V for address), which describes their prepacked columns. Total plate numbers (1 m columns) for the polystyrene gels are usually in the 2000-5000 range.

In other forms of LC, bands widen as t_R increases. This is not usually the case in gel chromatography, where band widths generally remain approximately constant throughout the chromatogram. The reason is that molecular size decreases sharply with t_R in gel chromatography, and plate heights H increase with molecular size (Appendix I). These two effects combine to roughly cancel the normal increased broadening of bands as elution proceeds.

10.3 MOBILE PHASES

In gel chromatography, unlike the other LC methods, the mobile phase is not varied in order to control resolution. Rather, the mobile phase is chosen for low viscosity at the temperature of separation (i.e.,

for higher N values; Section 2.3), and for its ability
to dissolve the sample. A low-viscosity mobile phase
will have a boiling point that is about 25-50°C higher
than the temperature of the column. In the case of
difficultly soluble samples, the solvent is sometimes
selected to provide maximum sample solubility.

When a refractometer is used as detector, the solvent can be chosen for optimum refractive index. Thus,
if maximum detector sensitivity is required, the refractive index of the mobile phase should be as different as possible from that of the sample. In some
cases the detection of a compound of interest is interfered with by an overlapping band. This interference
of the overlapping band can sometimes be eliminated by
matching the refractive indices of mobile phase and
interfering compound. An example is discussed in Section 10.6.

The effect of the solvent on the various nonrigid
packings for gel chromatography must also be considered. Thus, in gel filtration, a salt should be added
to the solvent to maintain constant ionic strength.
Otherwise the passage of ionic samples through the bed
can lead to shrinkage of the gel and a decrease in column permeability. The various gels for GPC can tolerate a wide range of organic solvents, but there are
some exceptions; for example, acetone and alcohols
cannot be used with polystyrene packings. The manufacturers' literature should be consulted in each
case. Several commonly used solvents for gel chromatography are listed in Table 10.2, along with properties of interest.

10.4 OTHER SEPARATION VARIABLES

Temperature in gel chromatography can be increased for
difficultly soluble samples, or to decrease solvent
viscosity in gel filtration. For convenience, most

TABLE 10.2. SOLVENTS FOR GEL PERMEATION CHROMATOGRAPHY

Solvents	Physical Properties			Safety Properties		
	Boiling Point (°C)	Viscosity @ 20°C (cP)	Refractive Index @ 20°C	Flash Point (°C)	Acute LD_{50} (mg/kg)	Toxicity $L(ct)50$ ppm
Tetrahydrofuran	66	0.51^{25}°C	1.4070	14	Dangerous	200
1,2,4-Trichloro-benzene (TCB)[b]	213	1.89^{25}°C	1.5717	99	Dangerous	75
O-Dichlorobenzene[b]	180	1.26	1.5515	66	500	50
Toluene	110.6	0.59	1.4961	4	Dangerous	6000-200
N,N-Dimethyl For-mamide (DMF)[c]	153	0.90	1.4280	58	Dangerous	100-10
Methylene Chloride	40.1	0.30	1.4242	None	3000	500
Ethylene Dichloride	84	0.84	1.4443	13	770	50
N-Methyl Pyrroli-done[c]	202	1.65	1.47	95.4	Dangerous	None
m-Cresol	202.8	20.8	1.544	94	Dangerous	5
Benzene	80.1	0.652	1.5011	27	7000	25
Dimethylsulfoxide (DMSO)	189	--	1.4770	95.0	Dangerous	None
Perchloroethylene	121	--	1.505	None	Dangerous	100
o-Chlorophenol	175.6	4.11	1.5473^{40}	None	670	Dangerous
Carbon Tetrachlor-ide	76.8	0.969	1.4607	None	Dangerous	10
Water	100.0	1.00	1.33	None	--	--
Trifluoroethanol	73.6	1.20^{38}°C	1.291	40.6	240	4600
Chloroform	61.7	0.58	1.446	None	Dangerous	50

[a] Generally contains butylated hydroxy toluene at a few hundredths of a per-cent as a stabilizer.
[b] These solvents are usually used at 135°C. The use of an antioxidant is recommended: Ionol (Shell) 5 g/gal or Santonox R (Monsanto) 1.5 g/gal.
[c] Quite hydroscopic. Relatively large amounts of water (several percent) may drastically affect fractionation. May damage Styragel® columns with exclusion limits of 500 Å or less.
[d] Ordinarily contains 0.75% ethanol as stabilizer.

Reprinted from the Waters Associates' Liquid Chromatography Catalog, with permission.

gel chromatography separations are carried out at room
temperature (without thermostatting the column). However, some high-molecular-weight polyolefins and poly-
amides require temperatures of 100-135°C for their
separation by gel permeation, since these samples are
not sufficiently soluble at lower temperatures. If
the sample is sufficiently stable, gel filtration can
be carried out at higher temperatures (e.g., 60°C).
The resulting lowering of the viscosity of the aqueous
mobile phase provides improved column efficiency and
better resolution.

Sample size in gel chromatography is limited
mainly by sample viscosity, as well as by the usual
requirement that the total sample volume be sufficiently small (see ref. 12, and Section 2.3). A rough guide
is that the sample solution injected onto the column
should have a viscosity no greater than twice that of
the mobile phase. In the case of very high-molecular-
weight samples, this may mean sample concentrations of
less than 0.1%. Since viscosity decreases sharply with
temperature, more concentrated samples can be charged
at higher column temperatures. About 15 mg of sample
per 100 ml of column volume can be charged in typical
cases (more for lower molecular weight samples, less
in the case of higher molecular weights). However,
much larger samples can be charged when the compounds
to be separated are easily resolved (as in application
of Figure 10.11, p. 361).

Calibrating a column in gel chromatography is carried out in various ways. GPC columns are usually
calibrated in units of molecular length, or molecular
hydrodynamic diameter (Å). The reason is that molecular size in gel chromatography is in fact determined
by hydrodynamic volume, which is related to the maximum length of the extended molecule; see Figure 10.7.
At first glance this can be confusing, since access
to the gel pore would seem to be determined by minimum
molecular cross section, rather than by maximum

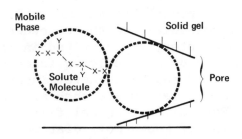

Figure 10.7   Molecular size in gel chromatography as a function of molecular length.

molecular length. Actually, this is the case for the separation of small molecules on the molecular sieves; for example, the retention of n-alkanes on Linde 5A molecular sieve, with exclusion of the wider (but shorter) isoalkanes. In gel chromatography, however, the molecules of interest are larger, and separation times are smaller. Therefore, molecules are unable to "snake" their way into pores, and the probability of entry increases as hydrodynamic diameter decreases. This is an oversimplified description of the gel exclusion process, but it does make clear a basic difference between separation on the molecular sieves and the LC gels. For more details, see refs. 3,4.

The calibration plots furnished by the suppliers of gels are usually only approximate. Therefore, it is desirable to prepare an exact calibration plot for each gel column that is to be used. To calibrate a column for GPC, polystyrene and polyglycol standards of known molecular length are used (e.g., Waters Associates). These narrow fractions cover the molecular size range of 50 Å and greater. Smaller molecular sizes are covered by the n-alkanes, whose molecular lengths L_m are given by

$$L_m = 2.5 + 1.25 \, n \quad \text{(in Å)}. \qquad (10.1)$$

n is the number of carbon atoms in the molecule. For example, L_m for n-pentane as sample is $2.5 + 5(1.25)$ = 8.75 Å. The molecular lengths of other compounds can also be calculated, as described in ref. 13.   In each case, the standard (a polymer fraction or n-alkane) is injected and V_R is determined; values of V_R versus molecular length are then plotted, as in Figure 10.3. For calibrating gel filtration columns, globular proteins of known molecular weight are sometimes used (e.g., Figure 10.4); see also discussion of ref. 14.   The exclusion volume in these cases can be measured using India ink or Blue Dextran 2000 (Pharmacia) as standards.

The relationship of molecular weight to length is often of interest, since this allows the conversion of a gel chromatogram (calibrated in length units) to a molecular-weight distribution in molecular-weight units. For large, linear polyethylene molecules, length (in Å) is approximately equal to 1.25 n (see Eq. 10.1), and molecular weight is equal to n times (monomer weight/2). In this case the monomer is ethylene (n = 2), and its molecular weight is 28. Therefore, for linear polyethylenes, the molecular weight is equal to $[(\text{monomer wt.}/2) \, n/1.25 \, n]$ times length, L_m (in Å) or

$$\text{mol. wt.} = 11 \, L_m \quad \text{(linear polyethylenes).} \qquad (10.2)$$

Similarly, for other substituted ethylene polymers,

$$\text{mol. wt.} = (\text{monomer wt.}/2.5) \, L_m. \qquad (10.2a)$$

For example, the monomer weight of polystyrene is 104, so that the molecular weight of polystyrene oligomers equals 41 times molecular length L_m. The proportionality factor between molecular weight and length (11 for polyethylenes, 41 for polystyrenes) is referred to as the Q-factor. The use of the Q-factor in calculating molecular-weight distributions is discussed further in Section 10.6.

Since the Q-factor of synthetic polymers normally varies between 11 and 41, an average value of 20 is often assumed (in the absence of information on sample structure) for estimating polymer molecular weights. Table 10.3 compares molecular weight versus length for several types of sample.

So-called universal calibration plots are sometimes used in molecular-weight-distribution analyses. These are plots of intrinsic viscosity times molecular weight, versus retention volume. Most samples fall on the same universal plot, regardless of their Q-factor values. For a further discussion, see refs. 3, 14. Quite recently a new approach to column calibration has been described, based on calculating the pore-size distribution of silica gels (15). This allows the prediction of calibration plots in the absence of actual data.

10.5 SPECIAL PROBLEMS

Limited peak capacity in gel chromatography was referred to earlier. This problem is further aggravated in the case of the nonrigid gels, because long columns and high pressures (for maximum N per unit time) cannot be used. One way around this problem is through the use of recycle chromatography. The basic apparatus for carrying out recycle chromatography is illustrated in Figure 10.8. This technique allows partly resolved bands to be sent back to the same column for further separation. Operation is begun in a normal mode, with carrier solvent flowing from the solvent reservoir through the pump, into the column, and through the detector to the fraction collector. When a partially separated band passes through the detector, the six-port valve (V2) is switched to allow these

TABLE 10.3   DEPENDENCE OF MOLECULAR WEIGHT ON LENGTH FOR DIFFERENT SAMPLES

Carbon Number (n-alkanes)	L_m (A)	Molecular Weight			
		Poly-ethylene[a]	Polyvinyl-chloride[b]	Polystyrene[c]	Globular Proteins[d]
5	8.8				
10	15				
20	27.5				
	100	1,100	2,500	4,100	7,000
	1000	11,000	25,000	41,000	200,000
	10^4	111,000	250,000	410,000	6×10^6
	10^5	1.1×10^6	2.5×10^6	4.1×10^6	1.8×10^8
	10^6	11×10^6	25×10^6	41×10^6	5×10^9

[a]Q-factor equal 11.
[b]Q-factor equal 25.
[c]Q-factor equal 41.
[d]Q-factor not applicable (molecular weight not proportional to length, because molecules are spherical rather than linear); see Electronucleonics application sheets.

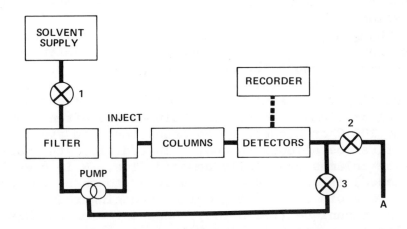

For RECYCLE — Valve 1 & 3 Open, Valve 2 Closed
For DRAWOFF — Valve 1 & 2 Open, Valve 3 Closed

Figure 10.8 Equipment schematic for recycle chromatography. Reprinted by permission of Waters, Associates.

bands to return to a low volume reciprocating pump and reenter the column for further separation. The process can be continued until separation of the bands in question is complete or until the recycled bands spread out to cover the entire column length (so that the two ends of the chromatogram begin to overlap). For a detailed discussion of recycle chromatography, see ref. 1 and recent applications sheets from Waters Associates. Recycle chromatography is finding increasing use in the other LC methods for very difficult preparative separations. This technique effectively yields longer column lengths, without a concommitant increase in column pressure (see Eq. II.1, Appendix II). When combined with automatic equipment for continuous, unattended recycling, recycle chromatography can yield tens of thousands of theoretical plates in a single overnight operation.

Changing columns for different samples is required more frequently in gel chromatography than in the other LC methods, because the column (not the solvent) controls separation. This can be inconvenient, and interchanging columns in gel chromatography often poses special problems. Thus, such columns are often operated at 100°C or higher, which means that the system must first be allowed to cool. In addition, the polystyrene columns are easily damaged (i.e., N is greatly decreased) if air bubbles are allowed to enter the column. To avoid this during column changing requires careful technique (see below). One solution to this problem is to use switching valves (Appendix V) which permit a choice of different columns merely by rotating the valve. However, such valves are moderately expensive, particularly when high temperatures and pressures are involved.

Gel columns with polystyrene packings are disconnected as follows. A syringe filled with solvent is first connected to the outlet of the last column before the detector, using a length of Teflon® tubing. The column is now disconnected, solvent is forced out the inlet end by pressing on the syringe, and the inlet cap is attached to the column. The syringe is disconnected, and the outlet cap is attached. Subsequent columns are removed in the same way. To reinstall a column, a similar procedure is used. A filled syringe is attached to the outlet end of the column, and the inlet end is reconnected while pressing on the syringe to force solvent from the column inlet. Solvent is then pumped into the column to force out any air entrained near the outlet end, before adding on the next column in series.

Adsorption effects in gel chromatography, particularly with the porous glass or silica gels, were noted in Section 10.2. Aside from the use of special packings (e.g., deactivated Porasil®, Aquapak®), various materials can be added to the mobile phase to suppress sample adsorption--by competing for active adsorption sites on the surface of the gel particle. In gel filtration, basic proteins and some neutral proteins tend to adsorb on porous glass. Such adsorption can be overcome in many cases by either raising the pH of the mobile phase, increasing its ionic strength, or by adding 0.05-0.1% polyethyleneglycol (e.g., Carbowax®, 20,000 mol. wt.) to the solvent. For large pore gels, a 100,000-mol. wt. Carbowax® should be used. Alternatively, the column can also be washed with a more concentrated solution of the polyglycol, reconditioned with pure mobile phase, and used for several separations (before retreatment is required). However, this can give a changing solvent baseline with refractometer detectors. It should be noted that adsorption effects are normally not observed for samples which do not permeate the gel. In gel permeation with the porous glasses, silanization of the glass before use (e. g., with refluxing hexamethyldisilazane; Pierce Chemicals) is recommended. For details on the treatment of porous glasses so as to avoid adsorption effects, see ref. 16.

Adsorption on the Sephadex® and Bio-Gel® packings has also been reported. Aromatic compounds can adsorb on Sephadex®, but this can be controlled by the proper choice of solvent (17). Aromatic compounds, acids and bases can adsorb on Bio-Gel®; the manufacturer's recommendations should be followed to minimize this effect.

Solvent degassing in GPC is a potentially serious problem, particularly in the case of the polystyrene packings. Such columns can be ruined by release of air bubbles within the column, and this effect is magnified by the use of separation temperatures above ambient. Therefore, it is strongly recommended to degas all solvents, by boiling the solvent for 5-10 min just before use.

False bands in gel chromatography can arise when polymeric samples are separated. This is illustrated in Figure 10.9, for the hypothetical separation of a monomodal polymer fraction (A). When a significant part of such a polymer fraction consists of oligomers which are totally excluded (i.e., have a molecular weight greater than the exclusion limit), these elute together as an apparent second peak at V_O. Thus, the presence of a separate polymer band at V_O should always be regarded with suspicion. In such cases, the sample should be reseparated on a column with a larger exclusion limit. If the band at V_O (from the first separation) disappears, then a false exclusion band was present. If the initial band is still seen in the second separation, a true bimodal molecular-weight distribution exists.

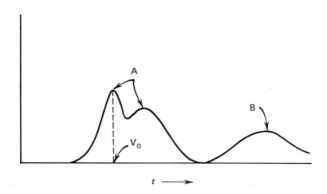

Figure 10.9   False band at the exclusion limit in gel chromatography.

Association of sample molecules in gel chromatography is an occasional problem. For example, ionic surfactants form micelles in some GPC solvents. The result is tailing bands whose V_R values change with sample size. Sample association can usually be controlled by using a mobile phase that is a better solvent for the sample; for example, a more polar solvent for ionic surfactants. Sample association can occasionally be used to advantage. For example, carboxylic acids dimerize in nonbasic solvents (e.g., benzene), but dissociate in basic solvents (e.g., THF). The V_R value of a carboxylic acid, therefore, varies with the solvent, since its apparent molecular size is doubled in nonbasic solvents. In such cases the solvent can be used to control separation, by adjusting the V_R values of carboxylic acids independently of other compounds in the sample. Association of solvent and sample molecules (e.g., alcohols plus ether solvents) can be used in similar fashion to control separation (see ref. 13), but the changes in V_R are usually small. Occasionally it may be advantageous to artificially increase molecular size through association, by adding a large associating compound to the solvent (e.g., an acid to complex a basic sample, alcohols for alcohol samples, etc.).

10.6 APPLICATIONS

A very large number of individual separations by gel chromatography are summarized in refs. 1-8. We will now look at some illustrative examples which show the unique capabilities of this general method, and we will briefly examine the factors involved in determining the molecular-weight distributions of polymers.

Gel chromatography is quite useful in providing initial information about unknown samples, particularly those samples which contain polymeric constituents.

Figure 10.10 shows such an example, the initial separation of a commercial adhesive formulation. Here the main part of the sample--the PVA polymer--is well resolved from surfactants present in lesser concentration. The two surfactants could each be recovered, and were then characterized by infrared spectrophotometry.

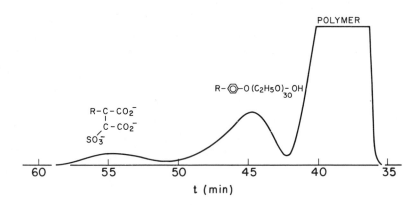

Figure 10.10   Separation of surfactants from polyvinyl acetate (PVA) polymer emulsion. Conditions: two 4-ft columns (3/8 in. i.d.), Poragel® A-3;   one 4-ft column, Styragel® 10^3 Å; one 4-ft column, Styragel® 10^4 Å; tetrahydrofuran solvent, room temperature.

Gel chromatography is also used to determine the concentrations of high-molecular-weight additives or reaction products in the presence of lower-molecular-weight components. Figure 10.11 illustrates the determination of a polymeric additive in lubricating oil, where the polymer band at V_R = 17 ml is well separated from the oil. Despite the marginal sensitivity of refractometer detectors, as little as a few hundredths of a percent of the polymer additive could be accurately determined in this system, because it was possible to charge relatively large amounts of sample (note

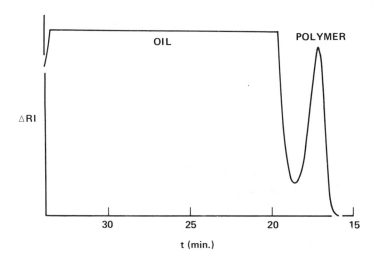

Figure 10.11   Determination of polymeric additive in
lubricating oil.   Conditions: Same as in Figure 10.10.

that the large lubricating oil band is off-scale).
Similarly, small amounts of low-molecular-weight compounds can be determined in a polymer matrix, as illustrated by Figure 10.10.

An essentially equivalent assay of a water-soluble
sample is illustrated in Figure 10.12.   Here it was
required to determine the total concentration of polyvinyl pyrrolidone (PVP) in aqueous solution which contained other constituents of low molecular weight.   It
was possible to use the semirigid gel Sephadex® G-15
as column packing, and to operate the column at pressures of up to 150 psi.   The high-molecular-weight PVP
appears at the exclusion limit, and is resolved from
the remaining sample constituents (ethanol plus glycerine) in less than 15 minutes.   Using peak height for
quantitation, with daily calibration, precisions of
$\pm 1\%$ (1σ) were possible in the determination of total

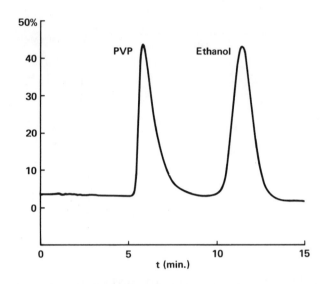

Conditions: 100 cm. x 2.8 mm Column of Sephadex G-15.
Operating Pressure 60 p.s.i
Solvent (Water) Flow Rate 0.5 ml/min.
Ambient Temperature
Sample Size 250μl.
Detector: Refractometer, 8 x 10^{-5} RI Units Full Scale.

Figure 10.12   Determination of polyvinylpyrrolidone
(PVP) in aqueous solutions.   Conditions: 100 cm x 2.8
mm column of Sephadex® G-15; operating pressure,   60
psi; solvent (water) flow rate, 0.5 ml/min; ambient
temperature; sample size, 250 μl; detector, refractometer, 8 x 10^{-5} units full scale RI.

PVP.
Figure 10.13 shows the value of gel chromatography
in the rapid analysis of relatively simple mixtures
which contain polymeric components.   Here it was desired to analyze a commercial hot-melt adhesive--a
blend of polymer, wax, and various resins.   Prior to
the application of GPC analysis, these competitor

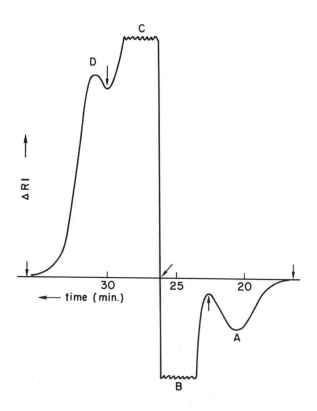

Figure 10.13   Separation and analysis of a hot melt
adhesive.   Conditions: same columns as in Figure
10.10; toluene solvent, 100°C.

samples were roughly characterized by infrared spec-
tral analysis of the total sample, followed by trial-
and-error blending experiments which attempted to dupli-
cate the properties of the sample of interest.   In the
example of Figure 10.13, a single GPC separation results in the separation and quantitative (gravimetric)
assay of the four components present (bands A-D).
About 200 mg of sample could be resolved in less than

an hour, with easy identification of the recovered
fractions (arrows in Figure 10.13 show cutpoints) by
infrared. Finally, the average molecular weight of
polymer (A) and wax (B), as inferred from the chromatogram, complete the total analysis of this particular
sample. In this way the GPC analysis was carried out
in less than a day, versus more than a month by the
old infrared/blending approach.

Figure 10.14 shows another analysis by GPC of a
competitor's hot-melt sample. In this case the initial separation yielded only two bands, and it was subsequently shown by infrared that the second band consists of unresolved wax plus resin. The refractive
index of the resin was greater than that of the solvent (toluene), while that of the wax was less. The
(net) wax band is sketched in as the dashed curve in
Figure 10.14. By repeating the separation with a new
solvent, a blend of chloro- and dichlorobenzene whose
refractive index was matched to that of the resin, it

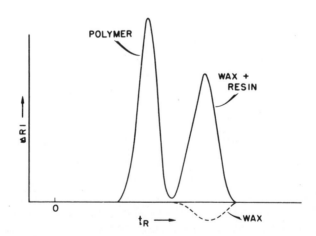

Figure 10.14 Separation and analysis of a hot melt
adhesive. Conditions: same as in Figure 10.13.

was subsequently possible to obtain a chromatogram in
which only the wax band appeared. This, in turn, permitted the quantitative analysis of the sample for its
wax concentration, and also allowed the molecular-
weight distribution of the wax to be determined.

A similar separation by gel filtration of an aqueous dye solution is shown in Figure 10.15. Here the

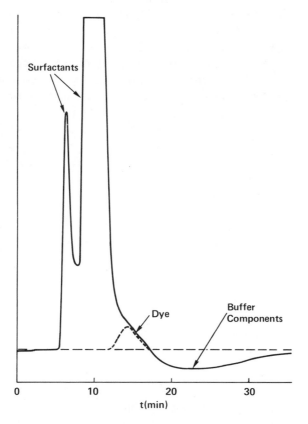

Figure 10.15 Separation of an aqueous dye solution
which contains higher-molecular-weight surfactants.
Conditions: same as in Figure 10.12, except for sample size (750 μl) and scale (1.28 x 10^{-3} RI units
full scale).

sample was suspected to contain one or more ionic surfactants, and it was desired to separate and identify
these compounds. Three bands including the dye were
recovered and submitted for infrared characterization.

Figure 10.16 shows the analysis of a polymer sample for low concentrations of phenolic surfactants.
The solid curve shows the chromatogram which resulted
from the use of a refractometer detector. In this
case the surfactant band at 42 min can be barely seen;
that is, a shoulder on the side of the polymer band.
It would be of little use to increase separation efficiency (i.e., N) in this analysis in order to improve
resolution. As is often the case in the separation of
polymeric samples, the overlap of these two bands is
mainly a result of polymer species (oligomers) whose
molecular size and V_R values are similar to those of
the compound of interest (in this case the surfactant).
That is, some of these polymer molecules elute at the

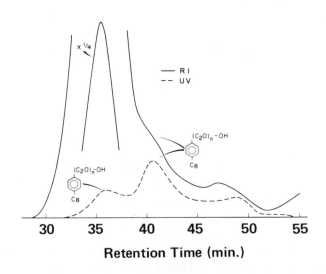

Figure 10.16 Use of dual detectors to follow separation of ethylene-vinyl acetate copolymer and surfactants. Conditions: same as in Figure 10.10.

same position in the chromatogram as surfactant molecules. However, by using a UV detector operating at
the wavelength maximum of the surfactant (280 nm), the
polymer response is eliminated, leaving only the chromatogram of the surfactant (dashed curve). Now the
concentration of the surfactant can be determined, as
well as its molecular weight (from its V_R value). Interestingly, <u>two</u> surfactant bands (at 36 and 42 ml)
are seen in the final UV chromatogram of Figure 10.16.
It was established that the band at 36 ml actually
corresponds to surfactant which has become chemically
bound to the polymer matrix. Thus, the final GPC analysis was also able to measure the fraction of free and
of polymer-bound surfactant.

A final example of separation by gel chromatography is shown in Figure 10.17, as an example of what
we can hope for in future. Here we see the very rapid

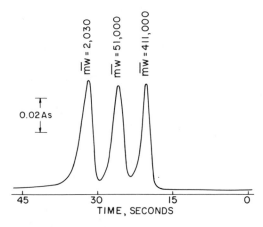

Figure 10.17 High-speed separation of polystyrene fractions. Column, 250 x 2.1 mm i.d. porous silica microspheres (~ 350 Å pores), 5-6 μ; carrier, tetrahydrofuran; temperature, 60°C; pressure, 1625 psi; flow rate,
1.0 ml/min; UV detector. J. J. Kirkland, J. Chromatog.
Sci., <u>10</u>, 593 (1972). Reprinted by permission of editor.

separation of three mono-disperse polymer fractions on
a porous silica packing of the Zorbax® type. Almost
all previous separations by gel chromatography have
been carried out on columns of much lower efficiency,
and the use of these newer packings will open the door
to much faster and/or better gel separations. The separation of proteins, nucleic acids, and other species
of biomedical interest with these Zorbax® packings
promises to be especially valuable, assuming that the
problem of sample adsorption can be solved.

The determination of polymer molecular-weight
distributions by GPC has been dealt with elsewhere
(see, e.g., ref. 2), and only a very simplified account will be given here. Beginning with a GPC chromatogram for a given polymer sample, as in the example
of Figure 1.4, the chromatogram is divided into small
segments; for example, 80-85 ml, 85-90 ml, ... 115-120
ml, 120-125 ml in the example of Figure 1.4. The area
of each of these segments of the chromatogram is determined, and converted into an equivalent weight percent
value. Similarly, the average molecular weight for
each segment is inferred from its average V_R value (and
a calibration plot). Finally, cumulative weight percent is plotted for the sample versus molecular weight.

In general, three problems are associated with
the determination of a molecular-weight distribution:

1. Determination of concentration versus detector response, as a function of V_R for the sample of
interest.
2. Determination of molecular weight versus V_R
for the sample of interest.
3. Correction of the initial chromatogram for
broadening as a result of the width of individual
oligomer bands.

In approximate calculations of molecular-weight distribution, we can assume that detector response per
unit concentration does not vary with V_R; that is, we

simply normalize the areas of the individual segments
to 100%. Similarly, we can assume a Q-factor of 20
(unless we know otherwise), thus yielding a relationship between V_R and molecular weight for the sample of
interest. Finally, we can ignore the effect of band
spreading on the chromatogram. With this approach,
approximate determinations of molecular-weight distribution are both simple and rapid.

Usually a somewhat more accurate approach to such
calculations is used. Typically, a representative sample of interest is fractionated by preparative GPC into
several narrow fractions, and each of the fractions is
rerun by analytical GPC. This calibration step estab-
lishes a correlation between relative detector response
(per unit sample concentration) and V_R. Then the average molecular weight of each of these fractions is independently determined by another method (e.g., vapor
phase osmometry, light scattering, ultracentrifugation,
etc.). The V_R values of the various fractions are then
plotted versus molecular weight, to give the necessary
calibration relationship. This is illustrated in Figure 10.18 for an asphalt sample (18), where it was desired to set up a method for determining the molecular-
weight distributions of different asphalts. The plot
of Figure 10.18 not only shows the desired relationship, but also plots these asphalt data as molecular
length (Å) versus V_R, assuming a Q-factor of 20. In
this case the resulting data fall on the same curve
(through the dark circles for polystyrene and polyglycol standards) that had been previously established
for this set of columns, thus confirming the validity
of the Q-factor approach in this case.

For some samples it was found (by comparing the
calculated average molecular weight with that found
for the whole sample by vapor phase osmometry [VPO])
that the Q-factor varied slightly from a value of 20.
Then the VPO measurement allowed the use of different
Q-factor values for each sample, yielding a more

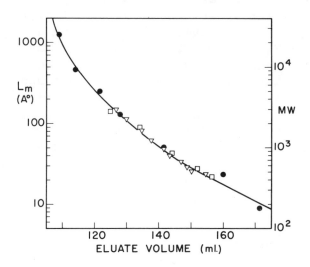

Figure 10.18   Determination of molecular-weight distribution of asphalts (16).   Calibration plot in molecular-weight units.   Conditions: same columns as Figure 10.10; 5 vol % methanol-chloroform solvent, room
temperature.   Reprinted from Analytical Chemistry (16).
Copyright 1969 by the American Chemical Society.   Reprinted by permission of the copyright owner.

accurate final molecular-weight distribution.

Correction of molecular-weight-distribution data
for band broadening is somewhat more complicated (see
review in ref. 3).   For samples with a Gaussian distribution, the baseline band width of the chromatogram
(V in ml) is related to the average band width of an
individual compound V_c as $V^2 = V_c^2 + V^{o^2}$; V^o is the
true (corrected) band width of the chromatogram.   This
allows a correction of the chromatogram for band broadening.   For asymmetric elution bands from a given
polymer sample, the band can sometimes by divided into
two halves, each of which is approximately Gaussian,
and each of which can be corrected in the same way as

above. The approximate constancy of V_c for different compounds in a given gel chromatogram simplifies the determination of V_c as a function of V_R.

Gel filtration has found limited application for the separation of inorganic species. A good review is provided by Anderson et al. (19).

REFERENCES

1.   K. J. Bombaugh, in Modern Practice of Liquid Chromatography, J. J. Kirkland, ed., Wiley-Interscience, 1971, Chapter 7.
2.   J. Cazes, Gel Permeation Chromatography, American Chemical Society Audio Course, Washington, D. C., 1971.
3.   K. H. Altgelt, Theory and Mechanism of Gel Permeation Chromatography, Advan. Chromatog., 7, 3 (1968).
4.   H. Determann, Gel Chromatography: Gel Filtration, Gel Permeation, Molecular Sieves, Springer Verlag, New York, 1968.
5.   L. Fischer, An Introduction to Gel Chromatography, North-Holland, London, 1969.
6.   G. Zweig and J. Sherma, Anal. Chem., 44, No. 5, pp. 47-49 (1972). (A review of the literature for 1970-72).
7.   Pharmacia, Sephadex®. Gel Filtration in Theory and Practice.
8.   Bio-Rad Laboratories, Gel Chromatography, 1971.
9.   D. H. Sachs and E. Painter, Science, 175, 781 (1972).
10.   W. A. Dark and R. J. Limpert, J. Chromatog. Sci., 11, 114 (1973).
11.   J. J. Kirkland, J. Chromatog. Sci., 10, 593 (1972).
12.   K. H. Altgelt, Separation Sci., 5, 777 (1970).
13.   J. G. Hendrickson and J. C. Moore, J. Polym. Sci., A1 (4), 167 (1966).

14. P. Andrews, Biochem. J., 96, 595 (1965).
15. M. E. Van Kreveld and N. van der Hoed, J. Chromatog., 83, 111 (1973).
16. Electro-nucleonics, Inc., Applications Sheets: Controlled-Pore Glass. Polyether Treatment & Silanizing Procedure.
17. C. A. Streuli, J. Chromatog., 56, 225 (1971).
18. L. R. Snyder, Anal. Chem., 41, 1223 (1969).
19. D. M. W. Anderson, I. C. M. Dean, and S. Hendrie, Talanta, 18, 365 (1971).

CHAPTER ELEVEN

SELECTING AND DEVELOPING
ONE OF THE FOUR LC METHODS

11.1 INTRODUCTION

Previous chapters have provided detailed discussions of
each of the four basic LC methods: partition, exclusion, adsorption, and ion exchange. This chapter will
discuss each of these various methods from a comparative standpoint. Some guidelines will be provided for
the selection of a particular LC method (or methods)
for a given separation problem. This choice is determined by a variety of factors, such as:

- The nature of the sample.
- The type of separation selectivity required.
- Experimental convenience.
- Experience with the method.

The following discussion should not be interpreted as
favoring one LC method over another, except in the case
that a given sample needs a particular LC approach.
In fact, the unique advantages of LC are the diversity
of applications for the different methods and the special techniques which can arise from various

modifications or combinations of the various LC methods. To attempt to solve every problem by only one
or two methods would seriously reduce the potential of
LC. Thus, for optimum application we need to have the
capability for all of the various LC methods.

Nature of the Sample

The general nature of the sample is a principal guide
to the initial selection of a given LC method. The
answer to questions like, "Is the sample low or high
molecular weight?", "Is it water soluble or not?",
"Ionic or nonionic?", "Polar or nonpolar?", "Is the
sample labile?", can result in our favoring one particular LC method and ruling out another. We can make
several generalizations about the various LC methods
with regard to areas in which they have the greatest
potential. Adsorption chromatography generally is most
suitable for oil-soluble samples or relatively nonpolar mixtures, while ion-exchange and liquid-partition
chromatography are usually best for water-soluble samples. Exclusion chromatography can be used for both
oil-soluble and water-soluble compounds, depending on
the type of packing and solvent used. However, there
are some important exceptions to these rough guidelines
regarding hydrophilic versus hydrophobic samples. For
instance, use of certain ion exchangers with nonaqueous
solvents permits the extension of this LC method to
oil-soluble samples (e.g., very weak acids or bases).
Similarly, adsorption chromatography has been successfully applied to the separation of very polar water-
soluble compounds using special techniques. However,
the original generalizations still stand as useful advice in the initial choice of an LC method. The preferred LC systems are most likely to offer immediate
success for typical separations problems.
     If the nature of the sample is such that it in-
volves ionic or ionizable compounds, or compounds

capable of interacting with ionic species (e.g., ligands and organic chelating agents), then ion-exchange
chromatography generally is the first choice. However,
gel filtration and liquid-partition chromatography are
both easily adaptable to ionic compounds. Adsorption
chromatography is generally less useful (although not
to be ruled out) since ionic compounds often give severe tailing (Section 2.7). Separation of ionic compounds by gel permeation chromatography can lead to
band tailing and variation of t_R with sample size,
since ionic species tend to form micelles in nonpolar
organic solvents.

Very low-molecular-weight compounds are usually
best separated by gas chromatography, rather than by
LC. The exception is in the separation of unstable
compounds which can be altered by the higher temperatures required by the gas chromatographic method. Very
high-molecular-weight compounds are especially suited
to exclusion chromatography. Certain high-molecular-
weight compounds of biological interest (e.g., proteins) have also been separated by ion exchange, but
this has involved special column packings and quite
slow separations. Compounds of intermediate molecular
weight (e.g., 300-2000) can be successfully handled by
all four LC methods.

The above considerations in choosing a satisfactory LC method are illustrated in Figure 11.1. Some
typical column packings needed for a particular separation are suggested, but other possibilities also
exist. The initial question we should ask ourselves
about a sample is, "What is the expected molecular
weight and polarity range?" If we know that the sample involves compounds with molecular weights of
greater than about 2000, we will usually choose a form
of exclusion chromatography for the separation. An
aqueous solvent (gel filtration chromatography) would
be appropriate for water-soluble samples (e.g., proteins). If the high-molecular-weight sample is soluble in organic solvents (e.g., polyvinyl chloride),

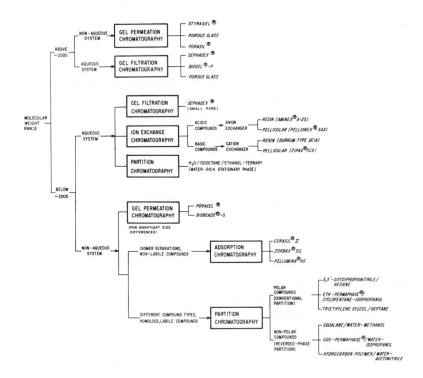

Figure 11.1   Selection of an LC method.

a nonaqueous system (gel permeation chromatography) would probably be used.

If the sample involves compounds having molecular weights of less than 2000, a greater choice of possible separation modes exists. For samples soluble in aqueous systems, we sometimes have the choice of three methods. Gel filtration might be the best approach for neutral compounds that have size differences (e.g., glucose oligosaccharides). If the compounds are water-soluble and ionic or ionizable, then ion-exchange chromatography is very likely the best method. Basic compounds (e.g., aromatic amines) should be separated on

cation exchangers, while acidic compounds (e.g., carboxylic acids) are separated with anion exchangers.
Water-soluble compounds, particularly nonionic and
weakly ionic materials with different type and numbers
of functional groups (e.g., substituted phenols), and
homologous series (e.g., polyethylene glycol) often
can be separated by partition chromatography. This
situation requires aqueous or partially aqueous stationary phases and immiscible, moderately polar mobile
phases.

If the sample compounds have molecular weights of
< 2000 and are soluble in nonaqueous systems, several
selection possibilities also exist. For mixtures consisting of compounds with significant size differences
(>10%; e.g., oligostyrenes), gel chromatography may
be favored, and small-pore chromatographic packings
would be used. If no such size differences exist with
the components of the sample, other approaches are necessary. For the separation of isomers and different
compound types, adsorption chromatography is often preferred (see Table 8.1). Various types of adsorbents
should be considered for separation selectivity.

Liquid-partition chromatography should be considered as the best separation method for organic-soluble
samples consisting of compounds of < 2000 molecular
weight and containing different compound types, homologs, or labile compounds (see Table 7.1). For the
separation of polar compounds, conventional LLC (polar
stationary phase and a relatively nonpolar mobile
phase) generally is employed, while the LLC separation
of nonpolar compounds may be accomplished with the use
of reversed-phase systems (nonpolar stationary phases
plus polar mobile phases).

Separation Selectivity

For very difficult separations involving highly similar compounds, or in very complex separations involving

mixtures with a large number of components of widely
different chemical types, use of very high-efficiency
systems may fail to provide the desired separation,
even in combination with gradient elution. In these
instances, precise control of separation selectivity
may offer the only chance for successful separation.
In addition, combinations of LC methods designed to
take advantage of the general selectivity of each method are common in the separation of very complex mixtures (Section 11.3). In any case, in the separation
of very similar substances or the separation of very
complex mixtures, a successful procedure requires some
understanding of the selectivity provided by different
LC methods. This understanding of the selectivity is
in turn based on knowledge of the mechanism of the retention process for each method. A discussion of each
of these two points, selectivity and mechanism, has
been presented in previous chapters for each of the
four LC methods. Some general indications of selectivity also have been given in connection with Figure
11.1. For additional information, see the bibliography
at the end of Chapters 7 through 10.

Experimental Convenience

Since there is a certain amount of skill and effort
needed to prepare and maintain a column and operate
equipment, experimental convenience is always a factor
which influences our selection of a particular method.
In this there are several considerations:

- Ease of column packing.
- Column efficiency.
- Column stability.
- Ease in determining experimental conditions.

The rigid solids used as supports in partition and

adsorption chromatography generally can be packed into columns with less skill (Section 6.4) than is required for the other packing materials. An exception is in the packing of columns with very small (< 20 μ) particles. Special equipment and skills are required for packing columns of these materials by the slurry technique; therefore, many users will probably want to purchase columns of these types.

Column efficiency can be a major factor in the selection of a method; however, columns for all of the LC methods can now be prepared with about the same efficiency, so this generally is not a major consideration. The stability of a column also may be a factor in the selection of a method. As previously mentioned, column stability is dependent on how well the columns for the various techniques are used. Nevertheless, in normal use columns for all of the four LC methods should last for at least 3-6 months, which means that many hundreds of separations are possible with a single column before it must be replaced.

The ease in determining satisfactory or optimum experimental conditions for a separation is important in our choosing of an LC method. Exclusion methods are most attractive, since the control of band migration rates is easily predicted in advance of the actual separations. In exclusion chromatography all of the sample components are eluted in a short time; this is not always the case with the rest of the LC methods. With the other LC methods, gradient elution or some equivalent technique may have to be used in separating mixtures containing compounds with a wide range of capacity factors. Selection of experimental conditions is probably most difficult for liquid-liquid chromatography. In some cases, paper chromatography can be used to determine an appropriate partitioning system for particular separations. Thin-layer chromatographic systems can be effectively used to pilot the conditions for separations by liquid-solid chromatography

(Section 8.1).

Experience with the Method

Our experience with a particular LC method may determine the LC approach which we use to solve a problem,
for it is tempting for us to use methods that we know
best. Actually, many problems can be solved this way,
since separations often can be accomplished with more
than one method. However, it should be stressed that
for a complete capability in LC, we must be able to
separate by any one (or combination) of the four LC
methods. Table 11.1 gives a list of suggested minimum
columns for complete LC capability in a laboratory.
These particular columns are not unique, but they do
represent a collection which will permit the separation of a large variety of mixtures.

TABLE 11.1.   SUGGESTED COLUMNS FOR COMPLETE LC CAPA-
BILITY[a]

Exclusion	Ion Exchange	LSC	LLC
Styragel® Series	Strong cation	Corasil®-II	ETH-Per-maphase®
Sephadex® Series	Strong anion	Zorbax®-SIL	ODS-Per-maphase®

[a]Satisfactory alternative column packings are given in
lists presented in the chapters for each of the LC
methods.

Other Considerations

There are several miscellaneous factors which are sometimes involved in the selection of an LC method. For

instance, in gel chromatography all compounds in the mixture elute, but only a limited number of compounds can be resolved in a single chromatogram. Thus, this approach is often inadequate for completely resolving complex mixtures. Gel chromatography is very useful for the initial examination of a completely unknown sample; in a short time one can learn a great deal about the nature of an unknown system. Separations are rapid, and the apparent complexity of the sample and the molecular-weight distribution of the components are revealed in a single gel chromatogram.

The possible reaction of the sample during LC separation is usually not a problem. Of the four LC methods, adsorption chromatography is most likely to lead to alteration or irreversible loss of the sample; this is characteristic of traditional adsorption column chromatography and in TLC, but less likely to be a problem in modern adsorption chromatography. The fast separation allows little opportunity for sample alteration to take place.

11.2   DEVELOPING A PARTICULAR SEPARATION

Each LC separation problem requires a different experimental approach. It is impossible, therefore, to devise a set of rigid guidelines for the development of experimental conditions in a particular separation. However, there are some general steps that should be followed in each method development.

Table 11.2 presents a brief systematic approach to an LC separations problem. After proper analytical sampling, the decisions involve choosing an LC method, selecting a specific column, establishing the proper operating conditions, and coping with any special consideration, such as the general elution problem.

At this point, it may be helpful to illustrate the steps involved in the development of an LC method

TABLE 11.2.   SYSTEMATIC APPROACH TO LC SEPARATIONS

Sampling

Select method	High-molecular-weight samples --exclusion, IEC (if ionic)
	Ionic, ionizable samples--IEC
	Separation by molecular weight --exclusion, LPC
	Isomer separation--LSC
	Separation of compounds with different functional groups --LSC, LLC
Select column	Control k' in exclusion
	Small H, large K°--select packing type
Select conditions	Control k'--solvent strength, etc.
	Control N--L and P, dp, particle type, flow rate
	Control α--solvent and stationary phase combination
	Low-viscosity mobile phase
Special considerations	General elution problem (gradient elution, other)
	Temperature

by recounting the history of an actual separations
problem (1). Although every parameter in the

systematic approach given in Table 11.2 was not experimentally determined, these variables generally were
taken into consideration. As is the case in many practical situations, not every decision was necessarily
the best one that could have been used. Nevertheless,
the desired separation goal was reached by following
the general guidelines of the systematic approach.

The object of the study was to develop a rapid,
precise quantitative procedure for mixtures of two moderately polar, closely analogous compounds. These two
proprietary heterocyclic compounds of approximately
250 molecular weight differed only by the presence of
a chlorine instead of a bromine group, the chlorine-
containing compound (compound A) being present at levels about 1/100 that of the bromine-substituted component (compound B). Attempts to analyze this system by
gas chromatography resulted in imprecise data. This
was due to the partial decomposition of the bromine-
containing compound at the high temperatures required
to gas chromatograph these materials.

Liquid-partition and adsorption chromatography
were initially considered for the separation of these
organic-soluble compounds. Gel chromatography was re-
jected since there was very little size difference.
In addition, high resolution (R_s = 1.25) was required
because of the large differences expected in the relative concentration of compounds A and B. Ion-exchange
chromatography was not considered to be a good possibility because of the poor solubility of these compounds in aqueous systems.

Initial attempts were made to separate these compounds by liquid-partition chromatography using the
commercially available bonded-phase packing, ETH-Per-
maphase®. This packing was chosen for three reasons.
First, a column of this type was already mounted in the
LC apparatus. Second, bonded-phase packings are preferred in this laboratory because of their convenience
(i.e., no precautions necessary to avoid loss of

stationary phase, etc.). Lastly, prepacked columns of
this material can be purchased, allowing convenient
transfer of analytical methods to other locations (e.
g., plants, formulators, governmental regulatory agen-
cies). Note that none of these considerations involved
the chemistry of the separation problem or any theoretical insights. However, such considerations are often
paramount as starting points for the solution of a separation problem.

Initially, selection of the carrier liquid for
the bonded-phase system was by trial-and-error. Although it was determined that a mobile phase of 1.5%
tetrahydrofuran/hexane gave optimum k' values, compounds A and B were unresolved (R_s < 0.4) because the
separation factor α was essentially unity. Various
modifying polar solvents were tested in conjunction
with hexane mobile phase to increase the separation
factor by employing different sample-solvent interactions. Mobile phases of 5% chloroform/hexane, and 1%
methanol/hexane both failed to improve resolution, although the components eluted in the optimum k' range.
Chloroform and methanol modifiers were selected for
these tests because their solubility parameters (Section 7.3) were different from the tetrahydrofuran modifier used in the initial attempt.

When these approaches failed to achieve desired
resolution, several different types of experimental
bonded-phase packings (5) containing various functional groups (e.g., nitrile, amino, ester) were studied
to find a stationary phase that would promote the desired selectivity. Again, all these systems were un-
successful in increasing the resolution of the subject
compounds.

It became apparent that the two compounds were
too much alike chemically to permit their separation
by liquid-partition chromatography. Therefore, it was
decided that adsorption chromatography might be better
suited for the separation of these two compounds,

which exhibit only a slight difference in functionality.

A summary of adsorbent systems subsequently investigated is given in Table 11.3. Corasil®-II was selected for initial study, since it was thought that, in this particular case, the relatively polar subject compounds would require the relatively low retentivity exhibited by a porous-layer bead adsorbent. Thin-layer chromatographic scouting tests (although not directly translatable to porous-layer bead adsorbents) suggested that a methanol/chloroform mobile phase might produce the desired selectivity. Using a mobile phase of 1% methanol in chloroform, a partial separation was obtained on the Corasil®-II column (attempt 1). Reducing the carrier velocity (attempt 2) resulted in a resolution increase, but much less than that required. This increase in resolution was a result of the improved efficiency of the column, as discussed in Chapter 3. Subsequent efforts to improve resolution were based on optimizing k', by altering the concentration of the polar modifier in the base mobile phase (attempts 3 and 5), and increasing α by changing the type of polar modifier (attempts 7 and 8). Attempts were also made to increase column efficiency N by decreasing carrier velocity (attempts 4 and 6). However, a maximum resolution of 1.0 was obtained as a result of all of these studies, and this separation was not sufficient for the desired application.

The totally porous silica adsorbent, Porasil®-A, was investigated using the best solvent pair found with the porous-layer bead adsorbent (attempt 9). However, there was a large decrease in column efficiency with this system, and resolution of the subject compounds failed. (In hindsight, this experiment was poorly conceived, since improved selectivity would not be expected for such a system; poor column efficiency and resolution could have been anticipated.)

While the methanol/chloroform mobile phase system

TABLE 11.3. SYSTEMS INVESTIGATED FOR SEPARATING COMPOUNDS A AND B BY ADSORPTION LC

Attempt Number	Adsorbent Column	Mobile Phase	k'	Flow Rate (ml/min)	Parameter Changed	R_s
1	Corasil®-II[a]	1% methanol in $CHCl_3$	1.8	1.5	---	0.5
2	Corasil®-II	1% methanol in $CHCl_3$	1.8	0.75	N	0.7
3	Corasil®-II	0.5% methanol in $CHCl_3$	3.6	0.75	k'	0.8
4	Corasil®-II	0.5% methanol in $CHCl_3$	3.6	0.30	N	1.0
5	Corasil®-II	0.2% methanol in $CHCl_3$	11	1.3	k'	0.8
6	Corasil®-II	0.2% methanol in $CHCl_3$	11	0.67	N	1.0
7	Corasil®-II	Acetonitrile	0.4	0.75	α	<0.4
8	Corasil®-II	1% dioxane in $CHCl_3$	4.7	0.75	α	0.8
9	Porasil®-Ab[b]	1% methanol in $CHCl_3$	4.3	0.3	k', N	<0.4
10	Zorbax®-SIL[c]	0.5% methanol in $CHCl_3$	12.2	1.2	N	1.25

[a] 37–50 μ; 1 m × 2.1 mm, i.d.; at k' = 3.6, and velocity 1.1 cm/sec, H = 1.3 mm.

[b] 37–75 μ; 0.5 m × 2.1 mm, i.d.; at k' = 4.3 and velocity 1.9 cm/sec, H = 4.4 mm.

[c] 8–9 μ; 0.25 m × 3.2 mm, i.d.; at k' = 12.2 and velocity 0.5 cm/sec, H = 0.12 mm.

provided some separation, none of the adsorbent columns tested had a sufficient number of theoretical
plates to permit the desired resolution. To obtain
larger N, a high-efficiency adsorbent column of Zorbax®
-SIL was used with the methanol/chloroform system
(attempt 10). This condition produced the desired R_s
= 1.25 for the subject compounds, and permitted the
development of a quantitative method.

11.3   SELECTING A COMBINATION OF DIFFERENT METHODS

Complex mixtures may require separation by more than
one LC method to obtain the desired information. By
combining different methods of LC in sequence, complex mixtures containing both similar and dissimilar
materials can be resolved. Since sample handling and
fraction collection are relatively simple in modern
LC, we can readily move from one separating system to
another using sequential LC operations. Figure 11.2
illustrates the technique of sequential analysis in

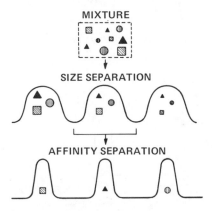

Figure 11.2   Sequential LC analysis.   J. N. Little,
Amer. Lab., 3, 59 (1971).   Reprinted by permission of
editor.

which the sample is first separated by gel chromatography (size). The various fractions are then collected for further separation by partition, adsorption, or ion-exchange chromatography. The techniques can be devised so that sufficient samples can be isolated in the final separation for positive identification by supplementary techniques, such as IR, NMR, and mass spectrometry.

Three different liquid chromatographic separations were performed sequentially to obtain complete information about a polyvinyl chloride sample containing several additives (2). Figure 11.3 shows a gel permeation chromatographic separation of the polymer using several series-connected Styragel® columns, selected for the very wide molecular weight range of 200 to 20×10^6. In the region of 17-24 counts (a measure of retention volume), the refractive index detector monitored the elution of the polyvinyl chloride polymer.

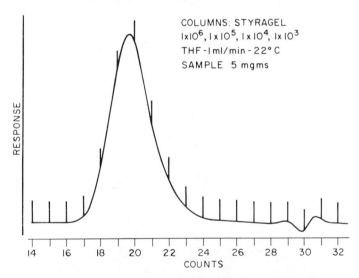

Figure 11.3 GPC separation of polyvinyl chloride sample. J. N. Little, Amer. Lab., 3, 59 (1971). Reprinted by permission of editor.

From the position and profile of this polyvinyl chloride peak, important information about the sample was
obtained: molecular-weight distribution of the polymer
and its average molecular-weight. The chromatogram
also shows a low-molecular-weight additive which eluted at about 29-31 counts (as a reverse peak with the
refractive index detector). The peak for this unknown
low-molecular-weight material was collected and subjected to the gel chromatographic separation shown in
Figure 11.4. For this second separation, the columns
were optimized for a molecular-weight range of 100-2000,
and Poragel® packings with small pores were employed.
The materials in the fraction collected from the first
separation produced this chromatogram, the first peak
being due to hydrocarbon, the second, phthalate, and
the last, benzoate, as determined by infrared spectrophotometry. These materials represent three different
types of plasticizers that were added to the polymer
to obtain certain desirable characteristics.

It was suspected that the second peak (phthalate)
consisted of more than one component. Therefore, this
peak was collected during the second gel chromatographic

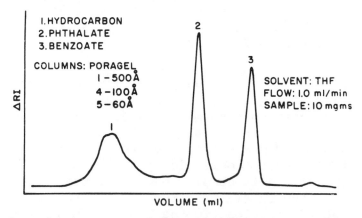

Figure 11.4   GPC separation of plasticizers from polyvinyl chloride.   J. N. Little, Amer. Lab., $\underline{3}$, 59
(1971).   Reprinted by permission of editor.

separation and rechromatographed by high-resolution adsorption chromatography, as shown in Figure 11.5. This separation indicated that peak No. 2 was actually composed of three phthalates: diisodecyl, diisooctyl, and dibutyl.

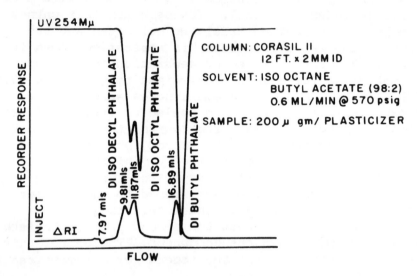

Figure 11.5 Phthalate plasticizers from polyvinyl chloride. J. N. Little, Amer. Lab., **3**, 59 (1971). Reprinted by permission of editor.

Exclusion chromatography also may be combined sequentially with gradient elution separations to characterize very complex mixtures (3). Figure 11.6 shows the chromatogram which is obtained on a total auto exhaust condensate sample by reversed-phase liquid-partition chromatographic system using a linear gradient. The highly complex nature of the sample is indicated by the raised baseline and severe overlapping of peaks at a retention time of about 30 min, resulting from unresolved components in the sample. To simplify this complex mixture, the sample was fractionated by

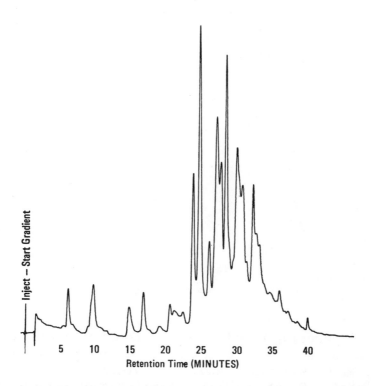

Figure 11.6   Gradient elution separation of total auto
exhaust condensates.   Conditions: linear gradient,
2%/min–20% methanol/80% water to 100% methanol; ODS-
Permaphase®, 60°C.   J. A. Schmit, R. A. Henry, R. C.
Williams, and J. F. Dieckman, J. Chromatog. Sci., 9,
645 (1971).   Reprinted by permission of editor.

gel chromatography with a series of Bio-bead®-S columns
selected for the resolution of low-molecular-weight
components (< 1000 molecular weight).   As illustrated
in Figure 11.7, this sample was divided into 2-ml fractions containing components of similar molecular size.
These fractions were then subjected to high-resolution
reversed-phase partition gradient chromatography for

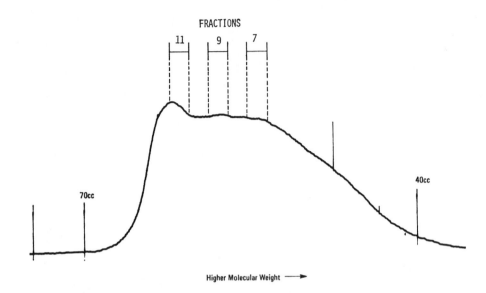

Figure 11.7 GPC separation of auto exhaust condensates. Operating conditions: Chloroform solvent,
SX-2, SX-8, Bio-bead® columns. J. A. Schmit, R. A.
Henry, R. C. Williams, and J. F. Dieckman, J. Chromatog. Sci., *9*, 645 (1971). Reprinted by permission
of editor.

characterization. Figure 11.8 shows the greatly simplified chromatogram which was obtained on GPC fraction No. 11. Figure 11.9 is the gradient elution
chromatogram of GPC fraction No. 7. The presence of
peaks throughout the chromatogram for this fraction
of closely sized material suggests that the components
contain different functional groups, since all components were approximately the same molecular size.
      These two examples show the versatility of LC in
analyzing very complex chemical mixtures. By combining
LC methods, optimum information can be obtained on
samples containing compounds differing greatly in

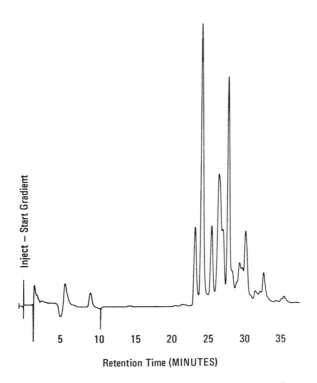

Figure 11.8 GPC fraction No. 11 from auto exhaust condensates. Operating conditions: linear gradient, 2%/
min–20% methanol/80% water to 100% methanol; ODS-Per-
maphase®, 60°C.   J. A. Schmit, R. A. Henry, R. C.
Williams, and J. F. Dieckman, J. Chromatog. Sci., 9,
645 (1971).   Reprinted by permission of editor.

functionality or covering a wide molecular weight
range.

11.4   THE USE OF DIFFERENT TECHNIQUES

Different LC techniques can be used in the solution of
a given problem.   Some insight on this point can be

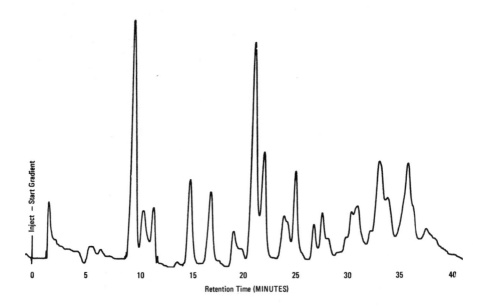

Figure 11.9   GPC fraction No. 7 from auto exhaust condensates. Operating conditions: linear gradient, 2%
/min-20% methanol/80% water to 100% methanol; ODS-
Permaphase®, 60°C.   J. A. Schmit, R. A. Henry, R. C.
Williams, and J. F. Dieckman, J. Chromatog. Sci., <u>9</u>,
645 (1971).   Reprinted by permission of editor.

obtained by reviewing the solution to a practical problem reported earlier (4).

    Dilute solutions of quinoline in dodecane had
been hydrogenated catalytically, and it was required
to determine what products were produced. Major interest centered on the aromatic, nitrogen-containing
products in the samples. The total concentration of
nitrogen in the hydrogenated products ranged from 1000
ppm down to less than 1 ppm. Figure 11.10 shows the
principal products of interest, although they were not
known at the start of the project. The first step was

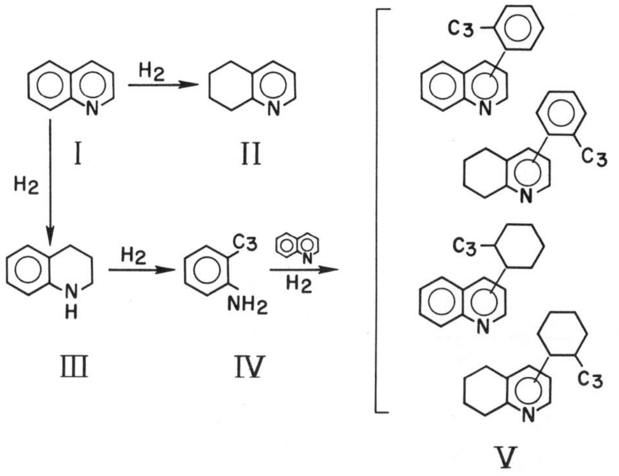

Figure 11.10   Products from the hydrogenation of quinoline in dilute solution.

the qualitative analysis of a typical product sample
to determine what compounds were present. Gradient
elution is well suited to initial exploratory separation for determining the compositional range of a sample. Figure 11.11 shows the chromatogram obtained
from the gradient elution of a typical product sample
by adsorption chromatography. The major components
(numbered) occur in the middle of the chromatogram,
with minor concentrations of hydrocarbons (i.e., nitro-
gen-free compounds) at the beginning, and minor amounts
of more polar compounds at the end. The major bands
shown in Figure 11.11 were readily isolated by scaling-
up this gradient elution separation, in much the same
manner described in Chapter 12. The identity of each
isolated peak was established by spectrophotometric
analysis, as indicated in Figure 11.11. The later
eluting (minor) components were also characterized,
but were found to be highly rearranged compounds of

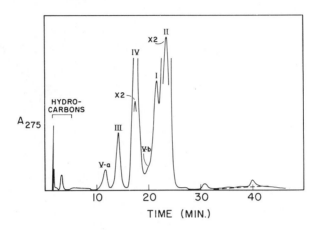

Figure 11.11   Gradient elution separation of hydrogenated quinoline sample on alumina. L. R. Snyder, J.
Chromatog. Sci., 7, 595 (1969). Reprinted by permission of editor.

little interest.   Band V-b in Figure 11.11 occurs as
an unresolved shoulder in the chromatogram.   Since
there was interest in the identity of this material,
the resolution of the LSC system was increased by using
a single solvent of optimum strength and a lower solvent flow rate (0.9 versus 7.8 ml/min).   In gradient
elution the elution of a given peak is a function of
the strength of the mobile phase (in the column effluent) at the time of elution.   Optimized single-solvent
separation of the same band will generally require a
solvent of somewhat lesser strength, and this approach
was used in the present separation.   The resulting separation of band V-b is shown in Figure 11.12.   Three
distinct bands are now observed between bands IV and I.
     After the nature of the products formed from the
hydrogenation of quinoline had been established, it
was desirable to analyze several hundred samples for
the compounds of interest: I, II, III, IV, V-a, and

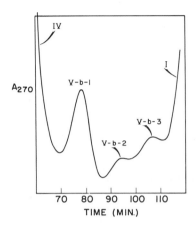

Figure 11.12   High-resolution separation of band V-b from hydrogenated quinoline. L. R. Snyder, J. Chromatog. Sci., _7_, 595 (1969). Reprinted by permission of editor.

V-b. A detailed breakdown of band V-b into its three components was considered unnecessary. As a result, it was possible to use a single-solvent system (since bands I-V elute within a narrow range) of lower resolution than that shown in Figure 11.12. This simplified the routine analysis of the samples and reduced the necessary time per sample. The resulting separation is shown in Figure 11.13. The good resolution of all bands of interest permitted convenient quantitation.

The various examples of this chapter show that the use of different LC techniques in combination with supplemental identification methods permits the solution of a wide variety of complex problems which cannot be approached by other methods.

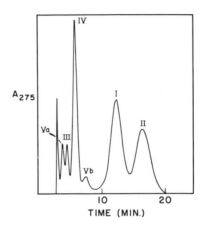

Figure 11.13 Medium-resolution separation of hydro-genated quinoline sample by normal elution from alumina. L. R. Snyder, J. Chromatog. Sci., $\underline{7}$, 595 (1969). Reprinted by permission of editor.

REFERENCES

1. J. J. DeStefano, unpublished study, 1972.
2. J. N. Little, Amer. Lab., $\underline{3}$, 59 (1971).
3. J. A. Schmit, R. A. Henry, R. C. Williams, and J. F. Dieckman, J. Chromatog. Sci., $\underline{9}$, 645 (1971).
4. L. R. Snyder, J. Chromatog. Sci., $\underline{7}$, 595 (1969).
5. J. J. Kirkland, U.S. Patent 3,722,181.

CHAPTER TWELVE

LARGE-SCALE SEPARATIONS

J. J. DE STEFANO

12.1 INTRODUCTION

Few systematic studies or applications have been published dealing with the large-scale (i.e., separations
with columns of larger internal diameter) capabilities
of modern LC, even though the major uses of classical
column chromatography are largely preparative in nature. This shortage of information concerning large-
scale LC has not been the result of a lack of interest
in using modern principles for preparative separations,
since the need for the isolation of larger amounts of
purified components for use as standards or for characterization, and so on, remains high. Rather, studies of large-scale liquid chromatography have been
delayed due to the major research efforts initially
being directed at improving the capabilities of the
technique for analysis.

Before discussing the principles and practices of
modern large-scale liquid chromatography, some definitions are in order. Large-scale columns are sometimes
evaluated in terms of <u>preparative efficiency</u>; that is,
the number of moles (or weight) of solute of a given
purity that can be isolated in a given period of time.
This definition is also sometimes referred to as

column throughput. Some examples of typical column
throughput are shown in Table 12.1, which assumes a
15-min separation of two compounds with a resolution
(R_S) of 1.25 and a collected purity of 99+%. Note the
differentiation made between a simple scale-up of an
analytical system, and a preparative separation. Some
workers regard the isolation of milligram amounts of
compounds using larger diameter columns as preparative
liquid chromatography. However, this type of isolation
might best be classified as a scale-up separation; most
column parameters, such as linear velocity, support
type and size, ratio of sample size to column cross-
sectional area, and so on, being the same as in analytical separations. On the other hand, preparative
isolations are performed with relatively larger quantities of sample in columns which are operated under
overloaded conditions. These conditions are needed to
achieve the goals of preparative chromatography - high
throughput with high purity. For this discussion the
term large-scale LC will include both preparative and
scale-up operations.

12.1. COLUMN THROUGHPUTS (R_S = 1.25, PURITY OF RECOV-
ERED FRACTIONS > 99%, 15 MIN SEPARATION)

	Superficially Porous Supports	Totally Porous Supports	Exclusion Gels
Analytical scale	0.1-0.2 mg	1-2 mg	10-20 mg
Semipreparative or scale-up	10-20 mg	100-200 mg	1-2 g
Preparative scale	--	0.5-1.0 g	> 10 g

Definition of preparative liquid chromatography
varies with the individual worker. The term "preparative LC" often has been used regardless of the amount
of sample that is isolated or of the conditions used
for the isolation. Table 12.2 summarizes the quantity
of sample that might be recovered by LC for particular
requirements. Each of these isolations has been called

TABLE 12.2.   HIGH-SPEED PREPARATIVE LIQUID CHROMATOG-
RAPHY SAMPLE QUANTITY REQUIREMENTS (1)

Objective	Sample Weight Required (mg)
Tentative identification by instrumental methods	<1
Positive identification by instrumental methods, including NMR. Confirmation of structure by chemical reactions	1–100
Positive identification and subsequent use in research or synthesis	>100

preparative LC since significant amounts of purified
material has been recovered in each case. However, the
first two cases can probably be achieved using scale-up
conditions as discussed above. The third situation
probably requires large diameter, overloaded columns
and thereby is properly classified as a preparative
separation. Thus, with the definitions discussed
above, it can be seen in Table 12.2 that analytical
chemists generally will be interested in scale-up separations where milligram quantities of purified materials are sufficient, whereas organic synthesis

chemists and biochemists will be more often inclined
toward preparative separations where greater amounts
of material can be isolated. Both will be interested
in large-scale separations involving larger internal
diameter columns.

    Because the goal of large-scale LC is to prepare
relatively large amounts of purified material, the
equipment and experimental conditions used can be somewhat different than those used for analysis. As shown
in Figure 12.1, the speed, resolution, and capacity of
the chromatographic system are related to each other
as are the three corners of an equilateral triangle.

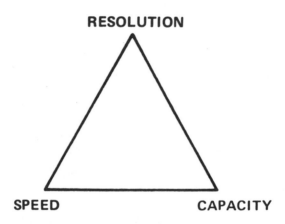

Figure 12.1   Schematic diagram of the goals in chromatography.

For a particular LC system, any one of the chromatographic characteristics can be improved at the expense
of the other two; or any two can be improved at the expense of the other one. In analytical LC, speed and
resolution are usually the desired characteristics;
hence, capacity is usually compromised (e.g., use of
superficially porous supports). In large-scale LC separations, capacity is the main objective, and speed

and resolution generally are sacrificed. Optimization
for capacity involves a different set of operating conditions than normally is employed for analytical LC,
and these differences should be considered before
attempting preparative separations.

12.2   STRATEGY FOR PREPARATIVE SEPARATIONS

The goal of both the scale-up and preparative modes of
large-scale liquid chromatography is the isolation of
pure compounds from mixtures. In the case of scale-up
separations, the maximum amount of material which can
be isolated is limited by the capacity of the column
for sample when operated in a non-overloaded condition.
On the other hand, for preparative separations of larger amounts of sample, the column is always operated in
an overloaded condition. Here, several strategies exist for maximizing column throughput (6).
     The situations most likely to be encountered when
performing preparative LC are demonstrated by the
three cases shown in Figure 12.2. The first case,
where the desired material is present as a single
major peak, can best be handled using the scheme outlined in Figure 12.3. First, an analytical separation
must be developed. Then the resolution should be increased (usually by modifying the composition of the
carrier) to provide room for the broadened peaks which
will occur under overload conditions. Next, the sample load should be increased to the <u>loading limit</u>
(that point at which the peaks threaten to overlap).
At this point, all of the desired component which was
injected can be collected as pure material. Further
overloading results in a yield loss for the desired
material, since a heart-cut (dotted lines) of the
product peak is required to obtain pure compound. On
the other hand, overloading the column with larger
sample will usually result in greater throughput since

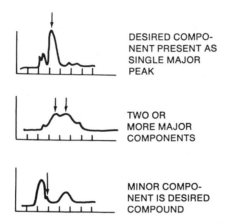

DESIRED COMPO-
NENT PRESENT AS
SINGLE MAJOR
PEAK

TWO OR
MORE MAJOR
COMPONENTS

MINOR COMPO-
NENT IS DESIRED
COMPOUND

Figure 12.2 Typical problems encountered in preparative liquid chromatography (6).   Reprinted with permission of International Scientific Communications.

the highest concentration of the solute occurs within
the heart-cut region.

In the second case, two major closely-eluting
components are to be isolated.   In this situation,
recycle techniques can be usefully employed (see Section 10.5).   As shown in Figure 12.4, by collecting
the front and back portions of the overloaded doub-
let peak, pure components can be immediately isolated.
The middle portion then can be recycled through the
column for additional accumulations of the two materials in the same manner.   Of course, resolution of
the two components also can be improved by increasing
α and/or N, as discussed in Chapter 3.

When the desired material is a minor component,
as in the third case, the component must first be en-
riched.   This enrichment is accomplished by overloading the column and collecting fractions in the area
of elution of the desired component.   These collected
fractions can then be pooled, concentrated, and

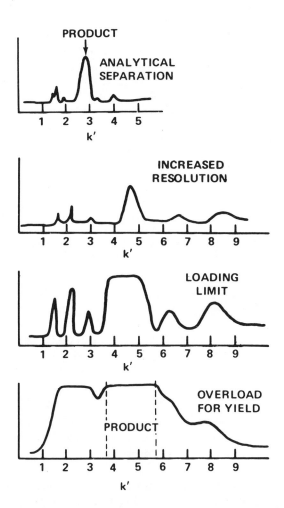

Figure 12.3   Scheme for maximizing throughput, single primary component case (6).   Reprinted with permission of International Scientific Communications.

reinjected.   The desired compound is now the major component of the injected sample and can be treated in the same manner as was discussed for the first case.

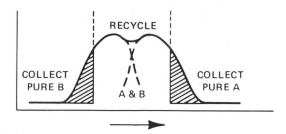

**Figure 12.4   Recycle approach for preparative isola-
tion of two closely-eluting, major components (6).
Reprinted with permission of International Scientific
Communications.**

12.3   EQUIPMENT

To perform modern large-scale liquid chromatography
effectively, some insights into the equipment requirements and the effects of various chromatographic parameters on column throughput are necessary. From the
equipment standpoint, the following components should
be optimized with specifications somewhat different
than those used in analytical LC: sample introduction
system, pumping system, detector, and collection
device.

Several studies have shown that sample introduction with a valve can be used instead of a syringe for
large-scale applications with no apparent loss in column performance. Most important, valves allow the
introduction of the much larger sample aliquots used
in large-scale LC without interrupting the flow of
mobile phase. Valves also have the added advantages
of being more convenient to use (e.g., stop-flow injections are not required, no rubber from septum can
enter the column, etc.) and of being automated, if

desired. On the other hand, the injection volume is
more conveniently varied when using a syringe (see
Section 4.6).

Pumps for large-scale LC systems should have a
high solvent delivery rate; between 10-50 ml/min is
usually satisfactory. Pumps capable of very high pressure generally are not required. A 1000-psi pressure
rating is usually adequate for most large-scale columns, since they are normally short and are packed
with relatively large-diameter particles that result
in relatively high permeability. Either reciprocating
or pneumatic amplifier-type pumps can be used. Damp-
ing the pulses from reciprocating pumping systems
usually is not critical, since no analytical inter-
pretation of the chromatogram is to be made. For the
same reason, the precision and accuracy of the carrier
delivery rate need not be as precise as for analytical
applications. Pneumatic amplifier pumps, because of
their higher pumping rates, can probably more easily
handle the large volumes of solvent required for large-
scale LC than can reciprocating pumps. However, the
relatively large volumes of compressed air required by
pneumatic pumps can be a minor disadvantage if a central supply is not available.

The detector for large-scale LC should be capable
of allowing large mobile phase flow rates with minimum
back pressure. In fact, the entire large-scale LC
system should be designed to provide minimum back pressure, because of the high solvent flow rates which must
be used. Low dead-volume tubing and fittings are not
as critical in large-scale systems because of the relatively large column volumes employed. However, good
flow characteristics must be maintained throughout the
system. While poorly swept areas can lead to tailing
peaks and poor column efficiency, this effect is less
critical in large-scale systems than in analytical
applications. Most commercially available flow-
through LC detectors have been primarily designed for

analytical applications; therefore, to provide for
minimum dead volume, these units are constructed of
narrow-bore tubing which may severely restrict the
flow of mobile phase. Some manufacturers (e.g., Waters
Associates) provide refractive index detectors with
larger bore tubing designed for the higher flow rates
used in large-scale applications. Another approach is
to use a stream splitter on the exit of the large-
diameter column. In this case, only a small flow of
the carrier liquid is put through the analytical detector which is supplied with narrow-bore tubing and the
flow of the carrier in the rest of the system is not
impeded.

High-sensitivity detectors generally are not required for large-scale applications. Solute concentrations are usually very high, making it difficult to
keep within the linear dynamic range of the detector.
The refractive index detector appears to be the most
suitable for use in large-scale applications. It is
more general than an ultraviolet detector and more
trouble-free than a flame ionization transport detector. Detectors for large-scale applications must be
nondestructive, since collection of the sample is the
prime goal. Currently available flame ionization detectors destroy only a small fraction of the total sample; hence are suitable for use with large-scale
columns.

A sample collection device must be employed to
collect the column effluent and should be capable of
handling large volumes of liquids. Collectors may be
manual or automatic; however, the latter are more convenient. The collection system should be equipped with
a switching valve which allows uncontaminated carrier
liquid to either be recycled back to the pump reservoir
or collected for reuse after suitable purification.

Equipment requirements for large-scale liquid chromatography are not as critical as for analytical systems. Lower demands on the pumping and detector

systems mean that a large-scale liquid chromatograph
can be developed with less sophisticated (and generally
lower cost) components than can an analytical instrument.   Several of the instrument companies and LC chromatographic suppliers are active in this LC area.   Recently, E. M. Laboratories (E. Merck; see Appendix V)
has begun marketing adsorbent LC columns specifically
designed for high capacity preparative applications.
These packaged glass columns are limited to about 100-
200 psi of pressure, but are particularly useful for
purifying products from organic syntheses, and the
like.

12.4   COLUMNS

Just as the instrument requirements are different for
large-scale and analytical chromatography, so too are
the techniques of preparative LC different from analysis.   Since the object of optimizing preparative separations is to maximize throughput, several chromatographic parameters will be examined for their effects
on column sample capacity.   The sample capacity of a
column is directly dependent on the types and amounts
of packing used.
    As described in Section 6.1, a totally porous sup-
port has a much higher surface area for chromatographic
interaction with the sample than has a superficially
porous support.   Therefore, the capacity of columns
packed with totally porous supports is markedly greater than for the same column packed with a superficially
porous packing.   Unless very large internal diameter
columns are employed (which may be prohibitive in cost),
it is difficult to isolate gram quantities of sample
using columns packed with superficially porous supports.
Although preparative LC involving large amounts of solute is not conveniently performed with superficially
porous supports, some excellent scale-up separations

410 Large-Scale Separations

involving milligrams of solute can be performed. It
is important to note, however, that preparative liquid
chromatography using overloaded columns usually is performed with totally porous supports.

The size of totally porous particles that should
be used for preparative liquid chromatography has not
been systematically studied. However, the 40-60-μ
particle size range appears to have the advantages of
being readily available, easily dry packed into columns,
and moderately priced. As found with narrow-bore analytical columns (Section 6.2), the plate heights of
larger-diameter columns continually improve as the
particle size of the packing is reduced. In addition,
it can be expected that under similar conditions of
separation time and column pressure, the efficiency of
large diameter columns would increase with decrease in
d_p -- at least down to the range of 5-10-μ particles.
However, two factors mitigate against using such small
particles in large-scale separations. First, small particles are more difficult to pack uniformly for small
H values (see Section 6.2). Second, the use of smaller
particles (constant column pressure and separation
time) normally requires a reduction in column length
due to decreased column permeability (Appendix VI).
However, this reduces column capacity and the amount
of sample that can be separated, other factors being
equal.

In addition to using totally porous packings, column capacity can be improved by increasing either column length or column internal diameter, since either
of these approaches provides a larger amount of column
packing for interaction with the solute. However, increasing the column length also decreases the column
permeability and requires more operating pressure. Increasing the column internal diameter has no significant effect on column permeability; thus, increased
column capacity is best achieved by increasing column
internal diameter rather than column length.

A misconception has existed concerning the effect of increasing internal diameter on the efficiency of modern LC columns. Several early studies showed that column efficiency decreased dramatically as the column internal diameter was increased. Since efficiency (and resolution) was thought to decrease as the column diameter was increased, application of larger diameter columns was discouraged. However, recent studies have shown that although the efficiency of dry-packed columns does show an initial decrease as the diameter is increased, beyond a certain diameter column efficiency actually improves. It should be noted, however, that in principle, if all LC columns could be homogeneously packed, column diameter should have no effect on column performance.

One reason that larger-diameter columns can be more efficient is that they can be constructed so that they are essentially of <u>infinite diameter</u>. Shortly after injection into a narrow-bore column, the sample band occupies essentially the entire cross section of the column. Since the carrier liquid moves faster near the walls than in the center of the column (as a result of packing inhomogeneities due to the immovability of the walls), solute molecules near the column wall move further down the column than do those molecules in the center. As the band travels through the column, this effect (eddy diffusion) results in a solute band that is considerably wider than when it was injected. On the other hand, a sample injected into the center of a much wider column can develop solute bands which reach the column exit <u>before</u> they reach the faster moving mobile phase in the region of the walls. The result is the so-called infinite diameter phenomenon proposed by Knox and Parcher (2).

The existence of the infinite-diameter column phenomenon in LC comes about because of the very slow diffusion of molecules in liquids. Since this diffusion is slow, stream splitting is the only major

mechanism by which sample molecules which have been
injected into the center of a column can migrate toward the walls. By taking this into account, Knox and
Parcher developed a relationship (Eq. 12.1) for predicting column and packing dimensions which will produce the infinite diameter condition. Subsequent
studies have proved this expression valid:

$$d_c > (2.4 \ d_pL)^{1/2} \qquad\qquad (12.1)$$

Figure 12.5 shows H versus u plots demonstrating
the infinite diameter effect in columns packed with a
totally porous packing, while Figure 12.6 shows this
same effect with columns packed with a superficially
porous support. As the diameter of the columns is increased from 2.1 to 4.76 mm, column efficiency worsens
due to the greater wall effects encountered in larger
diameter columns resulting from increased particle
size segregation. However, as the diameter is increased still further to 7.94 mm and 10.9 mm, the efficiency of the system is improved over results obtained
with a 2.1 mm column. These 7.94 mm and 10.9 mm i.d.
columns are infinite-diameter and so not affected by
wall effects. Thus it appears that column diameter
can be measurably increased without impairing (actually improving) the performance of the similar analytical scale system, providing the same ratio of sample
size to column cross-sectional area is maintained. It
should be noted, however, that when performing preparative separations, the infinite-diameter relationship
is no longer valid, due to the effects of overloading.
While larger-diameter columns can be used for scale-up
separations with equal (or greater) resolution than
analytical columns, loss in resolution is anticipated
when these columns are used in an overloaded condition
for preparative separations.
The infinite-diameter-column effect can be best
illustrated with photographs of an actual system.

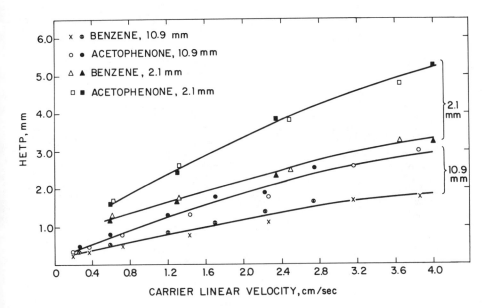

Figure 12.5 H versus mobile phase velocity curves for
500 x 2.1- and 10.9-mm i.d. columns packed with a totally porous support. Sample, 4 mg/ml benzene and 0.2
mg/ml acetophenone in CHCl$_3$; column packing, 35-75 μ
Porasil® A, 10% (w/w) H$_2$O-deactivated; carrier, 50%
H$_2$O-saturated spectrograde CHCl$_3$ (3). Reprinted by
permission of the publisher.

Figure 12.7 visually demonstrates the infinite diameter effect for a 30 x 3-cm i.d. column packed with
Davison Grade 62 silica gel using a chloroform mobile
phase containing sufficient benzene to match the refractive index of this solvent with that of silica
gel. This mobile phase makes the column essentially
transparent, and allowed the visualization of a sample
of an azo dye which was injected directly into the top
and center of the packing. The photographs show that

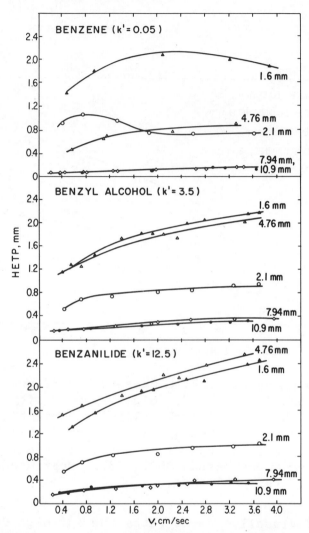

Figure 12.6   H versus mobile phase velocity curves for
500-mm-long columns packed with a superficially porous
support. Sample, 0.8 mg/ml benzene, 5 mg/ml benzyl
alcohol, and 0.2 mg/ml benzanilide in spectrograde hexane; column packing, 1% (w/w) triethylene glycol on
20-37 μ controlled surface porosity support; carrier,
spectrograde hexane saturated with triethylene glycol

(4).   Reprinted by permission of the publisher.

Figure 12.7   Demonstration of infinite-diameter-column effect.

this unretained dye never reaches the column wall before it is eluted from the column, and the solute band size grows only slightly as it moves through this infinite diameter column of rather coarse silica gel.
    Sample injection techniques greatly affect the performance of large-scale columns. This was demonstrated by injecting the dye solution 1 cm above the packing rather than directly into it (as was used for the column shown in Figure 12.7). When injected in

this manner, the sample band took the shape of the top
of the column bed (coned at the top due to the flow of
the carrier liquid) and the resultant band was badly
diffused. Sample introduction in preparative chromatography is important and needs to be more extensively
studied.

Many large-scale LC applications have been made
with columns of about 1/2-in. (1.1-cm) i.d. to 1-in.
(2.5-cm) i.d. Unpublished reports suggest that there
is a loss in performance with columns larger than 2.5-
cm i.d. Since there are no theoretical reasons why
columns larger than 2.5-cm i.d. should behave in this
fashion, this decrease in performance can probably be
attributed to column packing difficulties. Another
potential source of difficulty in using larger i.d.
columns is that the inlet system must be properly designed to ensure that the volume at the head of the
column is cleanly swept by the carrier liquid. It is
anticipated that proper packing techniques and inlet
design will eventually allow columns of larger than
2.5-cm i.d. to be utilized without loss in performance.

12.5 OPERATING PARAMETERS

The various chromatographic parameters will next be
examined for their effects on sample throughput, including sample volume, sample concentration, solute
weight, and carrier velocity. The first two of these
parameters, sample volume and sample concentration,
must be discussed for gel (exclusion) chromatographic
systems separately because the principles of separation are different from the other LC methods.

Column throughput is directly dependent on the
sample capacity of the chromatographic system. In
liquid-liquid, liquid-solid, and ion-exchange chromatography, the column capacity is largely dictated by
the stationary phase (i.e., the amount of stationary
liquid in LLC, the surface area in LSC, and the

number of exchange groups in IEC). In contrast, gel chromatography columns have no stationary phase to be overloaded; hence, capacity is normally governed by mobile phase effects. These mobile phase effects involve the ratio of injected sample volume to column void volume, and viscosity effects due to sample concentration. Since there are no active sites to be overloaded in gel chromatography, the sample capacity is generally considerably greater (about a factor of 10) than the same size column for one of the other LC methods.

As discussed in Chapter 10, the major limitation of gel chromatography is limited peak capacity. To overcome this limitation, the technique of <u>recycling</u> chromatography is sometimes employed (Section 10.5). The advantage of recycle chromatography is that it effectively provides a longer column for fractionation of the sample without increasing the pressure requirements of the system. In addition, recycling eliminates the need to prepare additional, expensive large-scale chromatographic columns. As indicated in Figure 12.8, relatively large sample weights can be used with this

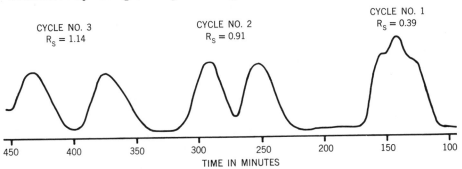

Figure 12.8 Separation of a polystyrene mixture using recycling chromatography. Column, 4 ft x 3/8 in. Styragel®, 2.5×10^4 Å; carrier, toluene; sample, 10 mg/ml polystyrene mixture 51K + 10.3K (1:1); injection volume, 350 ml (5). Reprinted by permission of the publisher.

approach, utilizing relatively short GPC columns, usually in conjunction with low dead-volume reciprocating
pumps. The use of a pneumatic amplifier pump for recycle chromatography has been reported recently (7).
Unfortunately, recycling is a less useful approach with
the other forms of LC because the time required for the
chromatogram is often much longer than the time required to elute the column void volume, as is the case
in gel chromatography. However, recycling techniques
sometimes can be used in preparative applications with
the other LC methods in certain instances (7).

The influence of sample volume on the column capacity of those chromatographic systems which are limited
by stationary phase effects has been investigated.
Figure 12.9 shows the effect of sample volume on the
efficiency of a 10.9-mm i.d. column packed with a totally porous support. After a small initial loss in

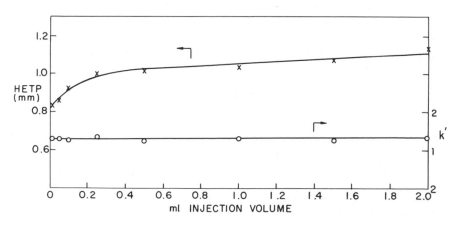

Figure 12.9   Effects of sample volume on H and k' with
totally porous support. Carrier and 10.9-mm i.d. column as in Figure 12.5; sample, varying concentrations
of aniline in CHCl$_3$; sample weight, random variation
from 0.02 to 0.12 mg; carrier velocity, 0.9 cm/sec (3).
Reprinted by permission of the publisher.

efficiency (probably due to less-than-optimum inlet design), no significant increase in HETP was observed with sample volumes up to 2.0 ml. On the other hand, as shown in Figure 12.10, a similar experiment using the same size column packed with a superficially porous support showed that only about 0.9 ml of sample could be injected without significant loss in column efficiency. This effect might be anticipated since the column of porous packing is less affected by extra-column band broadening effects (see Section 2.3); the void volume of this column is more than twice that of columns packed with the superficially porous material.

Sample volume has a more important effect on column performance when sample concentration is considered. Figure 12.11 compares the effects of sample

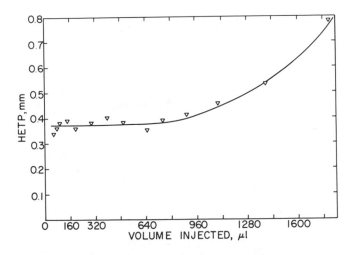

Figure 12.10   Effect of sample volume on HETP with a superficially porous support. Carrier and 10.9-mm i.d. column as in Figure 12.6; sample, 0.2 mg/ml 2,5-dichloroaniline in hexane (4). Reprinted by permission of the publisher.

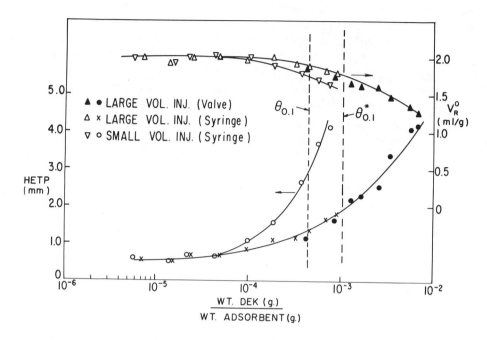

Figure 12.11   Effects of sample volume and concentration. Upper curves: solute equivalent retention volume, $V_r°$, versus weight of solute/weight of packing ratio. Lower curves: HETP versus weight of solute/weight of packing ratio; 10.9-mm i.d. column and carrier as in Figure 12.5; sample, various concentrations of diethylketone (k' = 2.8) in $CHCl_3$; carrier velocity, 0.5 cm/sec (3). Reprinted by permission of the publisher.

weight on the performance of a 10.9-mm i.d. column packed with a totally porous support when a small volume of a concentrated sample solution and a larger volume of a relatively dilute solution are injected. The two curves at the bottom of this figure show the effects of increasing sample weight on column efficiency. The upper of these two curves was obtained with small sample volumes and concentrated solutions, while the lower curve resulted from large sample volumes and dilute solutions. These data suggest that the column is less efficient when samples are introduced as concentrated solutions as compared to the same weight of sample applied to the column in a dilute solution. The two top curves of Figure 12.11 demonstrate that the use of large sample volumes and dilute solutions also increases the linear capacity of the column compared to that obtained when using small injection volumes of concentrated solute. This effect probably is the result of localized stationary phase overloading which is more prominent when a given sample weight is injected as a concentrated solution.

Column performance for a particular solute is governed by the injected weight of the single solute, rather than the total weight of injected sample, as illustrated in Figure 12.12. The lower curves represent the H of diethylketone (DEK) versus the sample weight for three solutions--one containing DEK only, another containing approximately equal weights of DEK and methylethylketone (MEK), and the third containing equal weights of DEK and acetone (DMK). The upper curve plots the H of DEK versus the injected weight of DEK only for these same solutions. Only in the latter case can all the data be described by a single curve. This sampling effect has not yet been studied under highly overloaded conditions.

Carrier velocity appears to have a reduced effect on resolution when the column is operated in overload condition, as illustrated in Figure 12.13. As anticipated by effects predicted by the resolution equation (Section 2.5), carrier velocity has a large effect on resolution in a non-overloaded column. With small sample weights both α and k' remain constant, but N is affected by carrier velocity. At larger sample weights, α and k' are reduced by overloading effects. In this case, affects on R_s due to changes in N from carrier velocity considerations apparently are

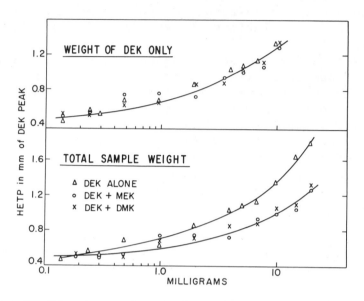

Figure 12.12 Comparison of component and total solute weight effects on H. Carrier and 10.9-mm i.d. column as in Figure 12.5; carrier velocity, 0.5 cm/sec; samples, 9.7 mg/ml diethylketone (k' = 2.8) in CHCl$_3$, 4.9 mg/ml diethylketone plus 5.3 mg/ml methylethylketone (k' = 3.6) in CHCl$_3$, and 4.9 mg/ml diethylketone plus 5.0 mg/ml acetone (k' = 4.9) in CHCl$_3$ (3). Reprinted by permission of the publisher.

relatively minor compared to changes in R$_s$ due to over-loading. Therefore, when scouting a system for maximum throughput, it appears that sample weight should be increased while operating the column at a low carrier velocity, until the resolution of the components of interest is slightly better than required. The carrier velocity may then be increased until the resolution is just adequate.

Existing knowledge of the various LC parameters affecting column throughput suggest that the following

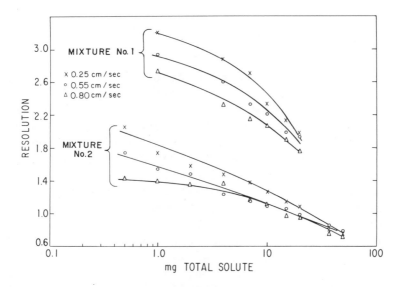

Figure 12.13   Effects of solute weight and carrier velocity on resolution.   Carrier and 10.9-mm i.d. column as in Figure 12.5; mixture No. 1, diethylketone and acetone; mixture No. 2, diethylketone and methylethylketone; sample concentrations as in Figure 12.12 (3).   Reprinted by permission of the publisher.

conditions should be used when attempting <u>preparative</u> liquid chromatography:

- Short, wide columns.
- Totally porous packings.
- Relatively high flow rates but low carrier velocities.
- Larger injection volumes and dilute sample solutions.
- Volatile carrier liquids of low viscosity.
- Valve injection.
- Insensitive detectors.

● Either reciprocating or pneumatic amplifier
  pumps.

12.6 APPLICATIONS

More use is currently being made of the scale-up technique than of preparative applications because of the
high resolution needed to isolate pure materials from
compounds of similar molecular structures. Because of
this need for high resolution (and apparently because
of the proprietary nature of most of the preparative
separations being performed today), few examples of
overloaded separations have yet been published. However, several scale-up separations have been reported
which utilize most of the principles discussed in this
chapter. Figure 12.14 shows the separation of two optically active compounds that were isolated to determine their optical activities. Initial measurements
after their separation by distillation and preparative-
scale gas chromatography were questionable because of
the possibility that the high temperatures required for
these isolation methods may have altered the optical
properties of the materials. No changes occurred with
the mild conditions used in the separation by liquid-
solid chromatography.
    The separation of the two aniline derivatives
shown in Figure 12.15 utilized a carrier velocity
change in mid-run to increase throughput by shorten-
ing separation time. With this mixture, throughput
could not be increased by further overloading of the
column due to the limited solubilities of the components in the carrier liquid.
    Gradient elution can be used to increase sample
throughput as shown in Figure 12.16. Use of the gradient technique not only shortens the separation time,
but also reduces band spreading during sample introduction. The low solvent strength of the initial carrier

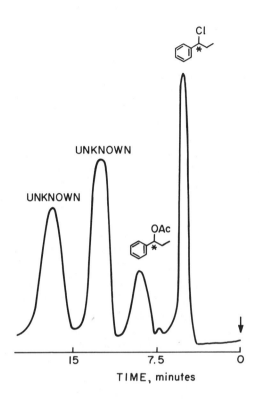

Figure 12.14   LSC separation of optically active compounds; 10.9-mm i.d. column as in Figure 12.5; carrier,
25% (v/v) spectrograde CHCl₃ in n-heptane, 50% H₂O-
saturated; injection volume, 1.0 ml; sample concentration, 25 mg/ml; carrier velocity, 1.0 cm/sec (3).
Reprinted by permission of the publisher.

solution results in the sample being collected at the
top of the column during its introduction. As the
solvent strength of the carrier is increased, the solute is migrated through the column as a sharp peak.

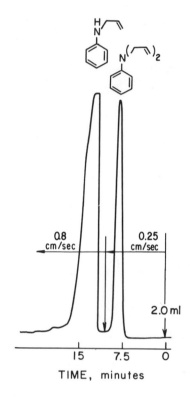

Figure 12.15   LSC separation of aniline derivatives;
10.9-mm i.d. column as in Figure 12.5; carrier, 0.5%
MeOH (v/v) in cyclopentane, 50% H_2O-saturated; sample
concentration, 50 mg/ml (3).   Reprinted by permission
of the publisher.

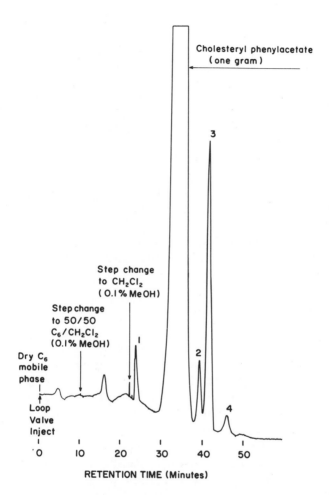

Figure 12.16 Purification of cholesteryl phenylace-
tate. Column, Spherosil® XOA 400 (two 500 x 23-mm
i.d. in series); carrier, step gradient, dry benzene
to CH$_2$Cl$_2$ (0.1% MeOH); flow rate, 30 ml/min; column
pressure, 1000 psi; column temperature, ambient; detector, UV photometer at 254 nm; sensitivity, 0.32 AUFS
(1). Reprinted with permission from Du Pont Instrument Products Division.

REFERENCES

1.   D. R. Baker, R. A. Henry, R. C. Williams, D. R. Hudson, and N. A. Parris, J. Chromatog., 83, 233 (1973).
2.   J. H. Knox and J. F. Parcher, Anal. Chem., 41, 1599 (1969).
3.   J. J. DeStefano and H. C. Beachell, J. Chromatog. Sci., 10, 654 (1972).
4.   J. J. DeStefano and H. C. Beachell, J. Chromatog. Sci., 8, 434 (1970).
5.   K. J. Bombaugh, in Modern Practice of Liquid Chromatography, J. J. Kirkland, ed., Wiley-Interscience, New York, 1971, p. 279.
6.   G. A. Fallick, American Laboratory, 5, 19 (1973).
7.   R. A. Henry, S. H. Byrne, and D. R. Hudson, Paper accepted for publication by J. Chromatog. Sci.

BIBLIOGRAPHY

K. J. Bombaugh, et al, J. Chromatog. Sci., 7, 42 (1969). (Discussion of recycle gel chromatography.)

J. J. DeStefano and H. C. Beachell, J. Chromatog. Sci., 8, 434 (1970); 10, 481 (1972); 10, 654 (1972). (Basic studies on the effects of various parameters on the performance of larger diameter columns.)

Du Pont Liquid Chromatography Applications Lab Report, Nos. 72-11, 73-03. Du Pont Instrument Products Division, Glasgow, Del. (Examples of preparative applications.)

F. E. Rickett, J. Chromatog., 66, 356 (1972). (Preparative separation by liquid-liquid chromatography.)

Waters Associates Bulletin AN-71-103. (Good synopsis of preparative LC principles.)

Waters Associates, Inc., Milford, Mass., Chromatography Notes, Vol. III, No. 2, June, 1973. (Demonstrates detrimental effects of high-sensitivity detectors.)

J. L. Waters, J. Chromatog., <u>55</u>, 213 (1971). (Discusses the uses of fractions separated by preparative LC.)

CHAPTER THIRTEEN

OTHER TOPICS

13.1 QUANTITATIVE LIQUID CHROMATOGRAPHY

An important aspect of modern LC is its ability to
analyze quantitatively a wide variety of materials.
The technique can be used both for the high-precision
assay of major components and for the analysis of
minor or trace components. Column chromatography was
generally difficult to quantitate prior to about 1968.
However, the use of modern equipment with continuous
detection, in conjunction with measurement and quantitative techniques developed for gas chromatography,
has made LC a highly accurate and precise analytical
technique. The precision of LC analytical results is
at present comparable to those obtainable by gas chromatography. LC does not show the variability associated with the gas chromatographic vaporization process;
therefore, with further developments in equipment and
techniques, we can expect that LC precision will eventually become superior.

Several possible sources of error are involved
in quantitating liquid chromatography:

- Sampling technique.
- Chromatographic separation.

431

- Detector response.
- Calibration and measurement techniques.

The ability to produce good quantitative data depends
directly on how well the original sample has been taken and whether or not it is a representative aliquot
of the material that is to be characterized. Sampling
also involves the accurate preparation of solutions
for LC analysis. There are no specific instructions
in this step, except that good analytical technique be
followed. If high-precision assays are required, solutions must be prepared in a manner that will allow the
level of precision which is desired in the final determination.

Errors from sample introduction are caused mainly
by imprecise injection volumes and by peak broadening.
Sample injection via syringe is a special problem when
high-pressure liquid systems are used (i.e., leakage
of the syringe), so that valve injection techniques
are preferred--particularly in routine applications.
Not only does valve injection minimize the effect of
leaks, but it provides closer control over delivery
volume (Section 4.6). The technique of sample injection also poses problems to the extent it affects band
shape and/or disturbs the packing at the column inlet.
However, fractionation of the sample during injection
(as in gas chromatography) is usually not a problem in
LC. Extra-column band broadening during sample injection usually causes peak tailing, which can lead to
band overlap and errors in peak-area measurements (1).

Inaccuracies can be caused by the chromatographic
process itself. Sample decomposition, although rare
in modern LC, can occur. However, decomposition of
sample components during the chromatographic process
can always be tested by chromatographing standards of
different known concentrations. Sample decomposition
should be suspected if a linear increase in peak area
is not obtained as a function of concentration. Errors

due to peak overlap and tailing are common and are usually best reduced by selection of a better column or
separating process. The desorption of tightly adsorbed
compounds due to solvent demixing may cause spurious
peaks in gradient elution analysis, particularly in
LSC (see Section 13.3). Therefore, for the most precise quantitative data, isocratic separations are preferred; gradient elution should only be used for quantitative analysis when absolutely necessary. For the
routine analysis of mixtures containing a wide range of
k' values, the coupled-column approach is preferred
(Section 13.3).

Variation in detector response can also cause
errors. As indicated in Section 5.1, detector sensitivity, noise, and linearity are important specifications, particularly for quantitative analysis. The
susceptibility of the detector to changes in temperature and column flow rate--and the level of the noise
associated with the detector--influence the errors involved in quantitative measurements; therefore, these
parameters must be maintained as stable as possible
for optimum precision. Because of the selectivity of
some detectors, certain peaks in some samples may be
missed in the quantitative measurement. Since detector response generally varies for different compounds,
most detectors must be calibrated with each compound
to be measured.

The techniques used in obtaining the quantitative
data can also cause errors in the analysis. For instance, if a recorder trace is used for the quantitative measurement, significant errors can result if the
recorder itself is not in correct adjustment. The procedures used for integrating the areas of chromatographic bands can also lead to errors in precision, as
discussed in greater detail below. The method of quantitation (i.e., peak height or area) can also influence
both the accuracy and the precision of analyses.

The Peak-Height Method

The simplest method of quantitation involves the measurement of peak heights. Calibrations are obtained by chromatographing a constant volume of several solutions containing different known concentrations of the compound (or compounds) of interest and measuring the heights of the peaks that are formed, as illustrated in Figure 13.1. Peak height is measured as a distance from the baseline to the peak maximum, as shown for peaks 1 and 3. Baseline drift is compensated by interpolation of the baseline between the start and finish of the peak. A standard calibration curve is obtained by plotting the concentration of the compound of interest versus the peak heights obtained. Unknowns are analyzed by chromatographing the same size aliquot of an unknown solution, and determining the peak height of the materials for which the calibration has been prepared. By referring to the standard calibration curve, peak heights can then be translated into

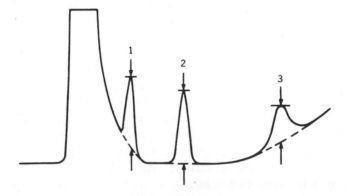

Figure 13.1 Peak-height measurement. N. Haddon, et al, _Basic Liquid Chromatography_, Varian Aerograph, Walnut Creek, Calif., 1971. Reprinted by permission of publisher.

the concentration of material in the unknown sample.

The peak-height method should only be used when peak heights change linearly with sample size. Serious errors will occur if the peak-height method is employed when peaks are distorted or if the column is overloaded. When using syringe injection, the peak-height method is usually no more precise than about ±5%, relative, since the results depend greatly on the precision with which a sample is introduced into the instrument. By carefully controlling instrumental parameters and making use of a sampling valve, it is sometimes possible to improve the long-term precision of the simple peak-height method to 1-2%; precision also is improved by using frequent calibrations to compensate for changes in instrumental parameters. However, the inherent simplicity of the peak-height method is much diminished by such additional manipulation.

The peak-height method is almost always used for trace analysis. In trace analysis, accuracy is usually at a premium; therefore, freedom from the interference of other components is of prime concern and high precision is not normally important. As discussed in Section 3.2, peak-height measurements are more accurate than peak-area measurements because peak heights are less interfered by neighboring, overlapping peaks. The peak-height method of quantitation also involves the least effort, and satisfactory quantitative methods can be obtained with quite modest investments of time and effort. It is often desirable to initially determine the approximate concentrations of components in a mixture by a simple peak-height method, before attempting to set up a more rigorous high-precision procedure for routine use.

The Peak-Area Method

The peak-area calibration procedure is generally

similar to that just described for peak height, except
that quantitation is based on area. Usually, the precision of this method is less influenced by changes in
instrumental parameters. In addition, improved results
are obtained with peaks whose shapes are not Gaussian.
However, results from the peak-area quantitative method are more affected by neighboring peaks. A good
rule of thumb is that peak area should be used for
better precision and peak height for maximum freedom
from possible interferences.

There are several techniques for measuring peak
area, as illustrated in Figure 13.2. Peak area may be
determined with a planimeter, a device which computes
the area of the peak by mechanically tracing its out-
line. The precision and accuracy of this tedious procedure is quite dependent on the skill of the operator.
Improved precisions can be obtained with this technique
by making repetitive tracings; however, this makes an
already time-consuming technique even more tedious.

A simple but very effective technique is to measure areas by the product of peak height, H, and the
width, W, at one-half the peak height. This technique
should only be used with symmetrical peaks or peaks
which have similar shapes. The precision of this measurement can be improved by increasing recorder chart
speed, so that the width of the peak can be determined
more precisely.

Another technique is to cut the peak out of the
recorder trace and weigh it. Unfortunately, this
approach destroys the recorder trace. The accuracy
of this approach depends on the constancy of the
weight of the chart paper and the care used in cutting
out the peak. By keeping the ratio of the peak heights
to the width at half-height in the range of 1 to 10,
the accuracy of the cutting process can be improved.
This method can be used satisfactorily for irregularly shaped peaks.

The triangulation technique involves the

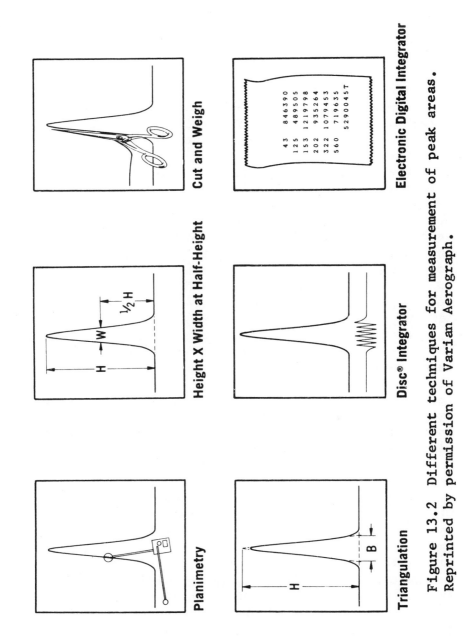

Planimetry

Height X Width at Half-Height

Cut and Weigh

Triangulation

Disc® Integrator

Electronic Digital Integrator

Figure 13.2   Different techniques for measurement of peak areas.
Reprinted by permission of Varian Aerograph.

437

calculation of the area by multiplying the peak height, H, by the width of its base, B, as determined by tangents constructed from the sides of the peaks so that they intersect the baseline. The areas found by triangulation are calculated by the formula: Area = (1/2)BH. This technique is particularly susceptible to operator error and demonstrates no advantages over the other area measuring methods.

Some recorders are equipped with ball and disc integrators which automatically produce tracings which indicate the area of the peak. This method provides a degree of automation, but the accuracy is still limited by the performance of the recording potentiometer. The ball and disc integrator often can be used with good accuracy for irregularly shaped peaks. However, adjustment for baseline drift and incompletely resolved peaks decreases the accuracy and increases the time of measuring the areas.

As indicated in Section 4.11, the electronic integrators that are now widely used in gas chromatography generally are also satisfactory for measuring peak areas in LC. The exception is negative peaks (e.g., from refractive index detectors), which cannot be handled by this method. The electronic digital integrator automatically measures peak areas and converts them into a numerical form which is printed out. Sophisticated versions of these devices also measure retention time and correct for baseline drift.

LC peak areas can also be conveniently measured with the computer systems which have been developed for gas chromatography. There are many different types of computer systems which can be used without change for LC analysis. Several systems are now in operation which service gas and liquid chromatography systems simultaneously. With the appropriate programming, most of these devices print out a complete report, including name of the compounds, retention times, peak areas, area correction factors, and the weight percent of the various sample components. A discussion of the detailed approaches to computer handling of LC data is beyond this presentation. However, a review of chromatographic automation is recommended for further reading (2).

In the experience of the author, the maximum precision one can expect of the various integration methods is given in Table 13.1. Since the type of data generated by LC is very similar to that found for gas chromatography, these area precision values are representative for both techniques. The data in Table 13.1 represent average values that can be obtained with the various integration methods. If the techniques are carried out with extreme care, the precision of these methods can be improved by as much as a factor of 2.

TABLE 13.1. AVERAGE PRECISION OF PEAK AREA MEASUREMENT TECHNIQUES (NO INTERNAL STANDARD)

Method	Relative Precision, 1σ (%)
Planimeter	3
Triangulation	3
Cut and weigh	2
H X (1/2) W	2
Ball and disc integration	1
Electronic digital integrator	0.5
Computer	0.25

The peak-area normalization techniques that are widely used in gas chromatography are generally not employed in LC. In this approach, the areas of all

of the peaks are summed, and the percent area for each
peak of the total is a measure of the concentration of
the unknown component. The reason that peak-area normalization does not apply in LC analyses is that the
response of LC detectors varies with different compounds. This variable response eliminates the possibility of this procedure unless response factors are
applied for each compound in the mixture.

Internal Standard Technique

A widely used technique of quantitation involves the
addition of an internal standard to compensate for
errors in the analytical measurement. In this approach,
a known compound is added to the unknown mixture at a
fixed concentration. This compound is used as an internal marker so that the precision of the analysis
does not depend on reproducing the size of the sample
introduced into the column. The internal standard
technique can be used for both peak-height and peak-
area ratio measurements. The compound/internal standard peak-height (or peak-area) ratio is used in the
calibration, rather than absolute values.

  Calibrations based on peak-height ratio are constructed by chromatographing appropriate aliquots of
calibration mixtures containing the compound (or compounds) of interest in various concentrations together
with a constant concentration of the internal standard
compound. The peak heights of the components of interest are determined, and the compound/internal standard
peak-height ratios are plotted against the concentration. This calibration plot is linear (if the calibration mixtures have been prepared properly) unless the
column is inadvertently overloaded or there is decomposition of the sample or internal standard. Measurements with a well-conducted peak-height ratio method
can be reproduced with a precision of 0.7-1%, relative
(1σ).

The peak-area-ratio measurement technique is most precise, since it compensates best for changes in instrumental parameters and variation in technique. However, this approach requires completely separated peaks, which may be difficult with some mixtures. For highest precision with the peak-area-ratio technique, a resolution of 1.25 is desired for the compounds of interest. Reproducibilities of < 0.5%, relative (1σ) are not uncommon with this approach. The method of calibration is identical to the peak-height-ratio method, except that measurements are based on peak area.

For highest precision, the selection of the internal standard is critical for both the peak-height- and peak-area-ratio methods. First, the internal standard peak should be separated from the other peaks in the mixture to be analyzed (with a resolution of at least $R_S = 1.25$ if peak-area-ratio measurements are to be made). The internal standard must be located in a "vacant" spot in the chromatogram of the unknown, but eluting as close as possible to the compounds to be measured. To determine the most feasible position for the internal standard, an appropriate range of unknown samples must first be qualitatively chromatographed. If more than one component is to be determined, it may be necessary to add more than one internal standard to the mixture to obtain highest precision. Optimum results are obtained when the concentration of the internal standard is adjusted so that the peak-height or peak-area ratio is approximately unity. The internal standard must not be present in the original sample, and it should be a compound which is stable and not reactive with sample components, column packing, or mobile phase. If possible, the internal standard should be commercially available in high purity.

It will be useful to show an example of a stable, high-precision LC method that was developed using the peak-area-ratio technique (3). Figure 13.3 shows a chromatogram containing an insecticide (methomyl) and

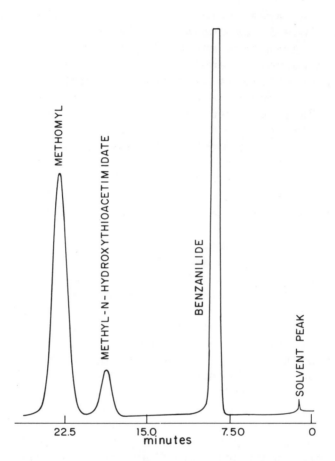

Figure 13.3   Typical chromatogram of a synthetic mixture containing Lannate® methomyl insecticide.   Column,
1 m x 2.1 mm i.d. 1% ODPN on Zipax® chromatographic
support; carrier, 7% chloroform in spectrograde n-
hexane; flow rate, 1.3 ml/min; detector sensitivity,
0.08 absorbance = full scale.   R. E. Leitch, J. Chromatog. Sci., 9, 531 (1971).   Reprinted by permission
of editor.

a precursor compound, together with benzanilide which
was employed as an internal standard.   This

conventional liquid-liquid chromatographic method was
used routinely for several years to measure samples of
technical and formulated methomyl. The peak-area-ratio
calibration used in this method is shown in Figure 13.4.
Note that this calibration plot is linear and extrapo-
lates through the origin. These data are typical of
the peak-area-ratio method for quantitative LC analysis; lack of linearity or extrapolation through the
origin in such cases usually is indicative of experimental difficulties.

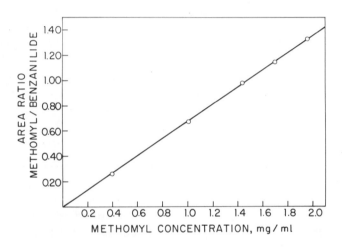

Figure 13.4 Typical calibration curve for methomyl
insecticide. R. E. Leitch, J. Chromatog. Sci., 9, 531
(1971). Reprinted by permission of editor.

Table 13.2 summarizes some LC analyses of experimental methomyl samples conducted over a period of
several months. The samples were analyzed in dupli-
cate and the method was conducted with a 2σ deviation
of 0.9%, absolute, over this period of time. Table
13.3 illustrates the reproducibility of retention
times obtained with this conventional liquid-liquid

TABLE 13.2. LC DETERMINATIONS OF EXPERIMENTAL SAMPLES
CONTAINING >90% S-METHYL-N- (METHYLCARBAMOYL)OXY -
THIOACETIMIDATE.

Date	Sample	% Methomyl[a]	
7/11/69	6441-43-C	98.6	98.5
7/18/69	6430-91-A	98.9	98.5
8/ 1/69	6430-82-A	92.0	92.5
8/ 1/69	6430-108-A	95.7	94.8
9/ 2/69	LL-390-17	95.4	95.0
9/ 5/69	6430-124-A	96.8	97.2
10/ 9/69	6430-136-A	93.8	94.0
10/ 9/69	6430-135-A	97.8	96.7
11/12/69	6430-157-A	93.4	94.3
11/19/69	6441-82-A	96.4	97.7
12/30/69	6441-87-A	97.4	97.5
12/30/69	6441-87-B	95.6	95.8
1/22/70	6441-89-A	93.1	93.6
1/22/70	6441-89-B	96.5	95.8

[a] 2σ deviation = 0.9% absolute.

chromatographic system. Over a 1-yr period, retention
times for the two compounds of interest varied only
about 5%.

Listed in the bibliography at the end of the chapter are references to some publications describing
some actual quantitative analyses by modern LC. These
papers should be studied for additional information in
setting up all types of quantitative LC procedures.

13.2 MORE ABOUT THE SOLVENT

Many organic liquids are available for use in liquid-
liquid or liquid-solid chromatography. Our choice of
a given liquid is governed by many considerations:

TABLE 13.3   RETENTION TIMES OVER A ONE-YEAR PERIOD

Date[a]	Methyl-N-Hydroxy-thioacetimidate	Methomyl
10/ 3/69	19.6	23.6
11/ 4/69	19.4	23.6
12/ 1/69	21.2	26.2
1/ 7/70	21.4	25.6
2/10/70	b	25.3
3/ 6/70	17.3	21.8
4/ 8/70	17.9	21.3
5/15/70	18.5	23.0
7/ 8/70	19.2	23.4
9/14/70	19.4	23.4
	Ave. 19.3 $\pm$ 1.4	Ave. 23.7 $\pm$ 1.5

[a]No mixture chromatograms were available for June and August, 1970.
[b]Not available.

cost, purity, viscosity (or boiling point), ability to dissolve a given sample, compatibility with the detector and column in use, solvent strength, and solvent selectivity. Each of these aspects of choosing a solvent has been discussed in Chapters 7 and 8. Here we will examine in greater detail the question of solvent strength and selectivity in liquid-liquid chromatography (LLC). Because this is a rather complicated subject, the beginner may wish to refer first to other discussions of this subject (e.g., Section 7.3 and especially ref. 4).

Because liquids differ in their selective interaction with various sample molecules (solutes), it is possible to vary k' and α over wide limits to fit the specific needs of a given separation problem. Thus, a knowledge of how solvent strength and selectivity depend upon the chemical nature of different liquids can be used to great practical advantage; it allows a rapid selection of the best mobile and/or stationary phase for any separation. Until recently, only rough qualitative rules were available for the prediction of solvent strength and selectivity in LLC, and the selection of optimum liquid phases was essentially by trial-and-error.

Many different workers have speculated on the basis of solvent strength and selectivity in LLC. Unfortunately, much of this past effort has not been suitable for practical application in the laboratory; in some cases the proposed treatments are impossibly complicated and require data that are normally unavailable, while other proposals are too approximate to be useful. Here we will review three recent approaches that appear to offer considerable practical utility. The first of these, by Snyder et al. (4), has been described briefly in Section 7.3. The second is based on the recent work of Rohrschneider (5), while the third is due to Huber and co-workers (6).

The Snyder, Rohrschneider, and Huber approaches are compared in Table 13.4. The first two treatments are intended for general application to any sample, including unknown samples that have not previously been studied. The Huber approach is applied to known solutes within a narrow range of possible chromatographic systems; for example, steroids as solutes, with iso-octane-ethanol-water solutions of varying composition as both mobile and stationary phases. The Snyder and Rohrschneider systems are approximate, but applicable over a wide range of possible experimental conditions. The Huber treatment can be quite precise, for a much narrower range of samples and conditions. The Snyder and Rohrschneider schemes are useful for the qualitative, rapid selection of optimum conditions for the separation of any sample. The Huber approach is

TABLE 13.4. CHARACTERISTICS OF THE SNYDER (4), ROHRSCHNEIDER (5), AND HUBER (6) SYSTEMS

Snyder	Rohrschneider	Huber
Based on pure compound data; does not require solubility or chromatographic data	Requires solubility data for various solvents and a few test solutes; necessary data largely available (Table 13.5)	Requires chromatographic data on all solutes and solvents to be used; necessary data available only for limited number of solutes and solvents
Applicable to any solute	Applicable to any solute	Applicable only to previously studied solutes
Applicable to any solvent whose physical properties have been described	Applicable to any solvent studied by Rohrschneider (Table 13.5)	Applicable (at present) only to solvents composed of varying proportions of isooctane, ethanol and water
Order-of-magnitude accuracy	Intermediate accuracy	Relatively precise ($\pm$5-10%)
Useful in optimizing separation of unknown mixtures	Useful in optimizing separation of unknown mixtures	Useful in optimizing separation of known mixtures, quite precisely
Easy to apply--no calculation required	Easy to apply--no calculation required	Calculation via Eq. 13.1 normally used
Not useful for predicting t_R values	Limited value in predicting t_R values	Quite useful for predicting t_R values

intended mainly for precisely optimizing the separation of some important class of related compounds, for
example, steroids and pesticides.

The Snyder and Rohrschneider systems are applied
in essentially the same way. We have already looked
briefly at the Snyder treatment, Section 7.3 (for details, see ref. 4), and we will concentrate here on
the Rohrschneider approach. On balance, the practic-
ing chromatographer may find the Rohrschneider data
somewhat more useful for LLC (but not for liquid-solid
chromatography; see Section 8.3). Data are available
for a wider range of solvents, and predictions of solvent strength and selectivity may prove more accurate.
Because of the relative complexity of the Huber scheme,
and its somewhat restricted applicability at present,
we will attempt only a brief review of this approach
at the end of this section. The reader is referred to
the original literature for details (6).

The Rohrschneider treatment begins with the classification of chromatographic liquids in terms of their
solubility for a number of test solutes: n-octane,
toluene, ethanol, methyl ethyl ketone, dioxane, and
nitromethane. These data are reproduced in Table 13.5,
expressed here as the distribution constants (K_g values) that would be observed in a corresponding gas-
liquid chromatographic system (solvent as stationary
phase). The larger is K_g, the lower is solute vapor
pressure, and the more strongly retained is the given
test solute in that solvent. Thus, for squalane as
solvent (No. 22 in Table 13.5), octane (K_g = 3512) is
retained more strongly than toluene (K_g = 1775), and
toluene is retained more strongly than nitromethane
(K_g = 84). That is, nonpolar solutes are preferentially retained on this nonpolar stationary phase (squalane). For dimethyl sulfoxide as stationary phase or
solvent, the order of retention is reversed: nitromethane (K_g = 13994), toluene (K_g = 1745), octane
(K_g = 121). Polar solutes are preferentially retained

TABLE 13.5. SOLUBILITY DATA OF ROHRSCHNEIDER, SIX SOLUTES AND 81 SOLVENTS (5). REPRINTED FROM (5) WITH PERMISSION OF THE PUBLISHER

Solvent		Octane	Toluene	Ethanol	M.E.-ketone	Dioxane	Nitromethane
					Kg Values for Test Solutes		
1.	Carbon disulfide	14,300	8,900	143	661	2,654	264
2.	Cyclohexane	13,500	3,900	80	355	1,137	126
3.	Triethylamine	13,300	4,450	690	677	1,692	458
4.	Ethyl ether	13,200	6,760	2,114	1,003	2,616	1,170
5.	Ethyl bromide	11,100	11,800	--	5,777	8,907	2,115
6.	n-Hexane	9,900	3,520	82	355	974	135
7.	Isooctane	9,800	2,680	61	310	886	118
8.	Tetrahydrofuran	8,870	10,800	2,168	2,136	6,141	5,379
9.	Isopropyl ether	8,800	4,020	719	786	1,725	719
10.	Toluene	8,705	--	273	1,339	3,956	1,452
11.	Benzene	8,310	7,930	320	1,880	6,597	1,994
12.	p-Xylene	7,620	5,120	268	1,260	3,632	1,191
13.	Chloroform	7,150	10,500	779	9,632	35,646	2,686
14.	Carbon tetrachloride	7,030	5,620	149	997	4,342	384
15.	Butyl ether	6,920	3,750	368	592	1,586	490
16.	Methylene chloride	6,302	11,100	632	--	3,368	4,994
17.	Decane	6,302	2,571	62	265	820	114
18.	Chlorobenzene	5,972	6,261	265	1,901	5,237	1,468
19.	Bromobenzene	4,808	5,762	248	1,600	4,812	1,327
20.	Fluorobenzene	4,567	7,245	336	2,339	6,361	2,460
21.	2,6-Lutidine	3,805	4,875	3,681	1,516	3,870	2,739
22.	Squalane	3,512	1,775	42	184	640	84
23.	Hexafluorobenzene	3,512	8,181	--	--	4,342	1,408
24.	Ethoxy benzene	3,043	4,609	337	1,371	3,708	1,812
25.	2-Picoline	2,945	5,174	4,705	2,035	4,685	3,778
26.	Ethylene chloride	2,852	7,459	559	4,442	--	4,369
27.	Ethyl acetate	2,852	5,020	1,279	--	4,946	5,179
28.	Iodobenzene	2,852	4,567	216	1,224	4,140	1,031

TABLE 13.5. (Continued)

Solvent	Octane	Toluene	Kg Values for Test Solutes Ethanol	M.E.-ketone	Dioxane	Nitromethane
29. Methyl ethyl ketone	2,535	5,121	1,878	--	4,812	6,994
30. Bis(2-ethoxyethyl)ether	2,466	4,260	1,384	1,174	3,235	4,369
31. Methoxybenzene	2,466	4,224	431	1,783	5,087	2,739
32. Octanol	2,339	1,895	2,168	639	1,518	411
33. Cyclohexanone	2,225	5,020	1,690	1,747	4,239	5,379
34. tert-Butanol	2,172	1,561	6,051	--	2,869	705
35. Tetramethylguanidine	2,027	4,738	19,715	1,600	4,046	7,362
36. Isopentanol	1,982	1,631	3,527	907	1,953	603
37. Pyridine	1,982	5,452	5,294	2,404	5,936	5,594
38. 1,4-Dioxane	1,900	5,070	1,878	1,729	--	5,594
39. Butanol	1,823	2,016	4,758	1,039	--	759
40. Isopropanol	1,688	1,566	--	1,318	2,195	858
41. Propanol	1,657	1,902	4,705	1,085	2,403	903
42. Phenyl ether	1,628	3,620	230	980	3,490	1,180
43. Acetone	1,599	4,121	--	2,838	5,237	8,800
44. Benzonitrile	1,519	4,260	926	2,307	5,936	4,994
45. Tetramethylurea	1,401	5,282	7,636	1,840	4,140	9,994
46. Benzyl ether	1,301	3,089	356	1,009	3,225	1,861
47. Acetophenone	1,282	3,899	1,054	1,646	4,946	4,112
48. Hexamethyl phosphoric acid triamide	1,264	4,447	38,539	1,747	3,632	15,724
49. Ethanol	1,264	2,221	--	1,490	2,695	1,499
50. Quinoline	1,247	3,424	2,918	1,416	3,708	2,911
51. Nitrobenzene	1,152	3,782	424	1,729	5,396	3,883
52. m-Cresol	891	2,270	12,284	24,091	99,028	2,973
53. N,N-Dimethylacetamide	776	4,155	9,211	2,035	4,685	13,202
54. Acetic acid	750	2,075	7,061	3,331	17,820	3,778
55. Nitroethane	714	3,782	834	2,935	7,422	8,377
56. Methanol	608	1,487	--	2,247	3,870	2,851
57. Benzyl alcohol	581	1,881	3,527	2,247	8,907	2,084

TABLE 13.5. (Continued)

Solvent		Octane	Toluene	Ethanol	M.E.-ketone	Dioxane	Nitromethane
					Kg Values for Test Solutes		
58.	Dimethylformamide	577	3,840	6,778	2,163	4,946	13,994
59.	Tricresyl phosphate	566	1,487	463	603	1,241	1,452
60.	Methoxyethanol	526	1,961	2,821	1,371	3,956	5,179
61.	Nonylphenoloxethylate	481	1,067	651	380	1,049	1,394
62.	N-Methyl-2-pyrrolidone	459	3,960	7,368	1,662	4,342	12,607
63.	Acetonitrile	418	2,666	1,761	2,935	8,097	13,327
64.	Aniline	356	2,240	1,837	3,685	13,707	5,594
65.	Methylformamide	285	1,528	5,647	1,516	3,632	5,594
66.	Cyanoethylmorpholine	235	1,532	936	853	2,280	3,494
67.	Butyrolactone	208	2,799	2,286	1,901	5,087	9,994
68.	Nitromethane	205	2,041	1,125	2,886	9,898	--
69.	Dodecafluoroheptanol	187	837	6,051	26,282	162,000	3,778
70.	Formyl morpholine	170	1,874	2,644	1,121	3,235	7,362
71.	Propylenecarbonate	168	1,769	1,205	1,543	4,451	8,745
72.	Dimethylsulfoxide	121	1,745	7,061	1,270	3,870	13,994
73.	Tetrafluoropropanol	119	880	8,560	28,437	71,298	8,229
74.	Tetrahydrothiophene-1,1-dioxide	61.0	1,248	1,407	997	3,870	7,362
75.	Triscyanoethoxyprpane	53.0	775	662	649	1,953	4,236
76.	Oxydipropionitrile	49.8	1,010	794	1,039	3,490	6,358
77.	Diethylene glycol	38.9	240	1,407	221	699	927
78.	Triethylene glycol	24.6	263	1,140	201	699	869
79.	Ethylene glycol		105	2,644	270	1,081	743
80.	Formamide		138	3,527	857	3,295	2,739
81.	Water			5,647	469	5,396	1,105

on this polar stationary phase.

These data of Rohrschneider are directly useful in LLC in two ways. First, the K_g values of the various test solutes characterize the selectivity of a given solvent. Differences in K_g between two solvents for these six test solutes indicate corresponding variations in selectivity. The larger are the differences in K_g, the greater is the variation in selectivity. Thus, in the above example, we see major differences between the nonpolar solvent squalane and the polar solvent dimethyl sulfoxide. More subtle differences can be recognized even for solvents of similar polarity; for example, cyclohexanone (No. 33 in Table 13.5) versus pyridine (No. 37). Thus, these two solvents exhibit similar selectivity (and K values) toward octane, toluene, and nitromethane, with preferential retention of methyl ethyl ketone and dioxane by pyridine, and markedly preferential retention of ethanol by pyridine.

A second feature of the data of Table 13.5 is that they can be used to calculate distribution constants K which are proportional to k' values (Eq. 2.5), for any of the six test solutes and any pair of the 81 solvents of Table 13.5. For a stationary phase liquid 1 (and K_g value K_1) and mobile phase liquid 2 (and K_g value K_2), the K value for the given solute and these two liquids as chromatographic phases is given as

$$K = \frac{K_1}{K_2} .$$

Thus, for ethanol as solute, hexane as mobile phase ($K_g = 82$), and dimethyl sulfoxide as stationary phase ($K_g = 7061$), K is calculated as

$$K = \frac{7061}{82} = 86;$$

that is, ethanol is preferentially retained by the polar stationary phase. These data of Table 13.5 can therefore provide many useful examples of the change in K and k' (from Eq. 2.5) for various solutes and pairs of LLC phases.

The data of Rohrschneider become easier to use if we express them in a different form, so as to eliminate the contributions of London or dispersion forces (Section 7.3) to K_g, and express the resulting corrected K_g values in logarithmic form (7). We then arrive at the characteristic solvent parameters of Table 13.6: P', I_{ap}, x_e, x_d, and x_n. Briefly, these solvent parameters can be characterized as follows:

P' A polarity index, which measures the polarity of the solvent. P' determines solvent strength, and is therefore similar to the parameter δ of Table 7.5.

I_{ap} An aromatic/paraffin index, which expresses the preferential retention of aromatic versus aliphatic compounds in that solvent (corrected for differences in P'). I_{ap} is roughly comparable to the parameter δ_d of Table 7.5.

x_e The fraction of P' that is contributed by interactions between the solvent and <u>ethanol</u> as solute. The quantity x_e is roughly comparable to the parameter δ_a of Table 7.5.

x_d The fraction of P' contributed by interactions between the solvent and <u>dioxane</u> as solute. The quantity x_d is roughly comparable to the parameter δ_h of Table 7.5.

x_n The fraction of P' contributed by interactions between the solvent and <u>nitromethane</u> as solute. The quantity x_n is roughly comparable to the parameter δ_o of Table 7.5.

Consider the solvent strength or polarity parameter P' first. Figure 13.5 shows a plot of calculated values of $K/10$ (roughly equal to k' for a 2.5% loading of stationary phase on support), for nitromethane as solute and oxydipropionitrile (ODPN No. 76) as stationary phase. The data points represent k' values for

TABLE 13.6. SOLVENT PARAMETERS DERIVED FROM ROHRSCHNEIDER DATA (5)[a]

Solvent	P'	I_{ap}	x_e (Ethanol)	x_d (Dioxane)	x_n (Nitromethane)
Squalane	-0.8	0.20			
Isooctane	-0.4	-0.10			
n-Decane	-0.3	0.08			
Cyclohexane	0.0	-0.15			
n-Hexane	0.0	0.00			
Carbon disulfide	1.0	0.11			
Carbon tetrachloride	1.7	0.15	0.30	0.38	0.32
Butyl ether	1.7	0.00	0.53	0.08	0.39
Triethyl amine	1.8	-0.22	0.61	0.07	0.32
Isopropyl ether	2.2	-0.15	0.54	0.11	0.35
Toluene	2.3	0.00	0.32	0.24	0.44
p-Xylene	2.4	-0.05	0.32	0.24	0.44
Chlorobenzene	2.7	0.11	0.24	0.34	0.42
Bromobenzene	2.7	0.18	0.24	0.34	0.42
Iodobenzene	2.7	0.30	0.24	0.36	0.40
Phenyl ether	2.8	0.43	0.25	0.33	0.42
Ethoxy benzene	2.9	0.23	0.27	0.29	0.44
Ethyl ether	2.9	-0.22	0.55	0.11	0.34
Benzene	3.0	0.04	0.29	0.28	0.43
Tricresyl phosphate	3.1	0.45	0.35	0.18	0.47
Ethyl bromide	3.1	0.04	0.32	0.28	0.40
n-Octanol	3.2	-0.10	0.61	0.14	0.25
Fluorobenzene	3.3	0.18	0.24	0.33	0.43
Benzyl ether	3.3	0.36	0.27	0.27	0.46
Methylene chloride	3.4	0.20	0.34	0.17	0.49
Methoxybenzene	3.5	0.18	0.28	0.31	0.41
i-Pentanol	3.6	-0.15	0.58	0.17	0.25
Ethylene chloride	3.7	0.32	0.36	0.19	0.45
Bis(2-ethoxyethyl) ether	3.9	0.08	0.35	0.19	0.46
t-Butanol	3.9	-0.30	0.55	0.23	0.22
n-Butanol	3.9	-0.10	0.53	0.21	0.26
n-Propanol	4.1	-0.15	0.54	0.19	0.27
Tetrahydrofuran	4.2	-0.15	0.41	0.19	0.40
2,6-Lutidine	4.3	-0.15	0.47	0.18	0.35
Ethyl acetate	4.3	0.00	0.34	0.25	0.42

TABLE 13.6. (Continued)

Solvent	P'	I_{ap}	x_e (Ethanol)	x_d (Dioxane)	x_n (Nitromethane)
i-Propanol	4.3	-0.30	0.54	0.20	0.26
Chloroform	4.4	-0.10	0.28	0.39	0.33
Acetophenone	4.4	0.20	0.33	0.27	0.40
Methyl ethyl ketone	4.5	0.00	0.36	0.17	0.47
Cyclohexanone	4.5	0.04	0.35	0.23	0.42
Nitrobenzene	4.5	0.20	0.30	0.27	0.43
Benzonitrile	4.6	0.11	0.35	0.26	0.39
Dioxane	4.8	0.08	0.38	0.21	0.41
2-Picoline	4.8	-0.10	0.51	0.19	0.30
Tetramethyl urea	5.0	0.18	0.46	0.14	0.40
Diethylene glycol	5.0	0.38	0.43	0.24	0.33
Triethylene glycol	5.1	0.59	0.43	0.24	0.33
Ethanol	5.2	-0.23	0.51	0.21	0.28
Quinoline	5.2	-0.05	0.40	0.27	0.33
Pyridine	5.3	-0.05	0.43	0.21	0.36
Nitroethane	5.3	0.23	0.31	0.27	0.42
Acetone	5.4	-0.10	0.36	0.24	0.40
Ethylene glycol	(5.4)		0.47	0.23	0.30
Benzyl alcohol	5.5	-0.05	0.42	0.28	0.30
Tetramethylguanidine	5.5	-0.23	0.52	0.11	0.37
Methoxyethanol	5.7	-0.05	0.39	0.25	0.36
Triscyanoethoxypropane	5.8	0.51	0.34	0.25	0.41
Propylene carbonate	6.0	0.30	0.31	0.28	0.41
Oxydipropionitrile	6.2	0.53	0.33	0.28	0.39
Aniline	6.2	0.00	0.34	0.30	0.36
Methyl formamide	6.2	-0.05	0.43	0.21	0.36
Acetic acid	6.2	-0.30	0.41	0.29	0.30
Acetonitrile	6.2	0.04	0.33	0.26	0.41
N,N-Dimethylacetamide	6.3	-0.10	0.43	0.20	0.37
Dimethyl formamide	6.4	0.00	0.41	0.21	0.38
Tetrahydrothiophene-1,1-dioxide	6.5	0.43	0.35	0.27	0.38
Dimethyl sulfoxide	6.5	0.28	0.35	0.27	0.38
N-Methyl-2-pyrolidone	6.5	0.08	0.41	0.21	0.28

TABLE 13.6. (Continued)

Solvent	P'	I_{ap}	x_e (Ethanol)	x_d (Dioxane)	x_n (Nitromethane)
Hexamethyl phosphoric acid triamide	6.6	-0.40	0.49	0.15	0.36
Methanol	6.6	0.3	0.51	0.19	0.30
Nitromethane	6.8	0.00	0.28	0.30	0.42
m-Cresol	7.0	-0.70	0.39	0.36	0.25
Formamide	(7.3)		0.40	0.28	0.32
Dodecafluoroheptanol	7.9	-0.70	0.35	0.40	0.25
Water	(9)		0.40	0.34	0.26
Tetrafluoropropanol	9.3	-1.00	0.36	0.34	0.30
Range		-1.0-+0.59	0.24-0.61	0.07-0.40	0.22-0.49

[a]A parameter x_m for methyl ethyl ketone can also be derived, but is found to correlate closely with x_e.

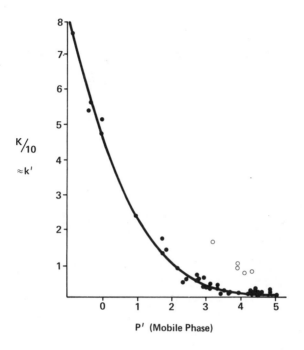

$^{K}/_{10}$

$\approx k'$

P' (Mobile Phase)

Figure 13.5 Dependence of solvent strength on P' in
normal liquid-liquid chromatography; calculated from
data of Table 13.5, assuming ODPN as stationary phase.
O alcohols as solvent (mobile phase).

nitromethane eluted by different mobile phase liquids,
of polarity P' (P' ≤ 5). In actual practice many of
these mobile phases would be miscible with ODPN, but
this difficulty is avoided if we assume that the ODPN
(or similar liquid) is chemically bonded to the support
without altering its solvent properties. Figure 13.5
clearly shows the expected decrease in k' with increasing solvent strength (polarity) P', thus establishing
the usefulness of this parameter. The major exceptions
or deviations in Figure 13.5 are for alcohols as solvents (open points), of which we will say more shortly.

In normal LLC with a polar stationary phase, solvent strength increases with P' as shown in Figure 13.5. In reversed-phase systems, on the other hand, solvent strength decreases for increasing P'. This has already been discussed in terms of solvent polarity in Section 7.3, and is illustrated in Figure 3.14. There, increasing solvent strength occurs for decreasing concentrations of water (i.e., for smaller P') in the methanol/water mobile phase.

The polarity P' of a solvent mixture can be estimated roughly from the P' values of the individual solvent components. For a binary mixture containing $\%_a$ volume percent of liquid A (for which $P' = P_a$), and $\%_b$ of liquid B ($P' = P_b$), the polarity P' of the mixture is

$$P' = 0.01 \; (\%_a \; P_a + \%_b \; P_b).$$

Thus, for a mixture of 20 vol % chloroform in benzene, we calculate P' equal to 0.01 (20 x 4.4 + 80 x 3.0), or 3.3. Since solvent mixtures are often used as mobile phases in LLC, this is an important (but approximate) relationship.

Having discussed solvent strength in terms of the Rohrschneider treatment, let us turn to solvent selectivity. The general aim in optimizing selectivity for a given problem is to maintain k' values within the optimum range--about 1 < k' < 10--while varying α for maximum resolution of all bands. Thus, after establishing what value of P' is best for a given separation, we will want to maintain P' constant while varying the selectivity parameters I_{ap}, x_e, x_d, and x_n. Let us consider an actual example. In Table 13.7 we have assumed ODPN as stationary phase, and have calculated K values for three solutes and three different mobile phases of similar P' value: ethanol, methyl ethyl ketone (MEK), and dioxane as solutes; triethyl amine (P' = 1.8), toluene (P' = 2.3), and carbon tetrachloride (P' = 1.7) as mobile phases. The parameters I_{ap}, x_e, and so on differ greatly among these three mobile phase

TABLE 13.7.   SOLVENT SELECTIVITY IN LIQUID-LIQUID CHRO-
MATOGRAPHY; ODPN STATIONARY PHASE[a] (CALCULATED FROM
DATA OF TABLE 13.5)

Solute	K (Different Mobile Phases)		
	Triethylamine	Toluene	Carbon Tetrachloride
Ethanol	1.15	2.91	5.32
Methylethyl ketone (MEK)	1.54	0.78	1.04
Dioxane	2.07	0.88	0.80
Mobile Phase Parameters			
P'	1.8	2.3	1.7
x_e	0.61	0.32	0.30
x_d	0.07	0.24	0.38
x_n	0.32	0.44	0.32
I_{ap}	0.22	0.00	0.15

[a] Chemically bonded to support to avoid miscibility.

solvents, so we can expect large differences in selectivity. As seen in Table 13.7 this is indeed the case;
in fact, three different elution sequences are found:

Mobile Phase	Relative k' values
Triethyl amine	Ethanol < MEK < dioxane
Toluene	MEK < dioxane < ethanol
Carbon tetrachloride	Dioxane < MEK < ethanol

For a drastic change in selectivity, as when $\alpha = 1$ for a pair of bands of interest, we must seek a new mobile phase with similar P' values but quite different values of I_{ap}, x_e, and so on. In this connection it is worth noting that the addition of a relatively nonpolar solvent (e.g., hexane) to a polar solvent (e.g., chloroform) reduces P' for the polar solvent, but should not affect the values of the selectivity parameters x_e, x_d, and x_n. This allows us to change P' over wide limits while holding selectivity constant. Some calculated examples are shown in Table 13.8, for $P' = 3$. Thus, for any value of P', a wide range in solvent selectivity is possible.

TABLE 13.8.   VARYING SOLVENT SELECTIVITY WHILE HOLDING
P' CONSTANT ($P' = 3$)

Solvent (volume %)	x_e	x_d	x_n
Benzene	0.29	0.28	0.43
11% hexane/methylene chloride	0.34	0.17	0.49
27% hexane/propanol	0.54	0.19	0.27
32% hexane/chloroform	0.28	0.39	0.33
33% hexane/tetramethyl urea	0.46	0.14	0.40
44% hexane/acetone	0.36	0.24	0.40
46% hexane/benzyl alcohol	0.42	0.28	0.30

Although we sometimes want a maximum change in selectivity, particularly for the separation of a very similar pair of compounds, in other cases we are looking for more subtle control over solvent selectivity. This is the case when several compounds of similar k' value are present in the sample, and we want a regular spacing of bands as in Figure 3.17. In this case we can anticipate that increase in one of the selectivity parameters (I_{ap}, x_e, etc.) will move some band centers toward t_o, and others toward larger values of t_R. Once these shifts have been related to the individual selectivity parameters (by actual experiment), the right values of I_{ap}, x_e, and so on for maximum resolution (most even spacing of band centers) can be estimated, and the solvent most closely matching these characteristics can be selected. This approach is obviously somewhat tedious, and would be justified only in setting up a routine separation-analysis system for application to many hundreds or thousands of samples.

Differences in solvent selectivity are also apparent in the data of Figure 13.5. Here the values of x_n for the alcohols (open circles) as mobile phases are noticeably lower than for the remaining solvents (P'< 5). This results in an apparent (modest) failure of the correlation of solvent strength (or k') with P' for the solute nitromethane. Similar minor failures of k' to correlate exactly with P' can be anticipated for actual samples encountered in the laboratory. However, large changes in P' should always result in changing k' in the expected direction. These minor failures of P' to correlate with solvent strength are much less important when using mixtures of polar solvents with nonpolar diluents (e.g., solutions of chloroform/hexane of varying composition). Therefore, such mixtures are often preferred for establishing optimum solvent strength, by observing the change in k' with change in composition of the solvent mixture--as in Figure 3.14.

The Huber scheme referred to earlier is based on an observed relationship between K values and parameters characteristic of the solute (a_1, a_2, a_3) and a given pair of LLC phases (x_1, x_2, x_3):

$$\log K = a_1 \, x_1 + a_2 \, x_2 + a_3 \, x_3 \, . \tag{13.1}$$

The values of the parameters a_1, a_2, x_1, and so on must at present be determined experimentally, and the solvent systems so far studied are restricted to water-poor mixtures of isooctane-ethanol-water as one phase, and water-rich mixtures of these same three solvents as the other phase. Solvent strength and selectivity in this particular ternary-solvent system can be varied over relatively wide limits by varying the proportions of the various solvents. For further details, see ref. 6.

Still another approach to describing the chromatographic properties of various solvents has been suggested recently by Guiochon and co-workers (8). These workers have related solvent strength to the interfacial tension between two liquid phases. At this point it is too early to assess the relative value of this approach versus the treatments summarized in Table 13.4.

13.3   TECHNIQUES FOR SOLVING THE GENERAL ELUTION PROBLEM

The general elution problem (GEP), which refers to the various difficulties presented by broad-range, multicomponent samples, was discussed in Section 3.4. Specialized equipment for carrying out gradient elution--one technique for dealing with the GEP--was covered in Section 4.5, and several examples of gradient elution separation have been presented in Chapters 7-9. Now we will examine some additional techniques for use with the GEP and also discuss the selection and application of one of these techniques for a given problem. For a more detailed discussion of the GEP and these related techniques, see refs. 9-11.

The general elution problem is illustrated in Figure 13.6A. Here isocratic (i.e., fixed-solvent or normal elution) of a multicomponent sample from a silica

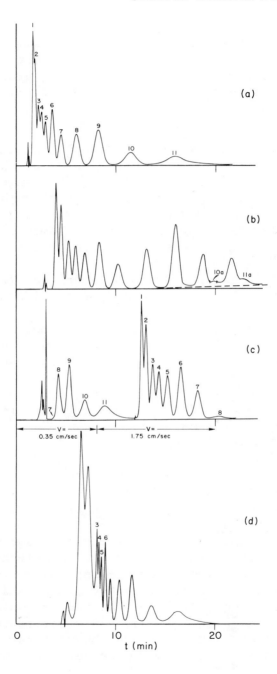

Figure 13.6 Separation of a synthetic mixture of substituted aromatics by liquid-solid chromatography on
silica (11). (a) Isocratic elution; (b) gradient elution; (c) coupled-column separation; (d) flow programming.

column is seen to give poor resolution at the front of
the chromatogram, and attenuated bands at the end of
the chromatogram. Because of the wide range in sample
k' values, most sample bands migrate through the column either too rapidly or too slowly. As a result,
average resolution per unit time is lower than would
be the case if all bands moved through the column with
optimum values of k'.
    In evaluating the various techniques for solving
the GEP, several factors should be considered:

- Average resolution per unit time.
- The range in k' values that can be handled.
- The complexity (and cost) of the necessary
  equipment.
- Possible limitations on the detector.
- Possible need for column regeneration follow-
  ing separation.
- Operational simplicity.

Keep in mind that the GEP only arises when the ratio
of k' values for the last band Z (k_z) and first band A
(k_a) is greater than about 10: $k_z/k_a > 10$.
    Gradient elution is carried out by continuously
varying the composition of the moving phase with time,
so as to provide a steady increase in solvent strength
throughout separation. This is done by mixing a weak
solvent A with a strong solvent B, in such a way that
the concentration of B is low at the beginning of separation, and high at the end. As a result the initial
bands in the chromatogram have their k' values increased, and are separated under near-optimum conditions.

By the time the final bands in the chromatogram have
moved appreciably along the column, solvent strength
has increased to the point where their k' values are
reduced into the optimum range. In this way all bands
migrate through the column with optimum average k' values (2 < k' < 5). The net result of such an operation
is illustrated in Figure 13.6b for the gradient elution
separation of the same sample as in Figure 13.6a.

In gradient elution (see refs. 9,10) the solvent
strength of the moving phase should change with elution
volume V in such a way that

$$\log k' = \underline{a} + \underline{b} \left(\frac{V}{V_m}\right) \tag{13.2}$$

or

$$\log k' = \underline{a} + \underline{b} \left(\frac{t}{t_o}\right). \tag{13.2a}$$

Here k' represents the value for any given band, and t
refers to the time after sample injection. The constant
$\underline{a}$ is determined by the initial band in the chromatogram:
that is, for that band, V or t will be small, so that
$\log k' \approx \underline{a}$. We want k' to fall in the optimum range of
values, so $\underline{a}$ should equal 0-0.3 (log 1 to log 2). This
means simply that the initial mobile phase composition
should be selected so as not to elute the first sample
bands too rapidly.

The constant $\underline{b}$ of Eq. 13.2 measures how rapidly
the composition or strength of the mobile phase is being changed. Large values of $\underline{b}$ mean a rapid change in
solvent composition or a steep gradient. Small values
of $\underline{b}$ mean a slow change in solvent composition or a
relatively flat gradient. The parameter $\underline{b}$ can be regarded as precisely analogous to 1/k' in isocratic elution. Thus for maximum resolution per unit time (10,
11), there is an optimum range of $\underline{b}$ values: 0.2 < $\underline{b}$ <
0.5, corresponding to 2 < k' < 5. If we desire rapid
elution of all sample bands and are prepared to

sacrifice resolution, then larger values of b (steeper gradients) can be used. Another useful feature of steep gradients is that all bands are eluted as much narrower bands, thus providing substantial increase in detection sensitivity.

In an optimum gradient formulation, for which Eq. 13.2 would then apply, the concentration of solvent B in the moving phase must generally increase exponentially with time. That is, a plot of the concentration of B versus t should be concave to the t axis. In liquid-solid chromatography, Eq. 13.2 is equivalent to a linear change in solvent strength $\epsilon°$ with time or elution volume (10). Table 13.9 provides an example of such a gradient for water-deactivated silica as adsorbent.

In ion exchange, with a salt gradient, k' is inversely proportional to salt concentration (9), so that the salt concentration should increase exponentially. Table 13.10 provides a representative example of such a gradient. Linear salt gradients are also often used, apparently with satisfactory results. A pH gradient in ion exchange should vary linearly for Eq. 13.2 to apply (9). Examples of such gradients are illustrated in Table 13.10.

In liquid-liquid chromatography, limited studies (12,13) with chemically bonded stationary phases suggest that log k' is linear with the volume percent B (over the range 20-90 vol % B). This implies that linear solvent composition gradients should be used in gradient elution from these columns (e.g., Table 13.11, and note Figure 4.11A). That is, the percent by volume of the stronger solvent component should increase linearly with time (while flow rate is maintained roughly constant).

As in other LC techniques, the viscosity of the moving phase should be kept as low as possible in gradient elution. Ideally this is accomplished by using nonviscous solvent components (i.e., A and B). However, during most of some separations the solvent will

TABLE 13.9. REPRESENTATIVE SOLVENT GRADIENT FOR LIQUID-
SOLID CHROMATOGRAPHY (WATER-DEACTIVATED SILICA)

t/t_o	$\epsilon°$	Solvent Composition[a]
0	0.00	pentane
2	0.05	3 vol % CH_2Cl_2/pentane
4	0.10	7 vol % CH_2Cl_2/pentane
6	0.15	15 vol % CH_2Cl_2/pentane
8	0.20	25 vol % CH_2Cl_2/pentane
10	0.25	45 vol % CH_2Cl_2/pentane
12	0.30	75 vol % CH_2Cl_2/pentane
14	0.35	3 vol % acetonitrile/CH_2Cl_2
16	0.40	12 vol % acetonitrile/CH_2Cl_2
18	0.45	50 vol % acetonitrile/CH_2Cl_2
20	0.50	100 vol % acetonitrile/CH_2Cl_2
22	0.55	4 vol % methanol/acetonitrile
24	0.60	12 vol % methanol/acetonitrile
26	0.65	25 vol % methanol/acetonitrile
28	0.70	60 vol % methanol/acetonitrile

[a]Note that the thermodynamic activity of water should
remain approximately constant throughout this solvent
series; e.g., 50% water-saturated solvents through t/t_o
equal 12. Also, for UV detectors at high sensitivity,
it may be necessary to balance the absorbance at the
wavelength measured (e.g., 254 nm), for constant baseline. This can be achieved by adding small amounts of
the compound BHT to the pentane, CH_2Cl_2, acetonitrile

and/or methanol from which the overall gradient is formulated.

TABLE 13.10.   REPRESENTATIVE GRADIENTS FOR ION-EXCHANGE CHROMATOGRAPHY[a]

t/t_o	Salt Gradient (NaCl)	pH Gradient[b] Cation Exchange	Anion Exchange
0	0.01 M	2.5	10.0
2	0.03 M	3.0	9.5
4	0.1 M	3.5	9.0
6	0.3 M	4.0	8.5
8	1.0 M	4.5	8.0
10	3 M	5.0	7.5
12		5.5	7.0
14		6.0	6.5

[a]These gradients might be applied independently or in series, but not concurrently.
[b]These gradients can be continued to higher or lower pH values, until elution of sample from the column is complete.

consist of a dilute solution of B, when there is an exponential increase in the concentration of B with time. In these cases solvent B can be somewhat more viscous (e.g., $0.5 < \eta < 2.0$ cP), with a sacrifice in separation efficiency only toward the end of the chromatogram.

Gradient elution often begins with pure solvent A, which means that concentrations of B are initially zero and remain very low for a significant part of the separation. If the solvent strength of B is very much greater than that of pure A (e.g., A equals pentane, B equals methanol, in liquid-solid chromatography), there will be a tendency for complete retention

TABLE 13.11. REPRESENTATIVE GRADIENTS FOR LIQUID-
LIQUID CHROMATOGRAPHY

t/t_o	Reverse-Phase[a]	Normal[b]
0	Water	hexane
2	10 vol % methanol-water	20 vol % CH_2Cl_2[c]-hexane
4	20 vol % methanol-water	40 vol % CH_2Cl_2-hexane
6	30 vol % methanol-water	60 vol % CH_2Cl_2-hexane
8	40 vol % methanol-water	80 vol % CH_2Cl_2-hexane
10	50 vol % methanol-water	100 vol % CH_2Cl_2-hexane
12	60 vol % methanol-water	
14	70 vol % methanol-water	
16	80 vol % methanol-water	
18	90 vol % methanol-water	
20	Methanol	

[a]E.g., Permaphase®-ODS, MicroPak®-CH, or other nonpolar chemically bonded phase.
[b]E.g., MicroPak®-NH2, Permaphase®-ETH, or other polar
chemically bonded stationary phase.
[c]Containing 1.25 vol % isopropanol (12).

of B by the stationary phase. The moving phase leaving the column will then consist of pure A for some
time, rather than exhibiting a continuing increase in
the concentration of B (as required by Eq. 13.2). This
phenomenon (solvent demixing, see Section 8.3) has a
number of undesirable consequences: (a) saturation and
deactivation of the front end of the column, (b) elimination of the solvent gradient in the remainder of the
column, and (c) displacement of unresolved sample bands
immediately ahead of the deactivated portion of the
column. The sum of these effects results in generally
poor separation.
    To avoid solvent demixing, gradients should not

be used in which the solvent strengths of pure A and
pure B are greatly different (e.g., pentane and methanol, as above). If for some reason the use of such
gradients is required, elution should begin with solutions that have appreciable concentrations of B (e.g.,
1-5%), and the column should be pre-equilibrated with
this solution. Where a wide range of solvent strengths
must be covered by a gradient, as in the case of samples with a very wide range in k' values, it is necessary to use more than two gradient components. For example, in Table 13.9 elution is begun with pentane (A)
and methylene chloride (B) as gradient components--the
gradient running from pure A to pure B. At the end of
this initial gradient, we switch to a new gradient
based on methylene chloride (B) and acetonitrile (C).
Finally, the gradient is concluded with the pair acetonitrile (C) and methanol (D). In this way we span the
solvent strength range from pure pentane to pure methanol, but greatly reduce the risk of solvent demixing
effects.

The main advantage of gradient elution--compared
to other LC techniques--is that it provides maximum
resolution per unit time. It is the only technique
that can provide adequate resolution for samples with
k' values k_z/k_a > 1000. This is shown in Figure 13.7
where comparative resolution (in terms of N_{eff}) is
plotted versus the sample k' range k_z/k_a. It is important to note that this improved resolution for complex
samples is reflected in all parts of the chromatogram--
not just at the front end. This effect can be seen in
Figure 13.6b, where the better resolution at the end
of the chromatogram (relative to isocratic elution in
A) results in the resolution of two additional bands,
10a and 11a. Gradient elution also provides maximum
enhancement of band heights for strongly retained compounds, with improved detection of these bands (which
also contributes to the appearance of bands 10a and
11a in Figure 13.6b. For these various reasons,

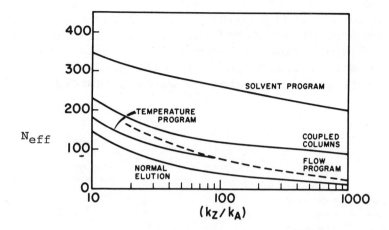

Table 13.7 Comparative resolution of various elution
techniques versus range in sample k' values (k_Z/k_a).
Number of effective plates N_{eff} on a time-pressure
normalized basis (11).

gradient elution is presently the most widely used technique for solving the GEP.

The disadvantages of gradient elution are that the
equipment required is usually complex and/or expensive;
cheaper units can be relatively inconvenient to operate. In addition the column must be regenerated following separation. This is usually accomplished by
running a reverse gradient through the column (e.g.,
ref. 14), particularly if the first solvent (A) cannot
remove strongly retained gradient components (C, D,
etc.). Another disadvantage of gradient elution is
that solvent composition must be matched to the detector to provide a stable baseline. The latter requirement means that refractive index and other bulk detection devices are essentially impractical for gradient
elution, but UV (see footnote a, Table 13.9) and wire-
transport detectors can be used. Finally, gradient

elution cannot be used in liquid-liquid chromatography, except for bonded-phase columns.

An interesting application of gradient elution is the procedure of Scott and Kucera (15), illustrated by the model separation of Figure 13.8. These workers use a single silica column and a solvent gradient of very wide polarity (heptane through water), thereby eluting

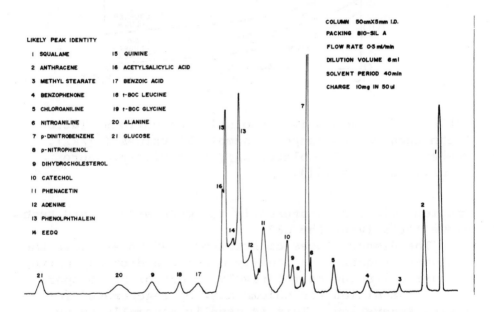

LIKELY PEAK IDENTITY

I	SQUALANE	I5	QUININE
2	ANTHRACENE	I6	ACETYLSALICYLIC ACID
3	METHYL STEARATE	I7	BENZOIC ACID
4	BENZOPHENONE	I8	t-BOC LEUCINE
5	CHLOROANILINE	I9	t-BOC GLYCINE
6	NITROANILINE	20	ALANINE
7	p-DINITROBENZENE	2I	GLUCOSE
8	p-NITROPHENOL		
9	DIHYDROCHOLESTEROL		
I0	CATECHOL		
II	PHENACETIN		
I2	ADENINE		
I3	PHENOLPHTHALEIN		
I4	EEDQ		

COLUMN 50cmX5mm I.D.

PACKING BIO-SIL A

FLOW RATE 0·5 ml/min

DILUTION VOLUME 6 ml

SOLVENT PERIOD 40 min

CHARGE IOmg IN 50 µl

Figure 13.8 Wide polarity range gradient elution, according to Scott and Kucera (15); separation of a synthetic mixture on silica (7-hr separation time). Reprinted from <u>Analytical Chemistry</u> with permission of the American Chemical Society.

both nonpolar and extremely polar compounds in one operation. Thus in the example of Figure 13.8 both squalane and glucose are separated in a single run. The major advantage claimed for this procedure (i.e., wide-

range gradient elution from silica) is that unknown
samples can be handled without varying experimental
conditions to fit each sample. This of course considerably simplifies the application of LC to an unknown
sample. Presumably samples that are not completely
resolved in the first separation are then handled on
an individual basis, using the initial chromatogram as
a guide to further action. Unfortunately, the use of
this technique to maximum advantage requires a universal detector of the moving-wire type (Section 5.4).
These detectors are at present marginal in terms of
sensitivity, and inconvenient to use. Another potential limitation of this technique is its questionable
suitability for all samples. Very polar samples are
usually not well separated on silica, and in many cases
would be predicted to give badly tailing bands, peak
artifacts, etc. While gradient elution overcomes these
difficulties to a certain extent, further study is needed to resolve this possible objection. The particular
solvents chosen by Scott, et al, are also subject to
question, as may be seen by contrasting their discussion with previous treatments of gradient elution (e.
g., 4,9,10) and the role of the solvent in liquid-solid
chromatography (e.g., 16).

Flow programming is a simple technique that can
be useful in dealing with the GEP, if the range in k'
values of sample components is small ($k_z/k_a < 20$). In
flow programming, the column pressure P (and flow rate)
is low during the initial part of the separation, which
increases N (Section 3.5) and improves resolution at
the front end of the chromatogram. Toward the latter
part of the separation P is increased, leading to the
rapid elution of compounds with larger k' values. Although column efficiency is less for later eluting compounds, resolution at the end of the chromatogram is
often not a problem. The application of flow programming to a sample is illustrated in Figure 13.6d. Apart
from the requirement that k_z/k_a be relatively small,

flow programming is best suited for samples that require little resolution at the end of the chromatogram.
An example would be a sample with several early-eluting bands (e.g., $0.5 < k' < 5$) and a single late-elu-
ting band (e.g., $k' = 10$).

Flow programming can be used with any type of column, but it should be kept in mind that this technique
is based on a change in N as a result of changing u.
When N is not very sensitive to changes in u (i.e.,
small values of n in Eq. 2.8), flow programming is less
useful. In such cases column length should be increased, the system run at maximum pressure (Section
3.5), and isocratic elution used.

The advantage of flow programming is simplicity
of equipment and operation, with an absence of experimental complications such as column regeneration and
variation of the detector baseline. Since refractometers are temperature sensitive, the use of flow programming with refractometer detectors requires careful
thermostatting of the detector. The disadvantages of
flow programming are: (1) it is not applicable to
broad range samples, (2) average resolution per unit
time is not much better than with isocratic elution
(Figure 13.7), (3) its application requires manual intervention by the operator--unless a timer is used to
control the pump output, and (4) it does not result in
any enhancement of band heights at the end of the chromatogram.

Coupled-columns is a recently proposed technique
(11; see also 17) that adjusts sample k' values through
the use of several interconnected columns; that is,
stationary phase programming. One possible arrangement
is shown in Figure 13.9. A short fore-column 1 is connected through a four-way valve to three secondary columns 2-4. The various column packings are selected
(see below) so as to provide small k' values in columns
1 and 4, large k' values in column 2, and intermediate
k' values in column 3. The operation of this coupled-

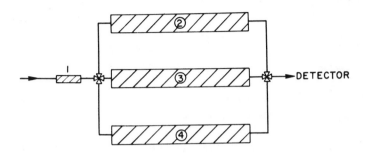

Figure 13.9   Column arrangement for carrying out coupled-column separation.

column set proceeds as follows.   The sample first en-
ters column 1 and is partially separated on this column
into three groups of compounds: fast, intermediate,
and slow-moving.   Because all k' values on column 1 are
reduced within some convenient upper limit (e.g., $k' =$
20), and column 1 is relatively short, little time is
required for this preliminary separation.   Column 1 is
initially connected through the four-way valve to column 2, and fast-moving compounds are sent to column 2
and held there temporarily.   After collection of this
initial group of compounds on column 2, the four-way
valve is switched to connect column 1 to 3, and the
intermediate group of compounds is collected on column
3.   Finally, the last, slow-moving group of compounds
from column 1 is collected on column 4.   At this point
elution is continued through columns 1 and 4, until
all slow-moving (large k' values) bands have left column 4.   The four-way valve is now switched to permit
elution of column 3, and then column 2.   In this way
each group of compounds is separated on a secondary
column which provides near-optimum values of k' during
elution.   As few as two secondary columns might be used
in coupled-column separation, or as many as four.   Figure 13.6c illustrates the application of coupled-column

separation with two secondary columns.

Coupled-column operation can be combined with flow
programming (18), as illustrated by the column arrangement of Figure 13.10 for liquid-solid chromatography.
An intermediate length of a weakly retentive column (I)
is connected via a three-way valve to a short length
of a more active column (II), which in turn is connected through a three-way valve to a longer column of the
same type (III). Sample is injected and elution begun
with all three columns in series. At time t_o for the
three-column set (i.e., when a nonretained band reaches the end of column III), the first three-way valve
is switched to allow elution of column I directly to
the detector. Pump pressure is maintained, so that the

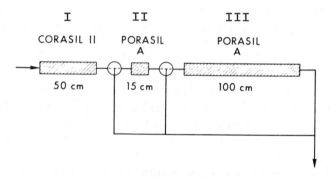

Figure 13.10   Column arrangement for carrying out
coupled-column liquid-solid chromatography with flow
programming (4); column diameter, 3 mm.

flow rate through column I is now much increased. In
this way strongly retained bands are quickly eluted
from column I. When no further elution from column I
is observed, the three-way valve is switched again to
allow serial elution of columns I and II to the detector (requiring change in the position of the second
three-way valve). When no further elution of sample

from column II is evident, the second three-way valve
is again changed to allow serial elution of columns I-
III through the detector. This technique is most conveniently applied with pumps of the constant pressure
type. Examples are shown in Figure 13.11 of the separation in this way of a model antioxidant mixture and
of an actual polymer extract.

The variation of column properties to allow for
different k' values in coupled columns (with the solvent held constant throughout) can be achieved in many
ways. Thus we can simply vary the concentration of
stationary phase (or the volume of stationary phase)
within the column, as by using porous particles for
strong retention and pellicular particles for weak retention (see Eq. 2.5). In the same way the relative
loading of stationary phase on particles in liquid-
liquid chromatography can be varied. In liquid-solid
chromatography we can use adsorbents that have been deactivated to varying extents (e.g., by added glycerine),
including the use of permanently deactivated adsorbents
(e.g., Vydac®, Sil®-X) in conjunction with conventional adsorbents (see Table 8.2 and related discussion).
We can also use adsorbents of differing surface area
(e.g., Porasil® series), since k' values are approximately proportional to the specific surface area of an
adsorbent (for surface areas less than 400 m^2/g). Finally in ion-exchange chromatography it should be possible to change k' by changing the number or concentration of ion-exchange groups in the resin.

The advantages of coupled-column operation are
that it avoids the equipment and operational complexity of gradient elution, since only a four-way (low volume) switching valve and a tee are required. Furthermore, since the same solvent is used throughout separation, column regeneration is not required, and any
column packing or detector can be used. Finally, the
advantages of coupled-column operation in terms of
resolution per unit time are almost as favorable as

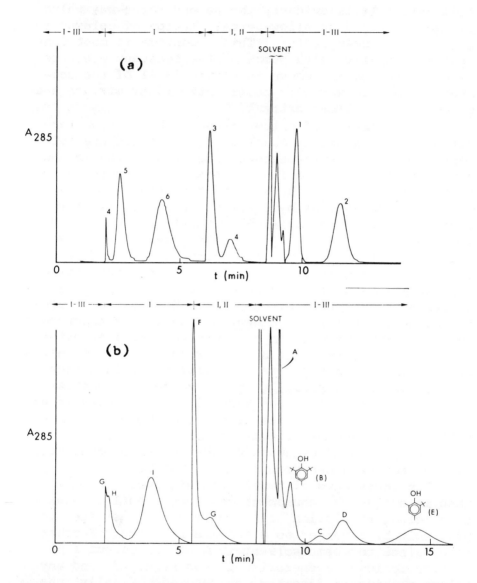

Figure 13.11   Separation of synthetic antioxidant mixture (a) and polymer extract (b) with column arrangement of Figure 13.10 (4).   Solvent: 17 vol % methylene chloride/pentane.

for gradient elution (cf., Figures 13.6B, C). The major disadvantage of this technique is that it is only useful for samples of moderate complexity (i.e., k_z/k_a < 1000). Whereas gradient elution is not well suited to the routine, repetitive analysis of similar samples, coupled columns are expected to be quite useful in such applications (particularly with a column arrangement as in Figure 13.9).

Temperature programming refers to an increase in column temperature during separation, which normally provides a decrease in all sample k' values as separation proceeds. Thus temperature programming is conceptually quite similar to gradient elution. Temperature programming has been little used in LC, mainly because the technique offers little in the way of improved separation, relative to isocratic elution (11). The problem with temperature programming in LC is that solvent viscosity changes during the separation, so that the initial viscosity is very much greater than desirable. This effectively degrades average column efficiency and counteracts the beneficial change in k' that occurs during temperature programming, as can be seen from the data of Figure 13.7. Further disadvantages of temperature programming include (a) applicability to only a narrow k' range in the sample (as a result of the limited temperature range available, before boiling of the solvent), and (b) the need for higher separation temperatures, which precludes the handling of thermally of thermally unstable compounds.

To summarize, when do we use a particular elution technique?

- For difficult separations, unknown samples and/or wide range samples, use gradient elution.

- For the repetitive analysis of samples with less than 1000-fold variation in k' values (normal elution), use coupled-columns.

- For the analysis of simple mixtures, where k'
  values vary by a factor of less than 20, use
  flow programming or normal elution.

- Do not rely on temperature programming, except
  in special cases.

REFERENCES

1.   J. J. Kirkland, Gas Chromatography 1972, S. G.
     Perry, ed., Applied Science Publishers, Ltd.,
     Essex, England, 1973.
2.   J. M. Gill, J. Chromatog. Sci., 7, 731 (1969).
3.   R. E. Leitch, J. Chromatog. Sci., 9, 531 (1971).
4.   L. R. Snyder, in Modern Practice of Liquid Chroma-
     tography, J. J. Kirkland, ed., Wiley-Interscience,
     New York, 1971, Chapter 4; R. A. Keller, B. L.
     Karger, and L. R. Snyder, in Gas Chromatography
     1970, R. Stark and S. G. Perry, eds., Institute
     of Petroleum, London, 1971, p.
5.   L. Rohrschneider, Anal. Chem., 45, 1241 (1973).
6.   J. F. K. Huber, E. T. Alderlieste, H. Harren, and
     H. Poppe, Anal. Chem., 45, 1337 (1973); J. F. K.
     Huber, C. A. M. Meijers, and J. A. R. J. Hulsman,
     Anal. Chem., 44, 111 (1972).
7.   L. R. Snyder, to be published.
8.   C. Eon, B. Novosel, and G. Guiochon, J. Chromatog.,
     83, 77 (1973).
9.   L. R. Snyder, Chromatog. Rev., 7, 1 (1965).
10. L. R. Snyder and D. L. Saunders, J. Chromatog.
     Sci., 7, 195 (1969).
11. L. R. Snyder, J. Chromatog. Sci., 8, 692 (1970).
12. R. E. Majors, private communication.
13. Du Pont Application Sheet, 820M9, Instrument
     Products Division.
14. R. E. Majors, Anal. Chem., 45, 755 (1973).
15. R. P. W. Scott and P. Kucera, Anal. Chem., 45,
     749 (1973).

16. L. R. Snyder, Principles of Adsorption Chromatog-
     raphy, Marcel Dekker, ed., New York, 1968, Chap-
     ter 8.
17. J. F. K. Huber and R. van der Linden, J. Chroma-
     tog., 83, 267 (1973).
18. L. R. Snyder, in Modern Practice of Liquid Chroma-
     tography, J. J. Kirkland, ed., Wiley-Interscience,
     New York, 1971, Chapter 6.

BIBLIOGRAPHY

H. Barth, E. Dallmeier, G. Courtois, H. E. Keller, and
   B. L. Karger, J. Chromatog., 83, 289 (1973). (Precision in modern LC with a dedicated computer.)
W. F. Beyer, Anal. Chem., 44, 1312 (1972). (Quantitative LC of sulfonylureas in pharmaceutical products.)
C. A. Burtis, J. Chromatog., 51, 183 (1970). (Determination of the base composition of RNA.)
M. Goedert and G. Guiochon, Analusis, 1, 443 (1972).
   (Propagation of random errors in peak-area measure-
   ments.)
G. Guiochon, M. Goedert, and L. Jacob, in Gas Chromatography 1970, R. Stock and S. G. Perry, eds., Institute of Petroleum, London, 1971, p. 160. (Detailed
   study of the measurement of peak areas.)
R. A. Henry and J. A. Schmit, Chromatographia, 3, 116
   (1970). (Analysis of analgesics.)
J. J. Kirkland, R. F. Holt, and H. L. Pease, J. Agr.
   Food Chem., 21, 368 (1973). (Determination of benomyl fungicide residues in soils and plant tissues.)
J. G. Koen and J. F. K. Huber, Anal. Chim. Acta, 51,
   303 (1970). (Analysis of parathion and methyl parathion residues.)
L. F. Krzeminski, B. L. Cox, P. N. Perrel, and R. A.
   Schlitz, J. Agr. Food Chem., 20, 970 (1972). (Deter-
   mination of methyl prednisolone residues in milk.)

482   Other Topics

C. A. M. Meijers, J. A. R. J. Hulsman, and J. F. K. Huber, Z. Anal. Chem., 261, 347 (1972). (Determination of cortisol in serum.)

J. Mellica and R. Strusz, J. Pharm. Sci., 61, 444 (1972). (Analysis of corticosteroid creams and ointments.)

J. J. Nelson, J. Chromatog. Sci., 11, 28 (1973). (Determination of food additives.)

L. J. Papa and L. P. Turner, J. Chromatog. Sci., 10, 747 (1972). (Determination of carbonyl compounds as their 2,4-dinitrophenylhydrazones.)

S. Ray and R. W. Frei, J. Chromatog., 71, 451 (1972). (Determination of polynuclear aza-heterocyclics.)

R. W. Ross, J. Pharm. Sci., 61, 1979 (1972). (Determination of barbiturates in pharmaceuticals.)

R. C. Williams, J. A. Schmit, and R. A. Henry, J. Chromatog. Sci., 10, 494 (1972). (Analysis of fat-soluble vitamins by gradient elution.)

LIST OF SYMBOLS

The following list is divided into commonly used and
less commonly used symbols, for convenience in referring to the symbols cited in this book. Units for
each symbol are given in parentheses.

Common Symbols

d_p Particle diameter of column packing
(cm in various equations containing
d_p, μ elsewhere).

d_c Column internal diameter (cm).

D Coefficient in Eq. 2.8; a measure of
column efficiency (proportional to H)
(cm).

D_m, D_s Solute diffusion coefficient in mobile
and stationary phases, respectively
(cm^2/sec).

F Mobile phase flow rate (ml/sec unless
otherwise specified).

h Reduced plate height, equal H/d_p
(dimensionless).

h_r, h_1, h_2	Heights of valley between two bands, band 1 and band 2, respectively (Figure 3.10).
H	Height equivalent of a theoretical plate, equal to L/N (cm).
k'	Solute capacity factor; equal to total amount of solute in stationary phase divided by total amount of solute in mobile phase (within column, at equilibrium) (Section 2.2).
K	Solute distribution constant (Eq. 2.5).
K'	Column permeability (Eq. II.2) (cgs units).
K_o	Distribution constant in gel chromatography, equal to $(V_R - V_o)/V_i$ (Section 10.2).
L	Column length (cm).
L_m	Molecular length (Section 10.4) (Å)
n	Coefficient in Eq. 2.8, determined as in Figure 6.9; required in use of Tables 3.2-3.5 (if not known, assume n equals 0.4).
N	Column plate number (Eq. 2.6).
P	Pressure drop across column (usually in psi, but dynes/cm^2 for equations involving P (1 atm = 14.7 psi = 1.01 x 10^6 dynes/cm^2).
R_S	Resolution function (Eq. 2.10).
R_1, R_2	Value of R_S in initial (1) and final (2) separations.
t	Time required for completion of separation, equal (approximately) to t_R

for last eluted band (Eq. 2.12) (sec). Also, the time of separation from sample injection (Eq. 13.1a).

t_R — Retention time of a given band (Figure 2.4) (sec).

t_w — Baseline band width in time units (Figure 2.4) (sec).

t_o — t_R value for mobile phase molecules injected as sample; also (except in gel chromatography), t_R for unretained sample components.

u — Mobile phase velocity (cm/sec).

V_m — Total volume of mobile phase within column (ml); equal to $t_o F$.

V_R — Solute retention volume (Eq. 2.4) (ml).

α — Separation factor (Eq. 2.11); equal to k' for second band divided by k' for first band.

ϵ° — Solvent strength parameter in liquid-solid chromatography (Table 8.3).

η — Solvent viscosity (Poise in equations; cP in tables).

$\theta_{0.1}$ — Column linear capacity; weight of sample/weight of column packing such that k' decreases by 10% from its constant value at small sample sizes (Figure 2.10).

ν — Reduced velocity, equal to $u\, d_p/D_m$.

Less Common Symbols

a, b — Constants in Eq. 13.1

a_1, a_2, a_3	Solute characterization factors; Eq. 13
C_d, C_e, C_m, C_s, C_{sm}	Plate height coefficients (Section 2.3).
d_f	Thickness of stationary phase layer (cm).
f	Total column porosity (Eq. II.2).
I_{ap}	Aromatic/paraffin index; Section 13.2.
k_1, k_2	Values of k' for bands 1 and 2.
k_a, k_z	Values of k' for first and last bands in a chromatogram (Section 13.3), using the same mobile phase and other experimental conditions.
K_g	Solute solubilities expressed as GC distribution constants; Table 13.5.
L_1, L_2	Column lengths L, in initial and final separation (cm) (Eq. 3.6).
N_{eff}	Column effective plate number (Eq. 3.1); value of N, corrected for effect of k' on resolution.
P'	Polarity index; Section 13.2.
P_1, P_2	Column pressure drop P in initial and final separations (psi) (Eq. 3.5).
R	Fraction of total sample molecules in the mobile phase, equal to $1/(1 + k')$.
$\underline{R}$	Corrected equivalent retention volume; equal to $V_R - V_m$ divided by weight of column packing (ml/g).
t_1, t_2	Values of t_R for bands 1 and 2 (Figure 2.6) (sec); also, times of separation, initial and finally (Eq. 3.5).

t_{w1}, t_{w2}	Values of t_w for bands 1 and 2 (Figure 2.6) (sec).
u_x	Velocity of sample molecules as they move through the column (Eq. 2.1) (cm/sec).
V_o, V_i	Volume of mobile phase (within column), respectively, outside and inside of gel particles (Figure 10.2) (ml).
V_i, V_j	Band broadening (ml) occurring in extra-column elements i and j (Eq. 2.9).
V_x	Volume of sample injected (ml).
V_s	Volume of stationary phase within column (ml).
V_w	True baseline volume of sample band (equal $t_w F$) (ml).
V_w'	Apparent baseline volume of sample band (Eq. 2.9)(ml).
V_q	Quinoline retention volume (Figure 8.8).
x	Peak-height ratio for two overlapping bands (Section 3.2).
x_e, x_n, x_d	Solvent selectivity parameters, Section 13.2.
$(X)_s$, $(X)_m$	Concentration of solute in stationary and mobile phases, respectively (moles/ml).
x_1, x_2, x_3	Solvent characterization factors; Eq. 13.1.
δ_d, δ_o, δ_a, δ_b	Solvent selectivity parameters; Section 13.2.
δ	Hildebrand solubility parameter; Section 7.3.

APPENDIX I

DEPENDENCE OF H ON EXPERIMENTAL CONDITIONS

According to Eq. 2.7a, a plot of h versus ν should give a single curve for different columns--regardless of differences in d_p, solvent or sample. This is the case for well-packed columns of porous particles (1, 2), or for pellicular particles (2,3). However, because the last term of Eq. 2.7a is zero for pellicular packings, two different h-ν plots are obtained for these two types of packings. This is illustrated in Figure I.1 for data taken from well-packed columns of particles in the 20-100 μ range. The lower dashed line for 480-μ glass beads represents a practical limit for an ideally packed column of pellicular particles. The actual solid curve is not very different, suggesting that further changes in column packing technique can yield only minor improvement (see ref. 2).

Experimental data for a given LC column can be compared with the curves of Figure I.1, to evaluate how good that column is. Generally it will be found that actual data fall somewhat above the appropriate curve in Figure I.1, because the latter curves are for very well-packed columns. When a column shows higher h values (relative to Figure I.1) at low values of ν (3 < ν < 30), this suggests a poorly packed column.

489

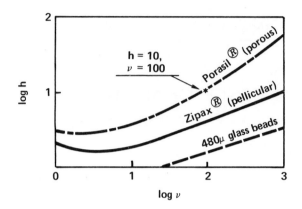

Figure I.1 Reduced plate height h versus velocity ν plots for well-packed columns (2). Typical* values of H and u for corresponding values of h and ν , and different size particles:

d_p	Value of H for h = 10 (mm)	Value of u for ν = 100 (cm/sec)
50	0.5	0.2
20	0.2	0.5
5	0.05	2.0
1	0.005	10.0

* Assumes $D_m = 10^{-5}$.

When a column agrees with the curves of Figure I.1 at low values of ν , but shows larger h values at high values of ν (30–1000), this suggests a defect in the column packing material (see discussion in ref. 2).

Figure I.1 also shows typical values of H and u for each reduced plot, as a function of particle size. The minimum in H which occurs at velocities of 0.01–

0.1 cm/sec ($\nu < 5$) represents the lowest desirable flow rate, and usually we will want to operate at velocities at least 3-10 times larger (see Section 3.5). In this region of ν values, term (ii) of Eq. 2.7a is negligible, and h is given as

$$h = \frac{1}{(1/C_e) + (1/C_m\nu)} + C_{sm} \ ,$$

or

$$H = \frac{1}{(1/C_e d_p) + (1/C_m u d_p^2/D_m)} + \frac{C_{sm} \, u \, d_p^2}{D_m}; \qquad (I.1)$$

that is, retaining only terms (i) and (iii) of Eq. 2.7. We see from Eq. I.1 that small H values are favored by small values of u and d_p, and large values of D_m. The sample diffusion coefficient D_m can in turn be approximated by the Wilke-Chang equation (4):

$$D_m = \frac{7.4 \times 10^{-8} \ (\psi_2 \, M_2)^{0.5} T}{\eta V_1^{0.6}} \ . \qquad (I.2)$$

Here M_2 is the molecular weight of the solvent, T is the temperature in °K, η is the solvent viscosity, and V_1 is the molar volume of the sample molecule. The association factor ψ_2 is unity for most solvents, 2.6 for water, 1.9 for methanol, and 1.5 for ethanol. Equation I.2 predicts that D_m is larger for less viscous solvents, for higher temperatures, and for small sample molecules. Therefore H, which decreases for larger D_m values, will be small for less viscous solvents, higher temperatures, and small sample molecules. Data on the dependence of H on η have been reported for liquid-solid chromatography (5,6), showing the expected increase of H with increasing solvent viscosity. At a u value of 1 cm/sec these data suggest that H is proportional to $\eta^{0.4}$; thus a doubling of η leads to a 20% increase in H. Because column permeability K is proportional to $1/\eta$ (Eq. 2.12a), the column performance factor $K^{(1-n)/2}/D$ of Eq. 3.2 is proportional to $\eta^{-0.8}$. This means (Eq. 3.2) that--other factors being equal--a doubling of solvent viscosity leads to a doubling of separation time.

Note for Eq. I.1 that the second term on the right is small for pellicular packings, so that at high values of u this equation becomes

$$H \approx C_e\, d_p \; . \tag{I.3}$$

That is, H approaches a limiting value which is now independent of u and D_m. In practice this is observed only rarely, and only over a limited range of u values. However, Eq. I.3 correctly predicts that the effect of D_m on H will be small for high values of u and pellicular packings, as has been observed experimentally (6). It also predicts that the dependence of H on solvent viscosity will be less at higher values of u, as observed in (7).

Other expressions for H as a function of u have been suggested (2,8), which combine the theoretical expression Eq. 7 with certain empirical observations. These relationships, in our opinion, have limited direct practical utility. They are essentially hybrids of Eqs. 2.7 and 2.8, and are mainly useful for detailed analysis of very precise H versus u data (as in fundamental studies of column performance).

REFERENCES

1.   R. E. Majors, J. Chromatog. Sci., 11, 88 (1973).
2.   G. J. Kennedy and J. H. Knox, J. Chromatog. Sci., 10, 549 (1972).
3.   J. J. Kirkland, J. Chromatog. Sci., 10, 592 (1972).
4.   J. C. Giddings, Dynamics of Chromatography, Marcel Dekker, New York, 1965, p. 233.
5.   L. R. Snyder, Anal. Chem., 39, 698 (1967).
6.   C. G. Horvath, B. A. Preiss, and S. R. Lipsky, Anal. Chem., 39, 1422 (1967).
7.   G. J. Krol, C. A. Mannan, F. Q. Gemmill, Jr., G.

E. Hicks, and B. T. Kho, J. Chromatog., <u>74</u>, 43 (1972).

8.   J. F. K. Huber, Chimia, Supplement, 1970 (Fifth International Symposium Separation Methods: column chromatography, Lausanne, 1969), p.24

APPENDIX II

FLOW RELATIONSHIPS IN LC

The quantity t_o is defined as L/u, and u is given by Darcy's law as

$$u = \frac{K\ P}{L} \qquad (II.1)$$

Here P is the pressure drop across the column, and K is a permeability constant for the column and the given set of experimental conditions. The quantity K is in turn given by the Kozeny-Carmen relationship (see below) for well-packed columns

$$K = \frac{d_p^2}{1000\ \eta\ f} \cdot \qquad (II.2)$$

Here η is the viscosity of the solvent (Poise) and f is the total column porosity, ranging from 0.4 for pellicular packings to 0.8 for typical porous packings. Inserting the above relationship into Eq. 2.12, followed by substitution into $t_o = L/u$, then gives

$$t_o = \frac{1000\ \eta\ f\ L^2}{P\ d_p^2} \cdot \qquad (II.3)$$

The specific permeability K° of a column is

defined as

$$K^O = \frac{u \, \eta \, L \, f}{P} .$$
                                                    (II.4)

The Kozeny-Carmen relationship (1) gives

$$K^O = \frac{d_p^2 f_o^3}{180 \, (1 - f_o)^2}$$
                                                    (II.5)

Here f_o is the interparticle porosity, equal to about 0.4 for well-packed columns. To a good approximation Eq. II.2 can then be written (2) as

$$K^O = \frac{d_p^2}{1000} .$$
                                                    (II.5a)

Kirkland (3) has reported values of K^O for a number of spherical LC supports; these are about twice as large as values predicted by Eq. II-5a, suggesting that spherical particles give generally greater column permeabilities.

REFERENCES

1. J. C. Giddings, Dynamics of Chromatography, Marcel Dekker, New York, 1965.
2. I. Halasz and E. Heine, in Progress in Gas Chromatography, J. H. Purnell, ed., Wiley-Interscience, New York, 1968, p. 153.
3. J. J. Kirkland, J. Chromatog. Sci., 10, 592 (1972).

APPENDIX III

RESOLUTION R_s AS A FUNCTION OF COLUMN PRESSURE P, SEPARATION TIME t, AND PROPERTIES OF THE COLUMN (K,D,n)

We begin with Eq. 3.1,

$$N_{eff} = \frac{N\ k'}{(1 + k')^2} \ . \qquad (3.1)$$

N can in turn be related to L and u through Eq. 2.6a and 2.8:

$$N = \frac{L}{H}$$

$$= \frac{L}{D\ u^n} \ .$$

The quantity u is next eliminated by means of Eq. II.1 (Appendix II):

$$N = \frac{L}{D\ (KP/L)^n}$$

$$= \frac{L^{1+n}}{D\ K^n\ P^n} \ . \qquad (III.1)$$

The time of separation t is given from Section 2.5:

$$t = t_0 (1 + k'); \hspace{2cm} \text{(III.2)}$$

t_0 equals L/u, where u is given by Eq. II.1, so that

$$t_0 = \frac{L^2}{K\,P} .$$

Eliminating t_0 between this relationship and Eq. II.2, and solving for L, gives

$$L = \frac{(t\,K\,P)^{1/2}}{(1 + k')^{1/2}} .$$

Inserting this expression for L into Eq. II.1 then gives Eq. 3.2. See ref. 1 for further details.

REFERENCE

1. L. R. Snyder, in Gas Chromatography 1970, R. Stark and S. G. Perry, eds., Institute of Petroleum, London, 1971, p. 81.

APPENDIX IV

RESOLUTION AS A FUNCTION OF N: VARIOUS OPTIONS FOR INCREASING R_s

It is assumed that α and k' will remain unchanged as we attempt to vary R_s from some initial value R_1 to some final value R_2. From Eq. 2.11, therefore,

$$\frac{R_2}{R_1} = \left[\frac{N_2}{N_1}\right]^{1/2} \qquad (IV.1)$$

Here, N_1 and N_2 refer to initial and final values of N. Combination of this relationship with Eqs. 2.6a and 2.8 gives

$$\frac{R_2}{R_1} = \left[\frac{L_2}{L_1}\right]^{1/2} \left[\frac{H_1}{H_2}\right]^{1/2} \qquad (IV.2)$$

$$= \left[\frac{L_2}{L_1}\right]^{1/2} \left[\frac{u_1}{u_2}\right]^{n/2} \qquad (IV.3)$$

Here, L_1 and L_2 refer to initial and final column lengths, u_1 and u_2 to initial and final solvent velocities. We next derive relationships for the required change in experimental conditions to give some required change in R_s.

Decrease P, hold L constant

Since $L_1 = L_2$, from Eq. IV.3 we get

$$\frac{R_2}{R_1} = \left[\frac{u_1}{u_2}\right]^{n/2} . \tag{IV.4}$$

From Eq. II.1 (L, K constant) we get

$$\frac{u_1}{u_2} = \frac{P_1}{P_2} ; \tag{IV.5}$$

P_1 and P_2 refer to initial and final column pressures. Equations IV.4 and IV.5 now yield

$$\frac{R_2}{R_1} = \left[\frac{P_1}{P_2}\right]^{n/2} , \tag{IV.6}$$

which gives P_2/P_1 as a function of the required change in R_s (Eq. 3.4). Separation time is given by (L/u) $(1 + k')$, so for L and k' constant

$$\frac{t_2}{t_1} = \frac{u_1}{u_2} . \tag{IV.7}$$

Here t_1 and t_2 refer to the initial and final separation times t. Equations IV.5 and IV.7 then give Eq. 3.5.

Increase L, hold P Constant:

Since P is held constant, as is K, Eq. II.1 can be written

$$\frac{u_1}{u_2} = \frac{L_2}{L_1} \tag{IV.8}$$

Equations IV.3 and IV.8 combine to give

$$\frac{R_2}{R_1} = \left[\frac{L_2}{L_1}\right]^{(1+n)/2} , \tag{IV.9}$$

which then yields Eq. 3.6.   Similarly, from $t_0 = L/u$ and Eq. II.2,

$$t = \frac{L}{u} (1 + k')$$

and, therefore

$$\frac{t_2}{t_1} = \frac{L_2}{L_1} \frac{u_1}{u_2} . \qquad \text{(IV.10)}$$

Equations IV.8 and IV.10 give Eq. 3.7.

Increase P and L, Hold t Constant

Since t is constant, from Eq. IV.10

$$\frac{L_1}{L_2} = \frac{u_1}{u_2} \qquad \text{(IV.11)}$$

Combining Eqs. IV.3 and IV.11 gives

$$\frac{R_2}{R_1} = \left[\frac{L_2}{L_1}\right]^{(1-n)/2} \qquad \text{(IV.12)}$$

or Eq. 3.8.   Equation 3.9 is then obtained from Eqs. II.1 and IV.11.   For further details and discussion, see ref. 1.

REFERENCE

1. L. R. Snyder, J. Chromatog. Sci., <u>10</u>, 369 (1972).

APPENDIX V

COMMERCIALLY AVAILABLE EQUIPMENTS FOR MODERN LC

The following tables are a listing of some equipment that is available for use in modern LC. The materials listed have specifically been designed for, or have been found to be useful for, high performance LC. Information on most of this equipment was abstracted from available literature and brochures. There are other designs and other suppliers which also undoubtedly have been, or could be, used successfully; therefore, this is only a partial listing of satisfactory equipment. These lists primarily refer to equipment available in the United States, although also listed are some foreign units that can be secured in the United States from local suppliers or obtained conveniently from foreign sources.

A. COMPLETE LC INSTRUMENTS

1. Research LC Instruments

Manufacturer	Detectors Available (see B)	Maximum Pressure of Pumping System (psi)	Type of Pumping System	Availability of Gradient Elution	Approximate Price Range ($)[a]
Chromatec (1)[c]	a,b	3450	E[d]	Yes	8,500[b]–10,900
Chromatronix (2)	a,b,d,h	7000	A	Yes	6,200–10,000
Du Pont (3)	a,b,e,f	4500	E	Yes	8,200–11,700
Hupe & Busch (4) (now Hewlett-Packard (5))	a,b,d,g,h,j	5000	B	Yes	12,800–18,000
Knauer (12)	a,b	NA	NA	NA	7,200
Meci (14)	a,d,h	4500	B	No	10,000
Packard (6)	b	3000	B	No	NA
Perkin-Elmer (7)	a,b,e	4000	C	Yes	8,000–10,800
		3000	C	Yes	13,800–16,200
Pye Unicam (8)	c	3565	B	Yes	13,200
Siemans (9)	g	5000	B	NA	NA
Varian Aerograph (10)	a,b,d,g	5000	C	Yes	8,000–12,000
Waters (11)	a,b,d	1000, 3000; 6000	A	Yes	6,000–11,000
Micromeritics (57)	a,b	6000	A	Yes	10,900

NA, not available.

aLower price: with UV detector (except as noted), but without recorder, column temperature control and gradient elution system; higher price reflects cost of two detectors (UV and RI) and column temperature control. For gradient elution, add $4000–8000, depending on instrument.

bIncludes column temperature control.

cSee G. for full names and addresses.

dSee C. for key to pump types.

A. COMPLETE LC INSTRUMENTS (CONTINUED)

2. Basic LC Instruments

Manufacturer	Detectors Available (see B)	Maximum Pressure of Pumping System (psi)	Type of Pumping System	Availability of Gradient Elution	Approximate Price Range ($)[a]
Chromatec (1)[b]	a,b	1000; 2000	A[c]	Yes	2600–5400
Chromatronix (2)	a,b,d,h	3000	D	No	2900–5800
Du Pont (3)	a,b,e	1500	D	No	2800–5700
Isco (19)	a,d,e	3000	C	No	2700–3300
Perkin-Elmer (7)	a,b,e	1000	A	Yes	3300–7400
Varian Aerograph (10)	a,b,d	1000	D	No	2100–6100
Waters (11)	a,b,d	1000	A	Yes	4500–7500
Molecular Separations (69)	a	2000	D	No	1900

[a]Lower price includes UV detector, but no recorder or column temperature control; higher price reflects cost of two detectors, column temperature control.
[b]See G. for full names and addresses.
[c]See C. for key to pump types.

B. LC DETECTORS (MODULAR UNITS)

Supplier	Type	Approximate Price (June, 1973)
Beckman (20)	i	$5000
Chromatec (1)	a	1350
	g	NA
Chromatronix (2)	a	1400; 2200
	d	1800
	h	1120
Du Pont (3)	a	1950; 3000
	b	2500
	e	3500–4500
	f	3000
	g	3800
Isco (19)	a	850
	d	1200
	e	1100–1400
Knauer (12)	a	2100
	b	2100
LDC (21)	a	1500
	b	2400
	e	2550–2700
	j	2200
LKB (22)	a	890–1440
	e	2900
Packard (6)	i	9400[a]
Perkin-Elmer (7)	a	1950
	b	2350
	e	2450
Pye Unicam (8)	c	8100

B.   LC DETECTORS   (MODULAR UNITS)   (CONTINUED)

Supplier	Type	Approximate Price (June, 1973)
Nuclear-Chicago (23)	i	6300[a]
Varian Aerograph (10)	a	1540
	b	2500
	d	1750
	g	5500
Waters (11)	b	2000
Wilks (24)	k	2600
Cecil (60)	g	2600
Gow-Mac (59)[b]	b	1400
Zeiss (61)	g	3100
Schoeffel (58)	g	3800

[a]Two channel, with two-pen recorder, without small
  volume cell.
[b]Based on Christiansen effect.

KEY TO DETECTORS

Code	Type
a	Ultraviolet photometer, 254 nm
b	Differential refractometer
c	Transport (wire, flame ionization, etc.)
d	Ultraviolet photometer, 280 nm
e	Multiwavelength photometer
f	Visible photometer
g	Spectrophotometer

KEY TO DETECTORS (CONTINUED)

Code	Type
h	Conductivity
i	Radiometric (scintillation)
j	Fluorescence
k	Infrared photometer

C. MODULAR PUMPS

Supplier	Model	Type	Maximum Pressure (psi)	Maximum Flow Rate (ml/min)	Approximate Price ($)
Chromatec (1)	3000-1	A	1000	10	950
	3000-2	A	2000	10	1350
	5000	E	6000	a	3450
Chromatronix (2)	CMP-1	A	500	2	1000
	CMP-2	A	500	2	1500[b]
	740	A	7000	20	4000
Du Pont (3)	843	D	1500	a	950
Haskel (52)	17082-3	E	6000	a	1100
Isco (19)	314	C	2000	3.3	1640
Lapp (53)	LS-10,-20,-30	B	1000-3000	4.5-70	400-500
Milton Roy (54)	196-31; -32	A	1000	3	410
	HDB-1-30-R	A	3000	2	440
Orlita (13)	DMP 1515	B	5000	5.5	2300
Varian Aerograph (10)	02-001468-01	C	5000	3.3	5700
	02-001231-00	D	1000	a	240
Waters (11)	6000	A	6000	10	3500
Whitey (48)	LP-10	B	5000	20	1200

[a]Not applicable, constant pressure system.

[b]With pulse compensation.

A, reciprocating, piston.
B, reciprocating, diaphragm.
C, positive displacement (driven syringe).
D, simple pneumatic.
E, pneumatic amplifier.

D. MODULAR GRADIENT ELUTION APPARATUS

Supplier	Type	Approximate Price ($)
Chromatec (1)	Single pump, high-pressure mixing	4650
Isco (19)	Two pumps, high-pressure mixing	3900
LKB (22)	Low-pressure mixing, no high-pressure pump; up to three solvents	2610
Varian Aerograph (10)	Two pumps, high-pressure mixing	13500
Waters (11)	Two pumps, high-pressure mixing	8400
Chromatronix (2)	Two pumps, high-pressure mixing	11000
Analabs (73)	Low-pressure mixing, no high-pressure pump; up to 20 solvents	3200

E. LC SAMPLING MICROVALVES[a]

Supplier	Sampling Type External Loop	Sampling Type Internal Cavity	Capacity (μl)	Maximum Pressure (psi)	Approximate Price ($)
Chromatronix (2)	X	X	2–1000	500	120–160
	X		20–500	3000	300
Disc (25)		X	5	5000[b]	290
	X		75–1000	5000[b]	250
		X	2–20	3500[c]	350
Du Pont (3)		X	0.5–5	5000	650
	X		30–250	1000	470
Valco (26)		X	0.2–3.5	1000	170
		X	0.2–3.5	2500	230
	X		30–1000	1000	200
	X		50–1000	2500	290
	X		5–1000	2000	250
Varian Aerograph (10)	X		50–200	3000	320
Waters (11)	X		2–10	1000	200
Hamilton (27)		X	0.5–5	5000	500

[a]Made with stainless steel and fluorocarbon; may be used with a wide variety of solvents.

[b]Delrin® transfer element. Not useful for some organic solvents.

[c]Fabricated entirely from stainless steel.

F. OTHER EQUIPMENT

Many of the materials listed below are also available from the suppliers of complete LC instruments. This list represents some of the other sources of materials which have been demonstrated to be particularly useful for high-performance liquid chromatography.

Item	Description	Supplier
1. Microsyringes	Series C-160, CG-130, B-110 Model HP-305 SGE Types A and B	Precision Sampling (18) Hamilton (24) Supelco (29)
2. Injection Ports	5000 psi, maximum 1000 psi, maximum	Precision Sampling (18) Chromatronix (2)
3. Septums	Silicone Silicone Silicone with Teflon® face, reinforced Neoprene with Teflon® face, reinforced Fluoroelastomer	Applied Science (28) Supelco (29) Canton Biomedical Products (15); Analabs (73) Canton Biomedical Products (15) Du Pont (3)
4. Blank columns, fittings, etc.		Chromatronix (2) LDC (21) Applied Science (28)
5. Pressure measuring devices	Diaphragms, gauges Diaphragms, gauges Diaphragms, gauges Diaphragms, gauges Transducers, strain gauge Transducers, strain gauge	Chromatronix (2) U.S. Gauge (31) Ashcroft/Dresser (56) Hoke (32) Statham (33) MB Electronics (34)
6. Pressure relief devices	Pop-valve type Pop-valve type Pop-valve type Electrical relay system	Chromatronix (2) Circle Seal (35) Nupro (36) United Electric (30)

F. OTHER EQUIPMENT (CONTINUED)

Item	Description	Supplier
7. Flow measuring devices	Flow tube	LDC (21)
8. Microfiltering equipment	"Swinney" holders, micro-filters, etc.	Millipore (37)
9. Column plugs	Porous metal frits Porous metal frits Porous metal frits Porous metal frits Porous Teflon® Teflon® wool	Du Pont (3) Mott Metallurgical (38) Applied Science (28) Pall Trinity (39) Fluoro-Plastics, Inc. (40) Du Pont (3); Supelco (24)
10. Line filters	Stainless steel	Nupro (36) Circle Seal (35) Hoke (32)
11. Recorders	See Chapter 4 for specs	Hewlett-Packard (5) Varian Aerograph (10) Esterline-Angus (41) Honeywell (42) Leeds & Northrup (43)
12. Digital electronic integrators	Use for high-speed LC reported	Hewlett-Packard (5) Varian Aerograph (10) Infotronics (45) Perkin-Elmer (17) Autolab (44)
13. Packed columns (in addi-tion to instrument manufacturers)	With certain packings	Reeve-Angel (46) Applied Science (28)
14. Constant-temperature water bath	Model FT	Haake Instruments (47)

F. OTHER EQUIPMENT (CONTINUED)

Item	Description	Supplier
15. Tubing	Precision-bore stainless steel; 316L superpressure tubing, mirror finish	Superior (49)
	Fluorocarbon	Chromatronix (2); Fluoro-Plastics (40); Franklin (55) Supelco (24)
	Glass-lined stainless steel (S.G.E.)	
16. Air regulators	For pneumatic pumps	Moore Products (50)
17. Switching valves	Teflon® - 500 psi	Chromatronix (2)
	Stainless steel--3000 psi	Chromatronix (2)
	Stainless steel--2000 psi	Valco (26)
18. Switching valves	Not low volume; useful for handling mobile phase system	Whitey (48) Hoke (32)
19. Compression fittings	Metal (Swagelok®)	Crawford (51)
	Metal (Gyrolok®)	Hoke (32)
	Fluorocarbon	Beckman (20)
	Fluorocarbon	Plasmatech (16)
20. Metal funnels	For packing columns	Pierce Chemicals (17)
21. Solvents	Purified especially for LC	Burdick and Jackson (74) Mallinckrodt (75)

G. LIST OF EQUIPMENT SUPPLIERS

1. Chromatec, Inc., 30 Main St., Ashland, Mass. 01721
2. Chromatronix, Inc., 2743 Ninth St., Berkeley, Cal. 94710
3. E. I. du Pont de Nemours & Co., Inc., Instrument Products Division, 1007 Market St., Wilmington, Del. 19898
4. Hupe & Busch, Gutenbergstrasse 6, 7501 Groetsingen, Karlsruhe, West Germany (now Hewlett-Packard (5))
5. Hewlett-Packard, Avondale Division, Rt. 41, Avondale, Pa. 19311
6. Packard Instrument Co., 2200 Warrenville Rd., Downers Grove, Ill. 60515
7. Perkin-Elmer Corp., Instrument Division, Main Ave., Norwalk, Conn. 06852
8. Pye Unicam Ltd., York St., Cambridge, England
9. Siemans AG, Rheinbruckenstrasse 50, 75 Karlsruhe 21, West Germany (U.S. - Siemens Corp., 186 Wood Ave., South, Iselin, N.J. 08830)
10. Varian Aerograph, 2700 Mitchell Dr., Walnut Creek, Cal. 94598
11. Waters Associates, Inc., 61 Fountain St., Framingham, Mass. 01701
12. KG Dr. Knauer, Holstweg 18, 1 Berlin 37 (West) Germany
13. Orlita KG, Max-Eyth-Strasse 10, 63 Giessen, Germany (in U.S., Federated Equipment and Supply Co., 2561 N. Clark St., Chicago, Ill. 60614)
14. Meci, 123 Bd. DeGrenelle - XVe, Paris, France
15. Canton Biomedical Products, P. O. Box 2017, Boulder, Colo. 80302
16. Plasmatech, 9812 Klingerman St., S. El Monte, Cal. 91733
17. Pierce Chemical Co., P. O. Box 117, Rockford, Ill. 61105
18. Precision Sampling Corp., 8275 W. El Cajon Dr., Baton Rouge, La. 70815

19. Instrumentation Specialties Co., 4700 Superior,
    Lincoln, Neb. 68504
20. Beckman Instruments, Inc., 2500 Harbor Blvd.,
    Fullerton, Cal. 92634
21. Laboratory Data Control, P. O. Box 10235, Inter-
    state Industrial Park, Rivera Beach, Fla. 33404
22. LKB - Produktur AB Fredsforsstigen 22-24, Foek,
    S-16125, Bromma, Stockholm, Sweden (U.S. - LKB
    Instruments, Inc., 12221 Parklawn Dr., Rockville,
    Md. 20852
23. Nuclear-Chicago, 2000 Nuclear Dr., Des Plaines,
    Ill. 60018
24. Wilks Scientific Corp., 140 Water St., South
    Norwalk, Conn. 06856
25. Disc Instruments, Inc., 2701 S. Halladay St.,
    Santa Ana, Cal. 92705
26. Valco Instruments Co., P. O. Box 19032, Houston,
    Texas 77024
27. Hamilton Co., 12440 E. Lambert Rd., Whittier,
    Cal. 90608
28. Applied Science Laboratories, Inc., P. O. Box 440,
    State College, Pa. 16801
29. Supelco, Inc., Supelco Park, Bellefonte, Pa.
    16823
30. United Electric Controls Co., 85 School St.,
    Watertown, Mass. 02172
31. Ametek/U.S. Gauge, P. O. Box 152, Sellersville,
    Pa., 18960
32. Hoke, Inc., Tenakill Park, Cresskill, N.J. 07626
33. Statham Instruments, Inc., 2230 Statham Blvd.,
    Oxnard, Cal. 93030
34. M. B. Electronics, P. O. Box 1825, New Haven,
    Conn. 06508
35. Circle Seal Corp., 1111 N. Brookhurst St., P. O.
    Box 3666, Anaheim, Cal. 92803
36. Nupro Co., 15635 Saranac Rd., Cleveland, Ohio
    44110
37. Millipore Corp., Ashby Rd., Bedford, Mass. 01730

38. Mott Metallurgical Corp., Farmington Ind. Park, P. O. Drawer L, Farmington, Conn.   06032
39. Pall Trinity Micro Corp., Rt. 281, Cortland, N.Y. 13045
40. Fluoro-Plastics, Inc., G and Verango Sts., Philadelphia, Pa.   19134
41. Esterline Angus, P. O. Box 24000, Indianapolis, Ind.   46224
42. Honeywell, Inc.,2701 4th Ave., S., Minneapolis, Minn.   55408
43. Leeds & Northrup Co., Sumneytown Pike, North Wales, Pa.   19454
44. Autolab, 655 Clyde Ave., Mountain View, Cal. 94040
45. Infotronics Corp., 2475 Broadway, Boulder, Colo. 80302
46. H. Reeve Angel, 9 Bridewell Place, Clifton, N.J. 07014
47. Haake Instruments, Inc., 244 Saddle River Rd., Saddle Brook, N.J.   07662
48. Whitey Research Tool Co., 884 E. 140th St., Cleveland, Ohio 44110
49. Superior Tube Co., P. O. Box 191, Norristown, Pa. 19404
50. Moore Products Co., Sumneytown Pike, Spring House, Pa.   19477
51. Crawford Fitting Co., 29500 Solon Rd., Solon, Ohio   44139
52. Haskel Engineering & Supply Co., 100 E. Graham Place, Burbank, Cal.   91502
53. Lapp Process Equipment, Leroy, N.Y.   14482
54. Milton Roy Co., 5000 Park St., N., St. Petersburg, Fla.   33733
55. Franklin Fiber-Laninex Corp., 129 Lafayette St., New York, N.Y.
56. Dresser Industrial Valve & Instrument Divn., Dresser Industries, Inc., 250 E. Main St., Stratford, Conn.   06497

57. Micromeritics Instrument Corp., 800 Goshen Springs Rd., Norcross, Ga. 30071
58. Schoeffel Instrument Corp., 24 Booker St., Westwood, N.J. 07675
59. Gow-Mac Instruments Co., 100 Kings Road, Madison, N.J. 07940
60. Cecil Instruments, Ltd., Cambridge, England
61. Carl Zeiss, Inc., 444 5th Ave., New York, N.Y. 10018
62. Bio-Rad Laboratories, 32nd and Griffin Ave., Richmond, Cal. 94804
63. E. Merck, 61 Darmstadt, Frankfurter Strasse 250, West Germany
64. E. M. Laboratories, 500 Executive Blvd., Elmsford, N.Y. 10523
65. Separations Group, 8738 Oakwood Ave., Hesperia, Cal. 92345
66. W. R. Grace & Co., Davison Chemical Division, Charles and Baltimore Sts., Baltimore, Md. 21203
67. Aluminum Co. of America, 1501 Alcoa Bldg., Pittsburgh, Pa. 15219
68. Durrum Chemical Co., 3950 Fabian Way, Palo Alto, Cal. 94303
69. Molecular Separations, Inc., Champion, Pa. 15622
70. Separation Technology, Inc., 48 First St., Cambridge, Mass. 02141
71. Electro-Nucleonics, Inc., 368 Passaic Ave., P. O. Box 803, Fairfield, N.J. 07006
72. Pharmacia Fine Chemicals, 800 Centennial Ave., Piscataway, N.J. 08854
73. Analabs, Inc., 80 Republic Dr., North Haven, Conn. 06473
74. Burdick & Jackson Labs, Inc., 1953 S. Harvey St., Muskegon, Mich. 49442
75. Mallinckrodt Chemical Works, Science Products Division, 2nd & Mallinckrodt Sts., St. Louis, Mo. 63160

APPENDIX VI

COMPARATIVE COLUMN PERFORMANCE

Given two columns, how can we tell which is better?
For two columns of the same size, packed with the same
material, there is little difficulty in making such
a distinction. The column that gives lower H values
(larger N values) is to be preferred. The question
becomes more difficult to answer, when we are faced
with columns whose packings differ with respect to particle size, particle shape, or the type of particle
(pellicular versus porous). In fact, even experts
often disagree over the criteria to be applied in such
cases.

To begin with, we should recognize that column
performance is a composite of several factors: efficiency, capacity, selectivity, convenience, and so on.
In some cases column capacity is of major importance,
and then totally porous particles will be favored over
pellicular particles. Some columns, specifically those
packed with very small particles ($d_p \leq 10\ \mu$), are capable of both high capacity <u>and</u> very large N values.
However, these columns with their small values of V_m
and large values of N are less convenient to use, because they place greater demands on experimental technique and the LC equipment; for example, V_i and V_x
(Eq. 2.9) must be quite small. Some workers have

519

assumed that the use of smaller particles must be compensated for by greater column pressures, but this is
not true. Instead column length can be reduced. Finally, differences in the stationary phase itself lead
to differences in separation selectivity or α values,
and hence to differences in performance for a given
sample.

The question of column capacity and selectivity
has been dealt with in Chapters 7-10. Here we will
focus on column efficiency, for a particular set of
experimental constraints (column pressure, separation
time, solvent composition, temperature, etc.). Therefore our interest will be concentrated on the N values
of different columns, for comparable experimental conditions. This discussion does not consider how convenient these various columns are to prepare or use.

To a first approximation, column performance is
adequately described by term (i) of Eq. 3.2; that is,
we can define a column performance factor PF,

$$PF = \frac{K^{(1-n)/2}}{D} .$$
(VI.1)

Here K is column permeability (Eq. II.1), D is the
value of H at a solvent velocity u = 1 cm/sec, and n
is the empirical column parameter from Eq. 2.8. Normally, $0.3 \leq n \leq 0.7$.

According to Eq. VI.1, column performance as measured by plates N (or effective plates N_{eff}) is greater for small values of H and large values of K. Keep
in mind that a column with a larger K value can be
used at lower pressures, other factors being equal,
which means that such a column provides performance
comparable to that of a less permeable column at higher pressures. And we have seen that higher column
pressures can generally be translated into a greater
number of plates per second. However, the role of K
is less important than that of H because N_{eff} increases
as (1/H) to the first power, but as K to a lower power
$(K^{0.2}-K^{0.3})$. Thus a decrease in H by a factor of 10

results in an increase in N_{eff} by tenfold. However,
an increase in K by tenfold leads (roughly) to only a
doubling of N_{eff}.

Consider next the significance of particle size d_p,
in comparing column efficiency. We have seen in Chapters 2 and 6 that small particles favor small values
of H, so that column efficiency tends to increase for
smaller particles. Column permeability, on the other
hand, decreases for smaller values of d_p (Eq. II.2).
The question is then: Are smaller particles desirable
for overall superior performance?

For particles above some minimum size (e.g., d_p
$\geq$ 5 μ), it appears that smaller particles are genuine-
ly better. Thus well-packed columns containing particles of differing average size usually give the same
reduced plots of h versus ν , where h = H/d_p and ν =
u d_p/D_m (e.g., Figure I.1). That is, H versus u data
for all such columns will follow a single function:

$$h = f(\nu),$$

or

$$\frac{H}{d_p} = f\left[\frac{u \ d_p}{D_m}\right]. \tag{VI.2}$$

If these same data can also be described in terms of
Eq. 2.8, H = D u^n, then it is seen that for D_m constant,

$$\frac{H}{d_p} = C \ (u \ d_p)^n . \tag{VI.2a}$$

where C is a constant for the different columns. For
Eq. VI.2a to hold for all values of d_p, it is now seen
that H must be related to d_p as

$$H = C' \ d_p^{\ 1+n} , \tag{VI.3}$$

as can be seen by substituting Eq. VI.3 into VI.2a.
The quantity C' is also a constant, for some given
value of u; therefore

$$D = C' \, d_p^{1+n} \qquad \text{(VI.4)}$$

for u = 1 cm/sec.

The quantity K can similarly be expressed as a function of K (Eq. II.2):

$$K = C'' \, d_p^2 \, . \qquad \text{(VI.5)}$$

Column performance is now related to d_p as

$$PF = \frac{K^{(1-n)/2}}{D}$$

$$= \frac{(C'' \, d_p^2)^{(1-n)/2}}{C' \, d_p^{1+n}}$$

$$= C''' \, d_p^{-2n} \qquad \text{(VI.6)}$$

Here C''' is a third constant, and column performance (effective plates/sec) is seen to increase as d_p^{-2n}; that is, since $0.3 \leq n \leq 0.7$, as $d_p^{-(0.6 \rightarrow 1.4)}$. Thus smaller particles are seen to definitely favor improved column performance. It should be noted that this conclusion is independent of the operating pressure of the column, since term (i) of Eq. 3.2 is independent of time and pressure. That is, <u>regardless of column pressure or separation time, small particles give generally better columns</u>, when the columns are well packed.

The latter conclusion can be accepted as generally correct, and in fact the highest number of plates per second are found for columns with $d_p \leq 10 \, \mu$ (e.g., refs. 1 and 2). However, certain reservations should be noted with respect to particle sizes smaller than $5 \, \mu$. First, while Eq. VI.2 is indeed obeyed for well-packed columns of particles in the 10-60 μ range (2,3), it is well known that the preparation of well-packed columns is more difficult with small particles than with large (Chapter 6). Therefore we should not be surprised to see Eq. VI.2 fail as d_p is increased below 5 μ, so that such columns may not show the expected continued increase in PF with decrease in d_p (Eq. VI.6).

Second, the application of the performance factor (Eq. VI.1) to columns of varying particle size, assumes that column length is decreased as Ld_p, when d_p decreases (to hold t_o and separation time constant). With columns of 10-μ particles, typical column lengths are 25-50 cm. Therefore, the corresponding column lengths for 1-μ particle columns would be only 2.5-5 cm, which represents a formidable problem in terms of extra-column dead volume.

The higher efficiency of pellicular columns over porous particles is apparent from Figure I.1. We have seen, however, that smaller particles generally imply smaller values of ν , other factors being equal. In Figure I.1 it is seen that the top two curves approach each other at small values of ν , which means that the advantage of pellicular particles becomes less for small values of d_p. In fact it seems unlikely that very small pellicular particles ($d_p < 20$ μ) will be found useful in LC, even if such particles were available (at present they are not). The reduced advantage of < 20-μ pellicular particles can also be seen in terms of other arguments (e.g., ref. 4).

REFERENCES

1. J. J. Kirkland, J. Chromatog. Sci., _10_, 593 (1972).
2. R. E. Majors, J. Chromatog. Sci., _11_, 88 (1973).
3. J. N. Done and J. H. Knox, J. Chromatog. Sci., _10_, 606 (1972).
4. L. R. Snyder, in Gas Chromatography 1970, R. Stock and S. G. Perry, eds., Institute of Petroleum, London, 1971, p. 81.